Technical Thermodynamics Workbook for Engineers

Achim Schmidt

Technical Thermodynamics Workbook for Engineers

Typical Tasks with Detailed Solutions

Springer

Achim Schmidt
Department Maschinenbau und Produktion
Hochschule Angewandte Wissenschaften
Hamburg, Germany

ISBN 978-3-031-50174-6 ISBN 978-3-031-50172-2 (eBook)
https://doi.org/10.1007/978-3-031-50172-2

This Springer imprint is published by the registered company Springer Nature Switzerland AG
The registered company address is: Gewerbestrasse 11, 6330 Cham, Switzerland

Paper in this product is recyclable.

Dedicated to Luisa.

Preface

After my textbook Technical Thermodynamics for Engineers—Fundamentals and Applications, cf. [1], has already been published in the second edition, I have decided to write an accompanying workbook aimed specifically at engineering students. I would like to give them the opportunity to apply the theoretical foundation in practical examples.

The aim of my textbook was to describe the theoretical thermodynamic correlations. The goal for me was to discuss the thermodyanmsic relationships to such an extent that at least my open questions were satisfactorily clarified. At the same time, example problems were solved and thermodynamic principles were explained by numerous illustrations. This led to the book being very extensive in terms of the number of pages. So there is a latent danger that it will deter rather than motivate!

In the last few semesters, the decision matured in me to write a supplementary book that conveys knowledge in a compact form and that can be used as a workbook. Whether one has understood the theory becomes apparent when solving concrete tasks. This book now contains over 170 exercises, some of which were covered in my lectures and exams: The comprehension questions in each chapter of this book are particularly suitable for checking the basic understanding of thermodynamics. The underlying theory is explained in sufficient detail. Even if the answers should be immediately clear to the reader, it is worth reading the explanations of the solutions, as I recapitulate theoretical correlations there. However, if further in-depth knowledge is required, I will refer to my textbook.

In addition to comprehension questions, the book contains numerous typical calculation problems. Here, too, the basics are summarised and solutions respectively solution strategies are discussed. These are designed to provide students with strategies and tools to solve complex tasks. All these exercises should help to deepen the physical knowledge. The greater the technical competence, the more confidently one can approach the exam! In the best case, the tasks help to find fun and enjoyment in thermodynamics!

I have tried to structure the tasks thematically, but it is not always possible to avoid overlaps between the topics.

As with my textbook, I appreciate constructive feedback on the workbook and hope that this book will be a useful tool for you when preparing for exams in *Technical Thermodynamics*.

Hamburg, Germany
Summer 2023

Achim Schmidt

Contents

Nomenclature

Roman Symbols

A	Area (m^2)
a	Cohesion pressure (Pa m^6 mol^{-2})
a	Mass fraction ashes (–)
a	Speed of sound (ms^{-1})
a	Acceleration (ms^{-2})
b	Molar co-volume (m^3 mol^{-1})
$\dot{B}_x$	Flux of anergy (W)
B_x	Anergy (J)
$B(T)$	Virial coefficient (m^3 mol^{-1})
b_x	Specific anergy (J kg^{-1})
C	Number of components (–)
C	Capacity (F)
C, c_M	Molar heat capacity (J $mol^{-1}K^{-1}$)
$\overline{C}$	Molar arithmetic mean heat capacity (J $mol^{-1}K^{-1}$)
$\overline{\overline{C}}$	Molar logarithmic mean heat capacity (J $mol^{-1}K^{-1}$)
C	Constant
$C(T)$	Virial coefficient (m^6 mol^{-2})
c	Mass fraction carbon (–)
c	Velocity (ms^{-1})
c_i	Molarity of a component i (mol m^{-3})
c	Specific heat capacity (J kg^{-1} K^{-1})
$\bar{c}$	Specific, averaged heat capacity (J kg^{-1} K^{-1})
$D(T)$	Virial coefficient (m^9 mol^{-3})
D, d	Diameter (m)
E	Energy (J)
E_x	Exergy (J)
$\dot{E}_x$	Flux of exergy (W)
$\Delta E_{x,V}$	Loss of exergy (J)

$\Delta\dot{E}_{x,V}$	Flux of loss of exergy (W)
E_{xm}	Molar specific absolute exergy (J mol^{-1})
$E_{xm,F}$	Molar specific absolute exergy of a fuel (J mol^{-1})
e_x	Specific exergy (J kg^{-1})
$\Delta e_{x,V}$	Specific loss of exergy (J kg^{-1})
F	Degree of freedom (–)
F	Force (N)
F	Helmholtz energy (J)
f	Specific Helmholtz energy (J kg^{-1})
G	Gibbs energy (J)
G_m	Molar specific Gibbs energy (J mol^{-1})
g	Specific Gibbs energy (J kg^{-1})
g	Gravitational acceleration, $g = 9.81$ m s^{-2} (m s^{-2})
H, h	Height in a gravity field (m)
H	Enthalpy (J)
H_m	Molar specific absolute enthalpy (J mol^{-1})
H_U	Mass specific lower heating value (J kg^{-1})
H_{UM}	Molar specific lower heating value (J mol^{-1})
H_{Uv}	Volume specific lower heating value (J m^{-3})
H_0	Mass specific higher heating value (J kg^{-1})
H_{0M}	Molar specific higher heating value (J mol^{-1})
H_{0v}	Volume specific higher heating value (J m^{-3})
$\Delta_B^0 H_m$	Molar specific enthalpy of formation at standard conditions (J mol^{-1})
$\Delta_R^0 H_m$	Molar specific enthalpy of reaction at standard conditions (J mol^{-1})
h	Mass fraction hydrogen (–)
h	Specific enthalpy (J kg^{-1})
Δh_v	Specific enthalpy of vaporisation (J kg^{-1})
Δh_m	Specific enthalpy of melting (J kg^{-1})
$\mathbf{I}$	Momentum (kg m s^{-1})
K	Revolutions per working stroke (–)
k	Heat transition coefficient (W m^{-2} K^{-1})
k_B	Boltzmann constant, $k_B = 1.3806 \times 10^{-23}$ J K^{-1} (J K^{-1})
k_F	Spring constant (N m^{-1})
L_{min}	Minimum molar-specific air need (–)
l_{min}	Minimum mass-specific air need (–)
M	Molar mass (kg mol^{-1})
M	Torque (N m)
m	Mass (kg)
$\dot{m}$	Mass flux (kg s^{-1})
$\dot{m}''$	Mass flux density (kg s^{-1} m^{-2})
Ma	Mach-number (–)
N_A	Avogadro constant, $N_A = 6.022045 \times 10^{23}$ mol^{-1} (mol^{-1})
n	Mass fraction nitrogen (–)
n	Molar quantity (mol)
n	Polytropic exponent (–)

n	Speed (s^{-1})
$\dot{n}$	Molar flux ($mol\ s^{-1}$)
$\mathbf{n}$	Normal
O_{min}	Minimum molar-specific oxygen need (–)
o_{min}	Minimum mass-specific oxygen need (–)
o	Mass fraction oxygen (–)
P	Number of phases (–)
P	Power (W)
p	Pressure (Pa)
p_i	Partial pressure of a component i (Pa)
Q	Electric charge (F)
Q	Heat or thermal energy (J)
$\dot{Q}$	Heat flux (W)
q	Specific heat ($J\ kg^{-1}$)
R	Individual gas constant ($J\ kg^{-1}\ K^{-1}$)
R, r	Radius (m)
R_M	General gas constant, $R_M = 8.3143\ J\ mol^{-1}\ K^{-1}$ ($J\ mol^{-1}\ K^{-1}$)
S	Entropy ($J\ K^{-1}$)
s	Specific entropy ($J\ kg^{-1}\ K^{-1}$)
S_i	Entropy generation ($J\ K^{-1}$)
s_i	Specific entropy generation ($J\ kg^{-1}\ K^{-1}$)
S_a	Entropy carried with heat ($J\ K^{-1}$)
s_a	Specific entropy carried with heat ($J\ kg^{-1}\ K^{-1}$)
s	Mass fraction sulphur (–)
s	Distance (m)
$\dot{S}_i$	Flux of entropy generation ($W\ K^{-1}$)
$\dot{S}_a$	Flux of entropy carried with heat ($W\ K^{-1}$)
$\Delta_R S_m$	Molar specific entropy of a reaction ($J\ mol^{-1}\ K^{-1}$)
S_m	Molar specific absolute entropy ($J\ mol^{-1}\ K^{-1}$)
T	Absolute thermodynamic temperature (K)
t	Time (s)
U	Internal energy (J)
U_m	Molar specific absolute internal energy ($J\ mol^{-1}$)
u	Specific internal energy ($J\ kg^{-1}$)
U	Voltage (V)
V	Volume (m^3)
$\dot{V}$	Volume flux ($m^3\ s^{-1}$)
v	Specific volume ($m^3\ kg^{-1}$)
V_m	Molar specific volume ($m^3\ mol^{-1}$)
W	Work (J)
w	Specific work ($J\ kg^{-1}$)
w	Mass fraction water (–)
x	Vapour ratio (–)
x	Water content (–)
x	Molar fraction (–)

x_i	Molar concentration of a component i (–)
x	Coordinate (m)
y	Coordinate (m)
Y	Pressure work (J)
y	Specific pressure work (J kg^{-1})
Z	Compressibility factor (–)
Z	Extensive state value
z	Specific state value, $z = \frac{Z}{m}$
z	Coordinate (m)
z	Distance (m)

Greek Symbols

α	Abbreviation
α	Angle (°)
α	Heat transfer coefficient (W m^{-2} K^{-1})
β	Isobaric volumetric thermal expansion coefficient (K^{-1})
δ_h	Isenthalpic throttle coefficient (K Pa^{-1})
δ_T	Isothermal throttle coefficient (m^3 kg^{-1})
ϵ	Compression ratio (–)
η	Abbreviation
η, ξ	Frictional constant (kg s^{-1})
η	Efficiency (–)
γ_i	Stoichiometric factor of a component i (–)
κ	Isentropic exponent (–)
λ	Air-fuel equivalence ratio (–)
μ	Chemical potential (J mol^{-1})
μ_i	Mass-specific exhaust gas composition of component i (–)
ν_i	Molar-specific exhaust gas composition of component i (–)
Ω	Flow function of a nozzle (–)
Ω	Statistical weight, measure of the probability (–)
ω	Acentric factor (–)
π	Pressure ratio (–)
Ψ	Dissipation (J)
$\dot{\Psi}$	Flux of dissipation (W)
ψ	Specific dissipation (J kg^{-1})
ψ'	Specific dissipation per length (J kg^{-1} m^{-1})
ψ	Relative saturation (–)
ψ_i	Volume ratio of a component i (–)
ρ	Density (kg m^{-3})
ρ_{m}	Molar density (mol m^{-3})
ρ_i	Partial density of a component i (kg m^{-3})

σ_i	Volume concentration of a component i (–)
ε	Coefficient of performance (–)
φ	Relative humidity (–)
ϑ	Celsius-temperature (°C)
ξ	Abbreviation
ξ_i	Mass concentration of a component i (–)

Subscripts

A	Air (wet)
a	Air (dry)
a	Outer
C	Cylinder
comp	Compression
cond.	Condenser
eff	Effective
EG	Exhaust gas
el	Electric
env	Environment
F	Fuel
fric.	Friction
G	Gas
gas	Gas
HP	High pressure
ice	Ice
in	Inlet
irrev	Irreversible
K	Control volume
kin	Kinetic
L	Liquid
liq	Liquid
LP	Low pressure
m	Mean
m	Melting
M,m	Molar
max	Maximum
Mech	Mechanic
min	Minimum
MP	Medium pressure
n	Narrowest cross section
Obj	Object
out	Outlet

P	Piston
p	$p = \text{const.}$
pot	Potential
R	Reservoir
ref	Reference state
rev	Reversible
S	Steel
s	Saturated
shift	Shifting
Source	Source term
spr	Spring
swing	Stroke
Sys	System
t	Technical
T	$T = \text{const.}$
th	Thermal
total	Total
V	Volume
v	$v = \text{const.}$
V,v	Vapour
W	Water

Acronyms

$'$	Saturated liquid state, saturated humid air
$''$	Saturated vapour state
δ	Process value
d	State value
1P	Single-phase
2P	Two-phase
C	Carnot
CM	Cold machine/fridge
cp	Critical point
HP	Heat pump
HP	High pressure
HT	Heat transfer
HVAC	Heating, ventilation and airconditioning technology
LP	Low pressure
Pr	Product
TE	Thermal engine
TP	Triple point

List of Figures

List of Tables

Chapter 1
Introduction and Basics

In this first chapter, the basic thermodynamic concepts are introduced and deepened in tasks. The focus is on the evaluation of thermodynamic equilibria. Principles from mechanics are adopted, for example, to determine the pressure of compressible fluids in thermodynamic equilibrium. Furthermore, the concept of process values, e.g. heat and work, is introduced. These can act on a system from the outside and change the state of the system regarded. It is shown how, for example, the work required to reduce the volume of a compressible fluid, i.e. the volume work of a closed system, can be calculated.

1.1 Problems

1.1 An ideal gas is filled in a vertical cylinder. Mass of the freely movable piston as well as ambient pressure are constant. Now, by a valve 20% of the mass of the gas are released to the environment. This takes place so slowly, that the temperature remains constant all time. Which statement is true when a new equilibrium is reached?

(a) The pressure inside the cylinder decreases by 20%.
(b) The specific volume remains constant.
(c) Such a change of state is impossible.
(d) None of the above statements is true.

1.2 Two tanks A and B are each filled with an ideal gas A resp. B:

- Tank A: ideal Gas A, $n_A = 2$ kmol, $T_A = 303$ K, $p_A = 2$ bar
- Tank B: ideal Gas B, $n_B = 2$ kmol, $T_B = 303$ K, $p_B = 2$ bar

A. Schmidt, *Technical Thermodynamics Workbook for Engineers*,
https://doi.org/10.1007/978-3-031-50172-2_1

What is true regarding the volumes of the tanks?

(a) $\frac{V_A}{V_B} = 1$
(b) $\frac{V_A}{V_B} = \frac{M_A}{M_B}$
(c) $\frac{V_A}{V_B} = \frac{R_A}{R_B}$
(d) None of the above statements is true.

1.3 The molar mass of an ideal gas is $44\,\frac{\text{kg}}{\text{kmol}}$. What is the gas constant of this gas?

(a) $188.96\,\frac{\text{J}}{\text{kg K}}$
(b) $178.16\,\frac{\text{J}}{\text{kg K}}$
(c) $168.23\,\frac{\text{J}}{\text{kg K}}$
(d) $158.71\,\frac{\text{J}}{\text{kg K}}$

1.4 In an industrial process, nitrogen is filled in a vessel with a constant volume. In state (1) the pressure is $p = 10\,\text{MPa}$ and the temperature is $T = 300\,\text{K}$. What is the pressure when the gas behaves ideally and is heated up to 500 K?

(a) 10 MPa
(b) 12.33 MPa
(c) 16.67 MPa
(d) 18.67 MPa

1.5 What is the molar volume of an ideal gas at 3 bar and 325 K?

(a) $1.31\,\frac{\text{m}^3}{\text{kmol}}$
(b) $5.12\,\frac{\text{m}^3}{\text{kmol}}$
(c) $9.01\,\frac{\text{m}^3}{\text{kmol}}$
(d) $22.4\,\frac{\text{m}^3}{\text{kmol}}$

1.6 The molar mass of an ideal gas is $44\,\frac{\text{kg}}{\text{kmol}}$. The gas is trapped in a volume of $0.5\,\text{m}^3$ at 20 bar and 400 K. What is the mass of the gas?

(a) 3.23 kg
(b) 13.23 kg
(c) 43.23 kg
(d) 113.23 kg

1.7 To what temperature does the carbon dioxide (ideal gas, 18 °C, 1.5 bar, $\kappa = 1.30$) contained in a bottle of beer cool when the cap suddenly pops open and the external air pressure is 1 bar. Approximately, let the change of state be isentropic.

(a) 4 °C
(b) 0 °C
(c) −4 °C
(d) −8 °C

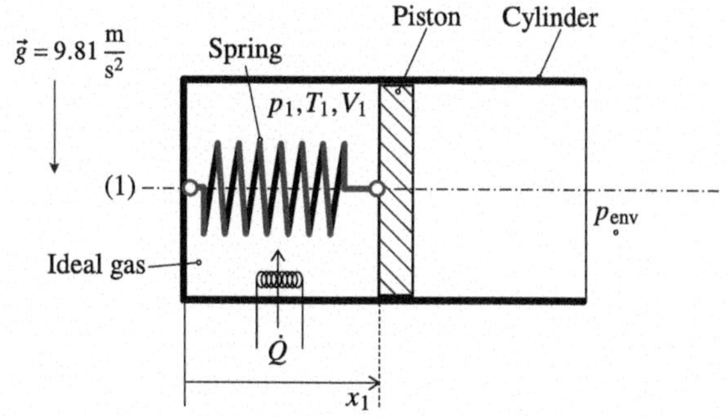

Fig. 1.1 Cylinder/piston/environment

1.8 A mass of $m = 4$ kg is to be lifted by $\Delta z = 2$ m in the earth's gravitational field ($g = 9.81\,\frac{m}{s^2}$).

(a) How much work must be done on the mass?
(b) How much work is to be done in total if the mass is additionally to be accelerated to $5\,\frac{m}{s}$?

1.9 The volume of an ideal gas with an initial density of $1.74\,\frac{kg}{m^3}$ under a constant pressure of 1 bar in a cylinder increases by 11% when it is heated by 20 K. What is the gas constant?

1.10 A horizontal cylinder is closed by a movable piston. The ideal gas inside the cylinder with a pressure $p_1 = 6$ bar and a volume $V_1 = 0.2\,m^3$ expands and then has a pressure of $p_2 = 3$ bar During expansion, the product of pressure and volume can be considered constant.

(a) Calculate the volume work $W_{V,12}$ of the gas.
(b) Determine the effective work $W_{eff,12}$ done by the piston rod when the ambient pressure is $p_{env} = 1$ bar and the movement of the piston can be considered frictionless.

1.11 An ideal gas ($R = 2077\,\frac{J}{kgK}$, $\kappa = 1.67$) is enclosed in a horizontal cylinder which is closed off by a freely moving piston. In the cylinder is a spring which obeys Hooke's law with a spring constant of $k_{Sp} = 1000\,\frac{N}{m}$. The spring is initially in an untensioned state, see Fig. 1.1. The diameter of the cylinder is $d = 0.1$ m. Ambient pressure is $p_{env} = 1$ bar. The initial state of the gas pressure is $p_1 = p_{env}$ and $T_1 = 300$ K. The initial volume is $V_1 = 0.002\,m^3$.

(a) What is the equilibrium pressure p_2 after the temperature of the gas has risen by 200 K due to the supply of heat?
(b) Now the transient heating process is to be investigated. Let the input variable be a heat flux $\dot{Q} = 2000$ W that is supplied to the system from outside. Develop a mathematical model that describes the temporal characteristic of the piston movement and a p, V-diagram. Include in your model dissipation within the

gas and the inertia of the piston mass ($m_P = 1$ kg). The inertia of the spring may be neglected. Let the cylinder wall and the piston be adiabatic. Let the gas and the spring ($m_{Sp} = 1$ kg, $c_{Sp} = 900\,\frac{\text{J}}{\text{kgK}}$) be perfectly thermally coupled and constantly in thermal equilibrium. Kinetic and potential energies of the gas are negligible.

1.2 Solutions

1.1 **(b)**
For a system in thermodynamic equilibrium, the balance of forces for a frictionless piston, cf. Fig. 1.2, results in

$$pA = p_{\text{env}}A + m_P g. \tag{1.1}$$

This leads to the gas pressure within the system,[1] i.e.

$$p = p_{\text{env}} + \frac{m_P g}{A}. \tag{1.2}$$

Since the environment and the piston are identical before and after gas extraction, the pressure in the system does not change either. Furthermore, the temperature is constant, so that the thermal equation of state applies to the specific volume:

$$v = \frac{RT}{p} = \text{const.} \tag{1.3}$$

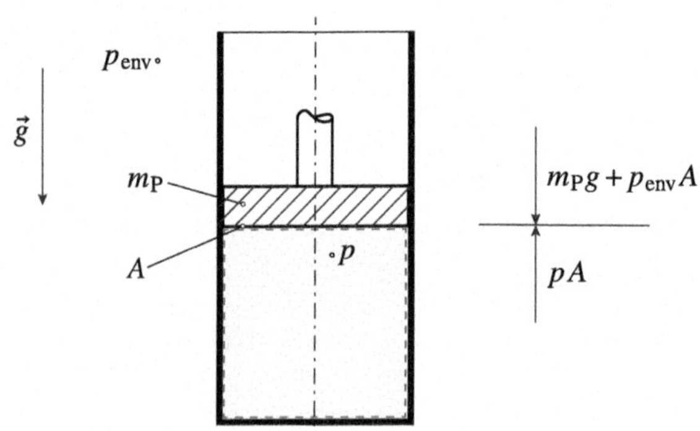

Fig. 1.2 Balance of forces

[1] A homogeneous system is considered, i.e. the field forces due to gravity are neglected. This is an acceptable simplification. However, on closer inspection, the pressure at the bottom of the vessel would be greater than immediately below the piston due to the gas column acting on it.

1.2 **(a)**
The thermal equation of state is

$$pV = mRT = n \underbrace{MR}_{=R_M} T = nR_M T. \tag{1.4}$$

Hence, the volumes follow

$$V_A = \frac{n_A R_M T_A}{p_A}. \tag{1.5}$$

and

$$V_B = \frac{n_B R_M T_B}{p_B} = \frac{n_A R_M T_A}{p_A}. \tag{1.6}$$

Since the molar quantities, pressure and temperature are identical for gases A and B, the volume must also be identical, i.e.

$$\frac{V_A}{V_B} = 1. \tag{1.7}$$

This finding is known as Avogadro's law, which states that the same amount of substance of different gases occupy the same volume at the same pressure and temperature.

1.3 **(a)**
The correlation between the general gas constant and the individual gas constant is as follows

$$R_M = RM = 8.3143 \frac{\text{kJ}}{\text{kmol K}} = \text{const.} \tag{1.8}$$

The individual gas constant can therefore be determined with the use of the molar mass, i.e.

$$R = \frac{R_M}{M} = 188.96 \frac{\text{J}}{\text{kg K}}. \tag{1.9}$$

1.4 **(c)**
The thermal equation of state for ideal gases applies, i.e.

$$pV = mRT. \tag{1.10}$$

For the process described, the mass, the volume and the gas constant remain unchanged. So the result is

$$\frac{p}{T} = \frac{mR}{V} = \text{const.} \tag{1.11}$$

Thus it follows that

$$\frac{p_1}{T_1} = \frac{p_2}{T_2}. \tag{1.12}$$

Finally, the pressure in state (2) is

$$p_2 = p_1 \frac{T_2}{T_1} = 16.67\,\text{MPa}. \tag{1.13}$$

1.5 **(c)**
The thermal equation of state in molar notation obeys

$$pV = nR_\text{M}T. \tag{1.14}$$

For the molar volume it follows

$$V_\text{m} = \frac{V}{n} = \frac{R_\text{M}T}{p} = 9.0072\,\frac{\text{m}^3}{\text{kmol}}. \tag{1.15}$$

1.6 **(b)**
The individual gas constant is

$$R = \frac{R_\text{M}}{M} = 188.9614\,\frac{\text{J}}{\text{kg K}}. \tag{1.16}$$

The mass yields

$$m = \frac{pV}{RT} = 13.2302\,\text{kg}. \tag{1.17}$$

1.7 **(d)**
For an isentropic change of state it applies for an ideal gas that

$$T_2 = T_1 \left(\frac{p_2}{p_1}\right)^{\frac{\kappa-1}{\kappa}} = 265.1431\,\text{K} \tag{1.18}$$

respectively

$$\vartheta_2 = -8.0069\,^\circ\text{C}. \tag{1.19}$$

1.8

(a) Mechanical work is required to increase the potential energy of the mass. It yields

$$W_{12} = E_{\text{pot},2} - E_{\text{pot},1} = mgz_2 - mgz_1 = mg\,(z_2 - z_1) = mg\,\Delta z = 78.48\,\text{J}. \tag{1.20}$$

(b) Now additional work is needed to increase the kinetic energy, i.e.

$$\begin{aligned} W_{12} &= E_{\text{pot},2} - E_{\text{pot},1} + E_{\text{kin},2} - E_{\text{kin},1} \\ &= mg\,(z_2 - z_1) + \frac{1}{2}mc_2^2 - \underbrace{\frac{1}{2}mc_1^2}_{=0} \\ &= mg\,\Delta z + \frac{1}{2}mc_2^2 \\ &= 128.48\,\text{J}. \end{aligned} \tag{1.21}$$

1.9 The thermal equation of state for state (1) obeys

$$p_1 v_1 = RT_1. \tag{1.22}$$

In state (2) it is

$$p_2 v_2 = RT_2 \tag{1.23}$$

respectively

$$\underbrace{1.11\,p_1}_{=p_2}\, v_1 = R\,\underbrace{(T_1 + 20\,\text{K})}_{=T_2}. \tag{1.24}$$

Subtracting Eqs. 1.24 and 1.22, one gets

$$0.11\,p_1 v_1 = R \cdot 20\,\text{K}. \tag{1.25}$$

Finally, with $v_1 = \rho_1^{-1}$ the gas constant yields

$$R = \frac{0.11\,p_1}{\rho_1 \cdot 20\,\text{K}} = 316.0920\,\frac{\text{J}}{\text{kg K}}. \tag{1.26}$$

1.10

(a) The following applies to the volume work $W_{\text{V},12}$

$$W_{\text{V},12} = -\int_1^2 p\,\text{d}V. \tag{1.27}$$

This means that to solve this integral, a correlation between p and V, i.e. $p = f\,(V)$ must be known. Such a correlation is given in the task description, i.e. the product of pressure and volume is constant[2]:

$$p_1 V_1 = p_2 V_2 = pV \tag{1.28}$$

[2] It is therefore an isothermal change of state, since $pV = mRT = \text{const.}$, so that if the mass m is constant, the temperature T of the gas must also be constant.

respectively

$$p = p(V) = p_1 \frac{V_1}{V}. \tag{1.29}$$

The volume in state (2) then follows

$$V_2 = \frac{p_1 V_1}{p_2} = 0.4\,\mathrm{m}^3. \tag{1.30}$$

By substitution of p in Eq. 1.27 with Eq. 1.29 one gets for the volume work

$$\begin{aligned} W_{\mathrm{V},12} &= -\int_1^2 p\,\mathrm{d}V = -\int_1^2 p_1 \frac{V_1}{V}\,\mathrm{d}V \\ &= -p_1 V_1 \int_1^2 \frac{1}{V}\,\mathrm{d}V \\ &= -p_1 V_1 \ln V\Big|_{V_1}^{V_2} \\ &= -p_1 V_1 (\ln V_2 - \ln V_1) \\ &= -p_1 V_1 \ln\frac{V_2}{V_1} = 83.1777\,\mathrm{kJ}. \end{aligned} \tag{1.31}$$

(b) As the gas expands, work is released to the piston. However, this work cannot be fully utilised because the expansion of the piston compresses the environment. This volume work on the environment therefore reduces the effectively usable work, i.e.

$$W_{\mathrm{eff},12} = \underbrace{W_{\mathrm{V},12}}_{<0} + \underbrace{p_{\mathrm{env}} (V_2 - V_1)}_{=W_{\mathrm{V,env}}} \tag{1.32}$$

The volume work due to the expansion of the gas has already been determined, so that finally follows

$$W_{\mathrm{eff},12} = -p_1 V_1 \ln\frac{V_2}{V_1} + p_{\mathrm{env}} (V_2 - V_1) = -63.1777\,\mathrm{kJ}. \tag{1.33}$$

It can be seen that for the expansion of the gas, the actual effective work is less than the volume work of the gas.

1.11

(a) The mass of the trapped ideal gas is

$$m = \frac{p_1 V_1}{R T_1} = 3.2098 \times 10^{-4}\,\mathrm{kg}. \tag{1.34}$$

The balance of forces in state (2) obeys

$$p_2 A = p_{\text{env}} A + k_{\text{Sp}} (x_2 - x_1) . \tag{1.35}$$

Based on geometric considerations and the thermal equation of state, it follows that

$$x_1 = \frac{V_1}{A} = 0.2546\,\text{m} \tag{1.36}$$

and

$$x_2 = \frac{V_2}{A} = \frac{mRT_2}{p_2 A} = \frac{p_1 V_1 T_2}{p_2 T_1 A} \quad \text{with} \quad T_2 = T_1 + \Delta T = 500\,\text{K}. \tag{1.37}$$

Hence, the balance of forces reads as

$$p_2 A = p_{\text{env}} A + k_{\text{Sp}} \left(\frac{p_1 V_1 T_2}{p_2 T_1 A} - x_1 \right) \quad \text{with} \quad A = \frac{\pi}{4} d^2 = 0.0079\,\text{m}^2. \tag{1.38}$$

Reformulating the equation gives

$$p_2^2 + p_2 \left(\frac{k_{\text{Sp}} x_1}{A} - p_{\text{env}} \right) - k_{\text{Sp}} \frac{p_1 V_1 T_2}{T_1 A^2} = 0 \tag{1.39}$$

The solution of the quadratic equation leads[3] to

$$p_2 = -\frac{1}{2} \left(\frac{k_{\text{Sp}} x_1}{A} - p_{\text{env}} \right) + \sqrt{\frac{1}{4} \left(\frac{k_{\text{Sp}} x_1}{A} - p_{\text{env}} \right)^2 + k_{\text{Sp}} \frac{p_1 V_1 T_2}{T_1 A^2}} = 1.1469\,\text{bar}. \tag{1.40}$$

(b) The partial energy equation for the work has to be extended since work is required to displace the spring, i.e. in differential notation it reads, cf. Fig. 1.5

$$\delta W = \delta W_{\text{V}} + \underbrace{\delta W_{\text{mech}}}_{=0} + \delta W_{\text{Sp}} + \delta \Psi. \tag{1.41}$$

The mechanical work refers, according to the system boundary,[4] to the gas. However, the kinetic as well as the potential energy of the gas shall be disregarded, i.e. $\delta W_{\text{mech}} = 0$.

$$\delta W = -p\,\text{d}V + k_{\text{Sp}} z\,\text{d}z + \delta \Psi = -pA\,\text{d}z + k_{\text{Sp}} z\,\text{d}z + \delta \Psi \tag{1.42}$$

[3] Only one solution of the quadratic equation makes physical sense. The second solution leads to a negative pressure.

[4] Therefore, the kinetic energy of the piston is not taken into account here.

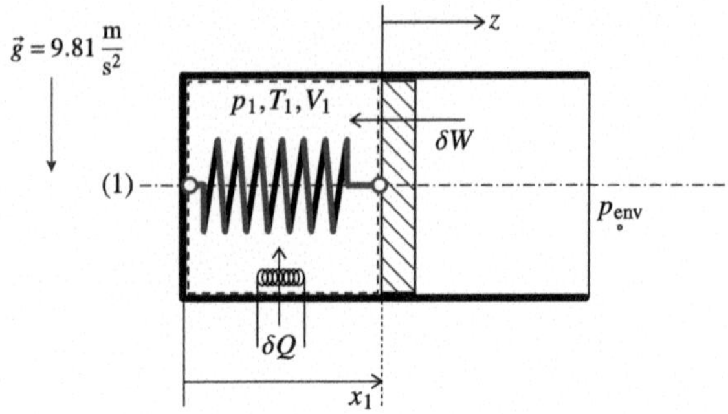

Fig. 1.3 System boundary

If one integrates the differential spring work, the known spring energy results in $0.5k_{Sp}\,\Delta z^2$.

The total work δW exchanged via the system boundary, cf. Fig. 1.3, is used to compress the environment[5] and to accelerate the piston, i.e.

$$\delta W = -p_{env}\,dV - m_P c_P\,dc_P = -p_{env} A\,dz - m_P c_P\,dc_P. \tag{1.43}$$

The coordinate system z is placed in the stress-free state of the spring see Fig. 1.3. A model approach is needed for dissipation. We assume that the dissipation is a function of the velocity of the piston and the distance travelled, i.e.

$$\delta\Psi \propto c\,dz > 0. \tag{1.44}$$

With a proportionality factor one obtains

$$\delta\Psi = \eta c\,dz = \eta\frac{dz}{dt}\,dz = \eta\frac{dz^2}{dt} > 0. \tag{1.45}$$

Finally, the partial energy equation is

$$-p_{env} A\,dz - m_P c_P\,dc_P = -pA\,dz + k_{Sp} z\,dz + \eta\frac{dz^2}{dt}. \tag{1.46}$$

With

$$c_P\,dc_P = \frac{dz}{dt}\,dc_P = dz\,\frac{dc_P}{dt} = dz\,a_P = dz\,\frac{d^2z}{dt^2} \tag{1.47}$$

one gets

$$-p_{env} A\,dz - m_P dz\,\frac{d^2z}{dt^2} = -pA\,dz + k_{Sp} z\,dz + \eta\frac{dz^2}{dt} \tag{1.48}$$

[5] Note that the ambient pressure remains constant though it is compressed by the cylinder.

respectively

$$\boxed{m_\mathrm{P}\,\frac{\mathrm{d}^2z}{\mathrm{d}t^2} = pA - p_\mathrm{env}A - k_\mathrm{Sp}z - \eta\frac{\mathrm{d}z}{\mathrm{d}t}} \tag{1.49}$$

For a very slow, quasi-static change of state, i.e.

$$\frac{\mathrm{d}^2z}{\mathrm{d}t^2} = 0 \quad \text{and} \quad \frac{\mathrm{d}z}{\mathrm{d}t} = 0 \tag{1.50}$$

the equation would simplify to[6]

$$pA = p_\mathrm{env}A + k_\mathrm{Sp}z. \tag{1.51}$$

Equation 1.49 is used in the following.
The first law of thermodynamics according to the system boundary in Fig. 1.3 obeys

$$\delta W + \delta Q = \mathrm{d}U + \mathrm{d}U_\mathrm{Sp} + \mathrm{d}E_\mathrm{Sp} = mc_v\,\mathrm{d}T + m_\mathrm{Sp}c_\mathrm{Sp}\,\mathrm{d}T_\mathrm{Sp} + k_\mathrm{Sp}z\,\mathrm{d}z. \tag{1.52}$$

Due to the thermal equilibrium between spring and gas it is $T = T_\mathrm{Sp}$, i.e.

$$\delta W + \delta Q = \left(mc_v + m_\mathrm{Sp}c_\mathrm{Sp}\right)\,\mathrm{d}T + k_\mathrm{Sp}z\,\mathrm{d}z. \tag{1.53}$$

With Eq. 1.42 it is

$$\underbrace{-pA\,\mathrm{d}z + k_\mathrm{Sp}z\,\mathrm{d}z + \delta\Psi}_{=\delta W} + \delta Q = \left(mc_v + m_\mathrm{Sp}c_\mathrm{Sp}\right)\,\mathrm{d}T + k_\mathrm{Sp}z\,\mathrm{d}z. \tag{1.54}$$

Rearranging leads to

$$-\,pA\,\mathrm{d}z + \eta\frac{\mathrm{d}z^2}{\mathrm{d}t} + \delta Q = \left(mc_v + m_\mathrm{Sp}c_\mathrm{Sp}\right)\,\mathrm{d}T. \tag{1.55}$$

Dividing by $\mathrm{d}t$ results in

$$-\,pA\,\frac{\mathrm{d}z}{\mathrm{d}t} + \eta\left(\frac{\mathrm{d}z}{\mathrm{d}t}\right)^2 + \underbrace{\frac{\delta Q}{\mathrm{d}t}}_{=\dot{Q}(t)} = \left(mc_v + m_\mathrm{Sp}c_\mathrm{Sp}\right)\,\frac{\mathrm{d}T}{\mathrm{d}t}\,. \tag{1.56}$$

Finally it is

[6] Compare with Eq. 1.35. This equation can be applied for the equilibrium states (1) and (2).

$$\boxed{\frac{dT}{dt}=\frac{1}{mc_v+m_{Sp}c_{Sp}}\left[-pA\frac{dz}{dt}+\eta\left(\frac{dz}{dt}\right)^2+\dot{Q}(t)\right]} \tag{1.57}$$

The supplied heat flux $\dot{Q}(t)$ is the cause of the change of state.

The correlation between pressure and temperature of the gas is given by the thermal equation of state, i.e.

$$pA(x_1+z)=pV=mRT \tag{1.58}$$

so that

$$\boxed{p=\frac{mRT}{V}=\frac{mRT}{A(x_1+z)}} \tag{1.59}$$

The following system of equations consisting of Eqs. 1.49, 1.57 and 1.59 can now be solved numerically:

$$\boxed{\frac{d^2z}{dt^2}=\frac{mRT}{m_P(x_1+z)}-p_{env}\frac{A}{m_P}-\frac{k_{Sp}}{m_P}z-\frac{\eta}{m_P}\frac{dz}{dt}} \tag{1.60}$$

$$\boxed{\frac{dT}{dt}=\frac{1}{mc_v+m_{Sp}c_{Sp}}\left[-\frac{mRT}{(x_1+z)}\frac{dz}{dt}+\eta\left(\frac{dz}{dt}\right)^2+\dot{Q}(t)\right]} \tag{1.61}$$

The results of the numerical calculation, i.e. the solution of Eqs. 1.60 and 1.61, are shown in the following Figs. 1.4, 1.5 and 1.6. Here, the process was started from the equilibrium state (1) and a temporally constant heat flux of $\dot{Q}=2000\,\mathrm{W}$ was supplied. Figure 1.4 shows the temperature rise of the system that is approximately linear and mainly dominated by the caloric effects of the gas and the spring inside the system. After about 90 seconds, the temperature has risen by 200 K and continues to rise because of the continued supply of heat. Figure 1.5 shows the velocity of the piston. A distinction is made between 2 cases: In case 1, a spring constant of $k_{Sp}=1000\,\frac{\mathrm{N}}{\mathrm{m}}$ is assumed, in the second case the spring constant is $k_{Sp}=0\,\frac{\mathrm{N}}{\mathrm{m}}$, i.e. the spring is thermally present but not connected to the piston. In part (b) of the Figure, which shows the first second after start-up, one can see the rapid increase in piston velocity due to the supplied heat. However, the velocity level is very low. If one looks at the range up to 100 seconds in part (a) of the figure, one can see that the piston velocity slows down due to the increasing spring displacement. Such a restoring force is not present for the case $k_{Sp}=0\,\frac{\mathrm{N}}{\mathrm{m}}$. Finally, the p, V-diagram is shown, see Fig. 1.6. Part (a) shows the full range, part (b) the start of the change of state. As before, cases 1 with $k_{Sp}=1000\,\frac{\mathrm{N}}{\mathrm{m}}$ and 2 with $k_{Sp}=0\,\frac{\mathrm{N}}{\mathrm{m}}$ are shown. It can be seen for case 2 that the change of state is approximately isobaric, but the pressure is slightly higher

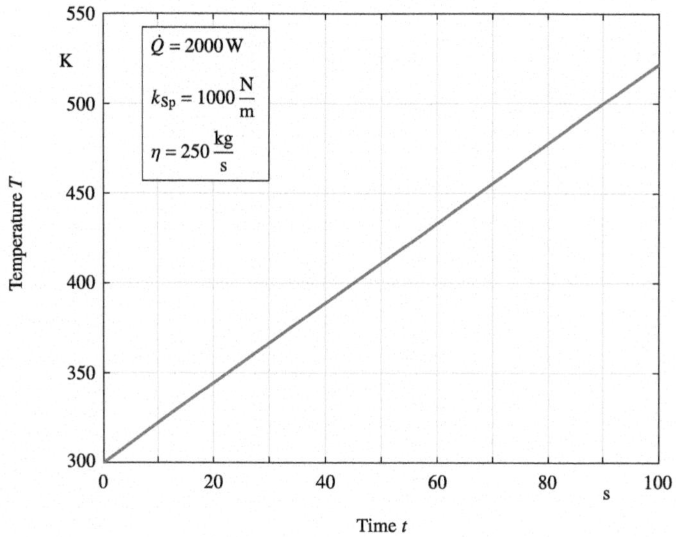

Fig. 1.4 Temperature rise

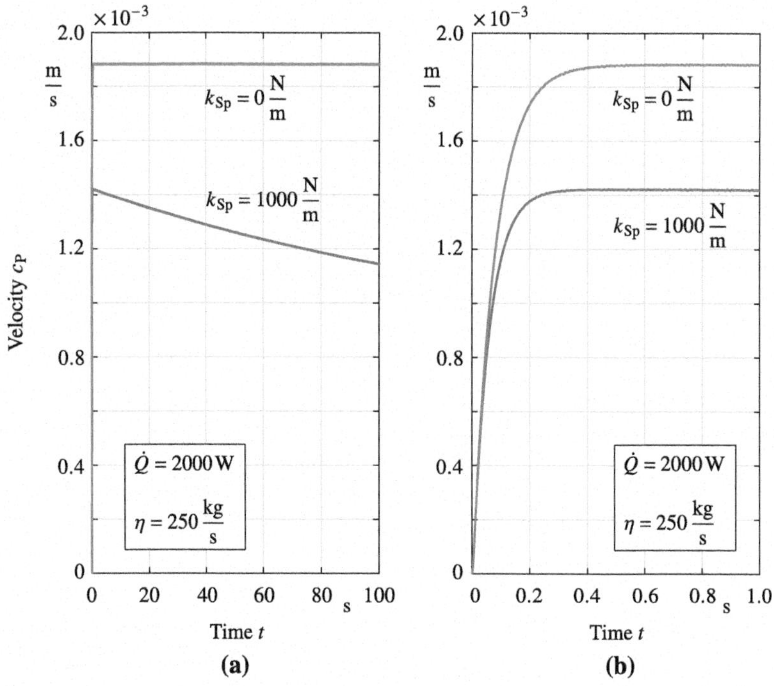

Fig. 1.5 Piston velocity: **a** Full range, **b** First second

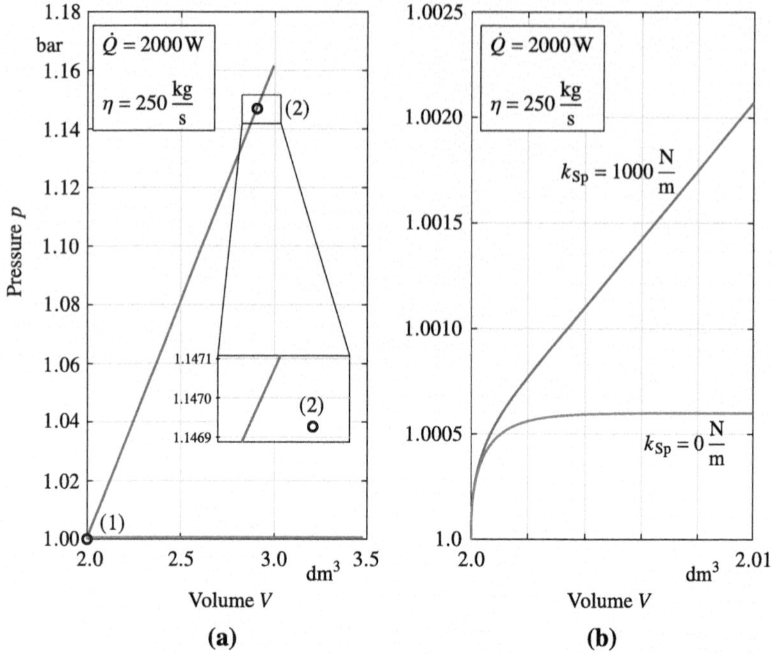

Fig. 1.6 p, V-diagram: **a** Full range, **b** Starting sequence

than the ambient pressure of $p_{\text{env}} = 1$ bar due to friction and low acceleration of the piston. In addition, the equilibrium state (2) from task part (a) is depicted in Fig. 1.6a. Here, too, it can be seen that in the dynamic case the pressure is slightly above (2) because the system is not in thermodynamic equilibrium due to the continued supply of heat.

Chapter 2
Thermodynamic Systems

The concept of a thermodynamic system is clarified in this chapter. A classification into closed systems, which do not allow mass exchange with the environment, and open systems, whose boundary is permeable to mass, is made. Basically, the tasks presented show which balances are required to quantify changes of state. In general, force balances and energy balances are carried out. It is shown that for the description of energetic transformation, the state variable entropy also needs to be balanced. Unlike energy, entropy is not a conservation quantity. It can be generated by imperfections, balancing processes and frictional processes, and therefore can have a source term. The concept of a stationary open system is introduced as well as transient systems whose state changes over time.

2.1 Problems

2.1 What can cause a transient balancing process within a fully closed system?

(a) A pressure resp. temperature imbalance.
(b) Supplied heat.
(c) A transient balancing process can not run under the given premises.
(d) None of the above statements is true.

2.2 A heat flux of 12 kW passes a solid wall stationary. The rate of entropy generation is 3.5987 $\frac{\text{W}}{\text{K}}$. The surface temperature on one side is 22 °C. What is the surface temperature of the wall on the other side?

(a) 5 °C
(b) 0 °C
(c) −2 °C
(d) −6 °C

A. Schmidt, *Technical Thermodynamics Workbook for Engineers*,
https://doi.org/10.1007/978-3-031-50172-2_2

Fig. 2.1 Cylinder/piston

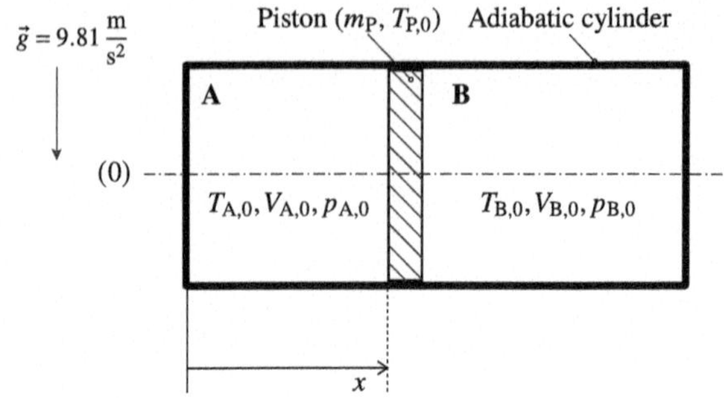

2.3 What is true of a fully closed system?

(a) The exergy in the system can decrease.
(b) The mass in the system must increase.
(c) Heat is exchanged with the environment.
(d) None of the above statements applies.

2.4 Which statement is true for a fully closed system?

(a) Due to balancing processes, exergy losses may occur internally.
(b) All state values strive towards a minimum.
(c) Such a system must be cooled.
(d) None of the above statements applies.

2.5 A vertical cylinder (diameter $d = 10\,\text{mm}$) is filled with an ideal gas. The mass of the freely movable piston is 5 kg. The constant of gravity is $g = 9.81\,\frac{\text{m}}{\text{s}^2}$, ambient pressure is 1 bar. What is the pressure of the gas in equilibrium state?

2.6 A freely moving piston divides a horizontal cylinder with the cross-sectional area $A = 0.5\,\text{m}^2$ into two chambers, each of which is filled with an ideal gas ($R = 2077\,\frac{\text{J}}{\text{kgK}}$, $\kappa = 1.67$), cf. Fig. 2.1. Ambient temperature is $T_{\text{env}} = 300\,\text{K}$. The piston ($m_\text{P} = 50\,\text{kg}$, $c_\text{P} = 900\,\frac{\text{J}}{\text{kgK}}$) between the two chambers is heat-permeable. The cylinder is adiabatic to the outside. The initial boundary conditions in state (0) are:

- Volume A: $T_{\text{A},0} = 370\,\text{K}$, $V_{\text{A},0} = 1\,\text{m}^3$, $p_{\text{A},0} = 1.2\,\text{bar}$
- Volume B: $T_{\text{B},0} = 300\,\text{K}$, $V_{\text{B},0} = 1.2\,\text{m}^3$, $p_{\text{B},0} = 1.2\,\text{bar}$
- Piston: $T_{\text{P},0} = 800\,\text{K}$

What is the position x_∞ of the piston after a long period of time?

2.7 A freely moving piston divides a horizontal cylinder with the cross-sectional area $A = 0.5\,\text{m}^2$ into two chambers, each of which is filled with an ideal gas ($R = 2077\,\frac{\text{J}}{\text{kgK}}$, $\kappa = 1.67$). In addition, chamber A contains a spring ($m_\text{Sp} = 40\,\text{kg}$, $c_\text{Sp} = 900\,\frac{\text{J}}{\text{kgK}}$) that obeys Hooke's law with a spring constant of $k_\text{Sp} = 1000\,\frac{\text{N}}{\text{m}}$, cf. Fig. 2.2. Ambient temperature is $T_{\text{env}} = 300\,\text{K}$. The piston ($m_\text{P} = 50\,\text{kg}$, $c_\text{P} = 900\,\frac{\text{J}}{\text{kgK}}$) between the two chambers is heat-permeable. The cylinder is adiabatic to the outside. The initial boundary conditions in state (0) are:

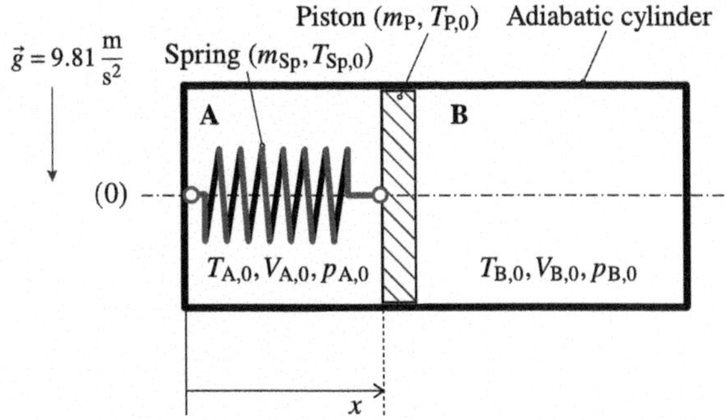

Fig. 2.2 Cylinder/piston/spring

- Volume A: $T_{A,0} = 300\,\text{K}$, $V_{A,0} = 1.6\,\text{m}^3$, $p_{A,0} = 1.2\,\text{bar}$
- Volume B: $T_{B,0} = 600\,\text{K}$, $V_{B,0} = 1.2\,\text{m}^3$, $p_{B,0} = 1.2\,\text{bar}$
- Piston: $T_{P,0} = 800\,\text{K}$
- Spring: $T_{Sp,0} = T_{A,0}$, initially in relaxed position

What is the position x_∞ of the piston after a long period of time?

2.8 An ideal gas is slowly compressed in a horizontal cylinder with a freely moving, perfectly sealed piston. The product of pressure and volume remains constant. What is true for such a change of state?

(a) The system is cooled.
(b) The change of state is adiabatic.
(c) It is an isochoric/isobaric change of state.
(d) None of the above statements applies.

2.2 Solutions

2.1 **(a)**
Pressure and temperature are both intensive state values. This means that they are drivers for changes of state, cf. Fig. 2.3:

- A temperature imbalance triggers a heat flux that persists until the system has settled into thermal equilibrium, i.e. a homogeneous temperature. Because of the then missing temperature gradient, the system cannot move back to its initial state.
- The same applies to a pressure imbalance within a system. Such a pressure imbalance is balanced until a force equilibrium is reached.

These balancing processes are always irreversible, i.e. they run in a direction that is determined by the second law of thermodynamics. Entropy is generated within the system[1]:

[1] Mass of the piston is disregarded.

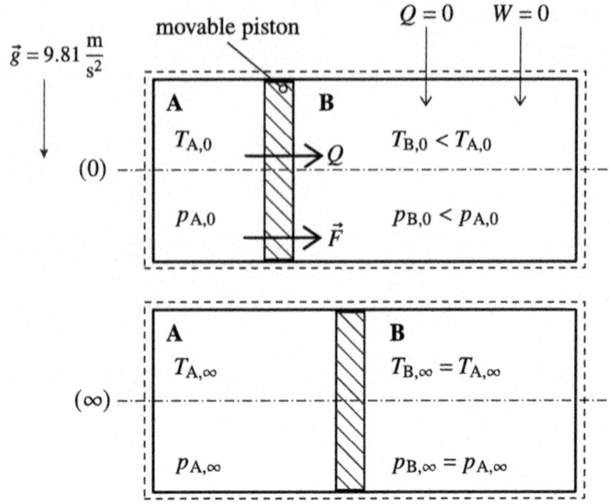

Fig. 2.3 Balancing processes in a fully closed system

$$S_i + \underbrace{S_a}_{=0} = S_\infty - S_0 = (S_\infty - S_0)_A + (S_\infty - S_0)_B \tag{2.1}$$

respectively

$$S_i = m_A \left(c_{p,A} \ln \frac{T_{A,\infty}}{T_{A,0}} - R_A \ln \frac{p_{A,\infty}}{p_{A,0}} \right) + m_B \left(c_{p,B} \ln \frac{T_{B,\infty}}{T_{B,0}} - R_B \ln \frac{p_{B,\infty}}{p_{B,0}} \right) \tag{2.2}$$

with $T_{A,\infty} = T_{B,\infty}$ and $p_{A,\infty} = p_{B,\infty}$. Once equilibrium has been reached, the system lacks the potential to move back on its own. Therefore, answer (a) is correct. Since the system is fully closed no heat is supplied, i.e. answer (b) is incorrect.

2.2 **(c)**
The wall represents a closed system, as no mass enters the wall and no mass exits. In the case described here, steady state is to be examined. The state variables, e.g. the temperature inside the wall, can vary locally, but they do not change with time, see Fig. 2.4a. This is referred to as local thermodynamic equilibrium, which is caused

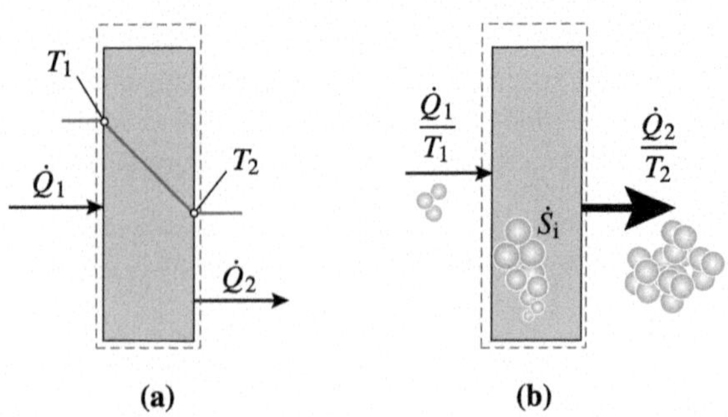

Fig. 2.4 **a** Energy balance resp. temperature profile, **b** Entropy balance

by the externally imposed heat flow. The state variables are spatially distributed; the system can be split into differential volume elements that are locally in thermodynamic equilibrium. For systems in steady state, the incoming energy flux must be balanced by the outgoing energy flux. Otherwise, the state of the internal energy and thus also the temperature would change with time. This would contradict the steady state. Therefore, the first law of thermodynamics obeys

$$\dot{Q}_1 = \dot{Q}_2 \stackrel{!}{=} \dot{Q}. \tag{2.3}$$

The entropy as a state value within the system must also be constant in time in steady state case. This means that the entering entropy flow must be exactly as large as the released entropy flow. The entropy flux entering includes the entropy flux supplied with the heat flux $\dot{Q}_1$ and the entropy produced within the system. The outgoing entropy flow correlates with the heat released from the wall $\dot{Q}_2$ at T_{env}, see Fig. 2.4b, i.e.

$$\frac{\dot{Q}}{T_1} + \dot{S}_{\mathrm{i}} = \frac{\dot{Q}}{T_2}. \tag{2.4}$$

For the temperature[2] T one therefore obtains

$$T_2 = \frac{\dot{Q}}{\frac{\dot{Q}}{T_1} + \dot{S}_{\mathrm{i}}} = 271.1497\,\mathrm{K}. \tag{2.5}$$

This means that the temperature is $\vartheta_2 = -2\,^\circ\mathrm{C}$, so answer (c) is correct.

2.3 **(a)**
A fully closed system means that neither mass nor energy can be exchanged across the system boundary. Therefore, answer option (b) is ruled out, since the mass in the system can only change if mass flows in or out across the system boundary. Answer option (c) is also ruled out, since a fully closed system is always adiabatic.

However, the system inside can be in thermodynamic disequilibrium, e.g. due to internal pressure/temperature or even concentration distribution. Such imbalances lead to balancing processes until, in the case of pressure differences, mechanical equilibrium is finally achieved. The same applies to temperature inhomogeneities, which balance out until thermal equilibrium is reached. In the case of concentration differences, i.e. a chemical potential is present, balancing takes place until a chemical equilibrium is reached. Such balancing processes are always irreversible, i.e. they continue until the mechanical/thermal/chemical potential disappears and can then no longer move back. Irreversible processes are characterised by entropy generation and thus also by exergetic loss. Therefore, answer (a) is correct.

[2] Note that the absolute temperature must be used for the entropy calculation.

2.4 **(a)**
The essential explanation has already been given in Problem 2.3. Due to irreversibility and the associated entropy generation, there is a loss of exergy. Therefore, answer (a) is correct.

Answer option (b) is incorrect because, for example, entropy generation in a possible balancing process causes an increase of the state variable entropy within the system. The entropy will therefore strive towards a maximum,[3] since in a fully closed system no entropy can be released to the outside via the system boundary.

2.5 To solve this task, an equilibrium of forces is applied to the piston, see Fig. 1.2, i.e.

$$pA = p_{\text{env}}A + m_{\text{P}}g \tag{2.6}$$

with

$$A = \frac{\pi}{4}d^2. \tag{2.7}$$

Solving for the pressure results in

$$p = p_{\text{env}} + \frac{m_{\text{P}}g}{A} = 7.245\,\text{bar}. \tag{2.8}$$

2.6 First we calculate the masses of the ideal gas. In chamber A there is

$$m_{\text{A}} = \frac{p_{\text{A},0}V_{\text{A},0}}{RT_{\text{A},0}} = 0.1562\,\text{kg} \tag{2.9}$$

and in chamber B

$$m_{\text{B}} = \frac{p_{\text{B},0}V_{\text{B},0}}{RT_{\text{B},0}} = 0.2311\,\text{kg}. \tag{2.10}$$

The specific heat capacities for the gas are

$$c_v = \frac{R}{\kappa - 1} = 3100\,\frac{\text{J}}{\text{kgK}} \tag{2.11}$$

and

$$c_p = R + c_v = 5177\,\frac{\text{J}}{\text{kgK}}. \tag{2.12}$$

To determine the temperature

$$T_\infty = T_{\text{A},\infty} = T_{\text{B},\infty} = T_{\text{P},\infty} \tag{2.13}$$

after a very long time, i.e. at equilibrium, the first law of thermodynamics is applied, cf. Fig. 2.5:

[3] Entropy rises in such a balancing process until equilibrium has been reached.

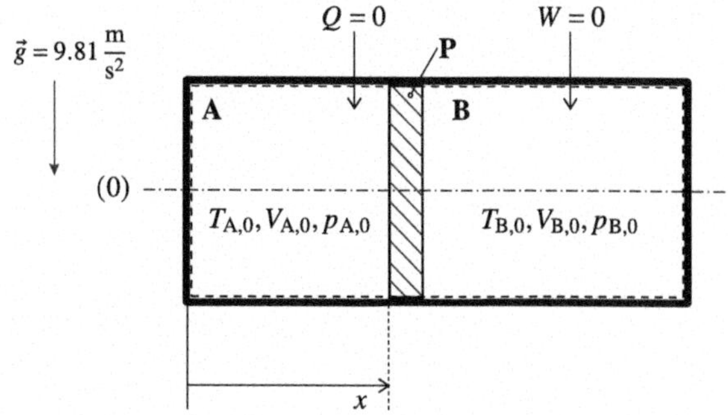

Fig. 2.5 System boundary

$$Q + W = 0 = \Delta U + \underbrace{\Delta E_{pot}}_{=0} = \Delta U_A + \Delta U_B + \Delta U_P. \tag{2.14}$$

Applying the caloric equation of state, it follows

$$0 = m_A c_v \left(T_\infty - T_{A,0}\right) + m_B c_v \left(T_\infty - T_{B,0}\right) + m_P c_P \left(T_\infty - T_{P,0}\right). \tag{2.15}$$

Solving for the temperature in thermodynamic equilibrium yields

$$T_\infty = \frac{m_A c_v T_{A,0} + m_B c_v T_{B,0} + m_P c_P T_{P,0}}{m_A c_v + m_B c_v + m_P c_P} = 787.7413\,\text{K}. \tag{2.16}$$

In thermodynamic equilibrium, a balance of forces must also rule at the piston, i.e.

$$p_{A,\infty} A = p_{B,\infty} A \tag{2.17}$$

respectively

$$p_{A,\infty} = p_{B,\infty}. \tag{2.18}$$

Applying the thermal equation of state for chamber A and B in thermodynamic equilibrium results in

$$\frac{m_A R T_\infty}{V_{A,\infty}} = \frac{m_B R T_\infty}{V_{B,\infty}} \tag{2.19}$$

with

$$V_{B,\infty} = V_0 - V_{A,\infty} = \underbrace{V_{A,0} + V_{B,0}}_{=V_0} - V_{A,\infty} \tag{2.20}$$

so that

$$\frac{m_A}{V_{A,\infty}} = \frac{m_B}{V_{A,0} + V_{B,0} - V_{A,\infty}}. \tag{2.21}$$

Solving for $V_{\mathrm{A},\infty}$ leads to

$$V_{\mathrm{A},\infty} = \frac{V_0}{\frac{m_\mathrm{B} R_\mathrm{B}}{m_\mathrm{A} R_\mathrm{A}} + 1} = 0.8871\,\mathrm{m}^3. \tag{2.22}$$

Hence, the pressure in equilibrium is

$$p_{\mathrm{B},\infty} = p_{\mathrm{A},\infty} = \frac{m_\mathrm{A} R T_\infty}{V_{\mathrm{A},\infty}} = 2.88\,\mathrm{bar}. \tag{2.23}$$

The position x_∞ follows accordingly

$$x_\infty = \frac{V_{\mathrm{A},\infty}}{A} = 1.7742\,\mathrm{m}. \tag{2.24}$$

Obviously, the determination of the equilibrium temperature was not necessary for the calculation of the rest position of the piston.

2.7 First we calculate the masses of the ideal gas with the thermal equation of state. In chamber A there is

$$m_\mathrm{A} = \frac{p_{\mathrm{A},0} V_{\mathrm{A},0}}{R T_{\mathrm{A},0}} = 0.3081\,\mathrm{kg} \tag{2.25}$$

and in chamber B

$$m_\mathrm{B} = \frac{p_{\mathrm{B},0} V_{\mathrm{B},0}}{R T_{\mathrm{B},0}} = 0.1156\,\mathrm{kg}. \tag{2.26}$$

The specific heat capacities for the gas are

$$c_v = \frac{R}{\kappa - 1} = 3100\,\frac{\mathrm{J}}{\mathrm{kgK}} \tag{2.27}$$

and

$$c_p = R + c_v = 5177\,\frac{\mathrm{J}}{\mathrm{kgK}}. \tag{2.28}$$

The initial position of the relaxed spring is

$$x_{\mathrm{Sp},0} = \frac{V_{\mathrm{A},0}}{A} = 3.2\,\mathrm{m}. \tag{2.29}$$

To determine the temperature

$$T_\infty = T_{\mathrm{A},\infty} = T_{\mathrm{B},\infty} = T_{\mathrm{P},\infty} = T_{\mathrm{Sp},\infty} \tag{2.30}$$

after a very long time, i.e. at equilibrium, the first law of thermodynamics is applied, cf. Fig. 2.6:

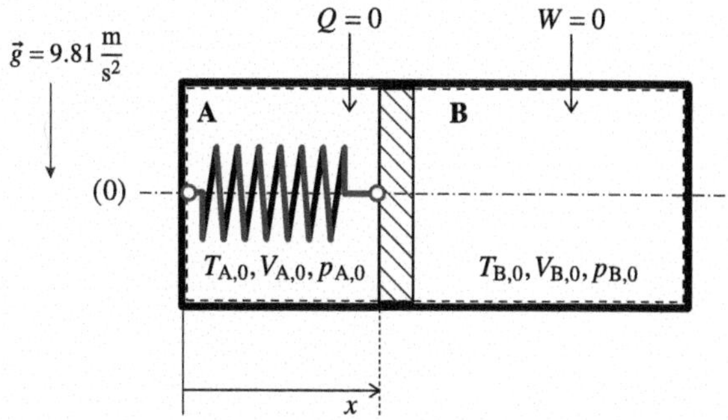

Fig. 2.6 System boundary

$$Q + W = 0 = \Delta U + \underbrace{\Delta E_{pot}}_{=0} + \Delta E_{Sp} = \Delta U_A + \Delta U_B + \Delta U_P + \Delta U_{Sp} + \Delta E_{Sp} \tag{2.31}$$

with the stored energy in the spring

$$\Delta E_{Sp} = \frac{1}{2} k_{Sp} \left(x_{Sp,0} - x_\infty\right)^2 . \tag{2.32}$$

Applying the caloric equation of state, it follows

$$0 = \frac{1}{2} k_{Sp} \left(x_{Sp,0} - x_\infty\right)^2 + m_A c_v \left(T_\infty - T_{A,0}\right) + m_B c_v \left(T_\infty - T_{B,0}\right) + \\ + m_P c_P \left(T_\infty - T_{P,0}\right) + m_{Sp} c_{Sp} \left(T_\infty - T_{Sp,0}\right) . \tag{2.33}$$

In thermodynamic equilibrium, a balance of forces must also rule at the piston, i.e.

$$p_{A,\infty} A + F_{Sp} = p_{B,\infty} A \tag{2.34}$$

respectively[4]

$$p_{A,\infty} A + k_{Sp} \underbrace{\left(x_{Sp,0} - x_\infty\right)}_{=\Delta x_\infty} = p_{B,\infty} A . \tag{2.35}$$

Applying the thermal equation of state for chamber A and B in thermodynamic equilibrium results in

$$\frac{m_A R T_\infty}{V_{A,\infty}} A + k_{Sp} \left(x_{Sp,0} - x_\infty\right) = \frac{m_B R T_\infty}{V_{B,\infty}} A \tag{2.36}$$

[4] If the spring is displaced from its rest position, a force vector results in the direction of the original rest position.

with

$$V_{B,\infty} = V_{total} - V_{A,\infty} = \underbrace{V_{A,0} + V_{B,0}}_{=V_{total}} - \underbrace{V_{A,\infty}}_{=Ax_\infty}. \quad (2.37)$$

This leads to

$$\frac{m_A R T_\infty}{V_{A,\infty}} A + k_{Sp}\left(x_{Sp,0} - x_\infty\right) = \frac{m_B R T_\infty}{V_{total} - V_{A,\infty}} A \quad \text{with} \quad V_{total} = 2.2\,\text{m}^3 \quad (2.38)$$

respectively

$$\frac{m_A R T_\infty}{x_\infty} + k_{Sp}\left(x_{Sp,0} - x_\infty\right) = \frac{m_B R T_\infty}{x_{total} - x_\infty} \quad \text{with} \quad x_{total} = \frac{V_{total}}{A} = 5.6\,\text{m}. \quad (2.39)$$

Solving Eqs. 2.33 and 2.39 numerically leads to

$$x_\infty = 4.0621\,\text{m} \quad (2.40)$$

and

$$T_\infty = 574.6464\,\text{K}. \quad (2.41)$$

Hence, the gas pressures in equilibrium are

$$p_{A,\infty} = \frac{m_A R T_\infty}{V_{A,\infty}} = 1.8108\,\text{bar} \quad \text{with} \quad V_{A,\infty} = 2.0310\,\text{m}^3 \quad (2.42)$$

and

$$p_{B,\infty} = \frac{m_B R T_\infty}{V_{B,\infty}} = 1.7935\,\text{bar} \quad \text{with} \quad V_{B,\infty} = 0.7690\,\text{m}^3. \quad (2.43)$$

2.8 **(a)**
An ideal gas is trapped in a closed system, i.e. the mass of the gas is constant and the thermal equation of state obeys

$$pV = mRT = \text{const.} \quad (2.44)$$

Since the product of pressure and volume is constant, the temperature must also be constant, i.e.

$$T = \text{const.} \quad (2.45)$$

The first law of thermodynamics for the closed system is as follows

$$Q_{12} + W_{12} = U_2 - U_1 = mc_v\left(T_2 - T_1\right) = 0. \quad (2.46)$$

respectivley

$$Q_{12} = -W_{12}. \tag{2.47}$$

As the gas is compressed, work is supplied, i.e. $W_{12} > 0$, so the amount of work supplied must be released as heat, so that

$$Q_{12} < 0. \tag{2.48}$$

Therefore, answer (a) is correct.

Chapter 3
Thermodynamic Balancing

After the classification of thermodynamic systems in the previous chapter, the necessary thermodynamic balances are now examined in more detail. Particular attention is paid to the strategic methodology for conducting balances. It is shown that the definition of a system boundary is the first essential step in calculating the effect of process values, i.e. heat and work, on the change of state of a system. The reader is shown how to conduct mass balances in case of open systems. For the energetic evaluation, the first law of thermodynamics of open and closed, stationary and transient systems is explained and practised with suitable tasks. The concept of dissipation is introduced and its influence on reversibility is discussed with the help of entropy balances.

3.1 Problems

3.1 Is it possible to decrease the entropy of a closed system?

(a) No, entropy can only increase.
(b) Yes, by cooling the system.
(c) Yes, by supply of technical work.
(d) None of the above statements is true.

3.2 An ideal gas is filled in a vertical cylinder. Mass of the freely movable piston as well as ambient pressure are constant. Now, by a valve 20 % of the mass of the gas are released to the environment. This takes place so slowly, that the temperature remains constant all time. Which statement is true when a new equilibrium is reached?

(a) The entropy inside the cylinder decreases by 20 %.
(b) The entropy inside the cylinder increases by 20 %.
(c) Such a change of state is impossible.
(d) None of the above statements is true.

A. Schmidt, *Technical Thermodynamics Workbook for Engineers*,
https://doi.org/10.1007/978-3-031-50172-2_3

3.3 A closed system with constant volume is considered. The change of state of the ideal gas shall be isothermal, the system is cooled. External energies are to be neglected. Which statement applies?

(a) Energy is dissipated inside the system.
(b) The change of state is reversible.
(c) It is a perpetual mobile of the 2nd kind.
(d) None of the above statements applies.

3.4 Which statement is true for an adiabatic turbine with kinetic/potential energies being ignored?

(a) The pressure rises.
(b) In case the change of state is reversible temperature will remain constant.
(c) Entropy needs to be released by heat.
(d) None of the above statements applies.

3.5 An incompressible liquid flows in a work-insulated, adiabatic horizontal pipe section. Which statement is true?

(a) The pressure can never rise.
(b) The pressure can increase if the flow is sufficiently decelerated, e.g. by cross-section expansion.
(c) Such a flow is not possible because it violates the first law of thermodynamics.
(d) None of the above statements applies.

3.6 A closed system with constant volume is considered. The change of state of the ideal gas inside the system shall be isothermal ($T = 1.2 \cdot T_{\text{env}}$). The system is cooled and shaft work is supplied. Outer energies are to be neglected. Which statement applies for the gas?

(a) Entropy inside the system rises.
(b) The change of state is reversible.
(c) The exergy loss equals 83.33 % of the shaft work.
(d) None of the above statements applies.

3.7 A closed system at rest with constant volume contains an ideal gas. Work of 20 kJ is supplied by a stirrer. In addition, heat of 60 kJ is released. Which statement is correct?

(a) Temperature will remain constant.
(b) The change of state is reversible.
(c) Exergy gets lost.
(d) None of the above statements applies.

3.8 Water (incompressible, $c_{\text{w}} = 4.19\,\frac{\text{kJ}}{\text{kgK}}$, $\rho = 1000\,\frac{\text{kg}}{\text{m}^3}$) is heated by 20 K in an open system. This increases the specific enthalpy by 83.7 $\frac{\text{kJ}}{\text{kg}}$. Which statement is correct?

(a) A pressure loss occurs within the system.
(b) Entropy is destroyed.
(c) Since the fluid is incompressible, the specific entropy needs to remain constant.
(d) None of the above statements applies.

3.9 An adiabatic tank ($V = 1\,\text{m}^3$) is filled with 1.4 kg of an ideal gas ($c_p = 1.004\,\frac{\text{kJ}}{\text{kgK}}$, $\kappa = 1.4$). Initial temperature is $\vartheta_1 = 20\,°\text{C}$. Now, by a stirrer 0.5 kW are supplied over 30 s. In state (2) a new equilibrium is reached. What is the thermodynamic mean temperature?

3.10 An ideal gas underlies an adiabatic, reversible compression in a horizontal cylinder. The following is known of the equilibrium states (1) and (2):

- State (1): $T_1 = 300\,\text{K}$, $p_1 = 1.013\,25\,\text{bar}$
- State (2): $p_2 = 5\,\text{bar}$

By what amount does the specific internal energy change in case $c_p = 0.519\,\frac{\text{kJ}}{\text{kgK}}$ and $c_v = 0.309\,\frac{\text{kJ}}{\text{kgK}}$?

3.11 An adiabatic tank ($V = 1\,\text{m}^3$=const.) is filled with 1.4 kg of an ideal gas ($c_p = 1.004\,\frac{\text{kJ}}{\text{kgK}}$, $\kappa = 1.4$). Initial temperature in equilibrium is $\vartheta_1 = 20\,°\text{C}$. Now, by an immersed electrical heating element 0.5 kW are supplied over 30 s. In state (2) a new equilibrium is reached. What is the temperature in state (2)?

3.12 In an adiabatic, horizontal cylinder (diameter $d = 12\,\text{cm}$) with a frictionless, freely movable, gas-tight piston, there is air (ideal, $c_p = 1.004\,\frac{\text{kJ}}{\text{kgK}}$, $\kappa = 1.4$) in chambers A and B, see Fig. 3.1. Kinetic energy of the gas shall be neglected. The mass of the piston is $m_\text{K} = 20\,\text{kg}$. At the beginning, the piston is fixed and the following parameters are given:

- Chamber A (adiabatic): $V_{\text{A},0} = 2\,\text{dm}^3$, $p_{\text{A},0} = 8\,\text{bar}$, $\vartheta_{\text{A},0} = \vartheta_0 = 20\,°\text{C}$
- Chamber B (adiabatic): $V_{\text{B},0} = 10\,\text{dm}^3$, $p_{\text{B},0} = 1\,\text{bar}$, $\vartheta_{\text{B},0} = \vartheta_0 = 20\,°\text{C}$

The fixation is now released. At which position x does the piston reach maximum velocity if the change of state is fully frictionless?

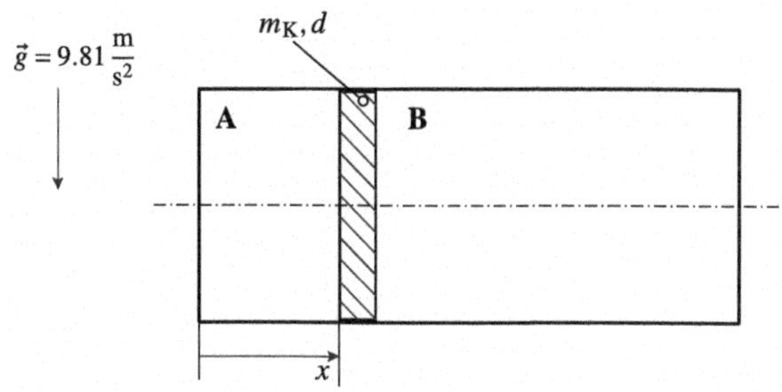

Fig. 3.1 Piston

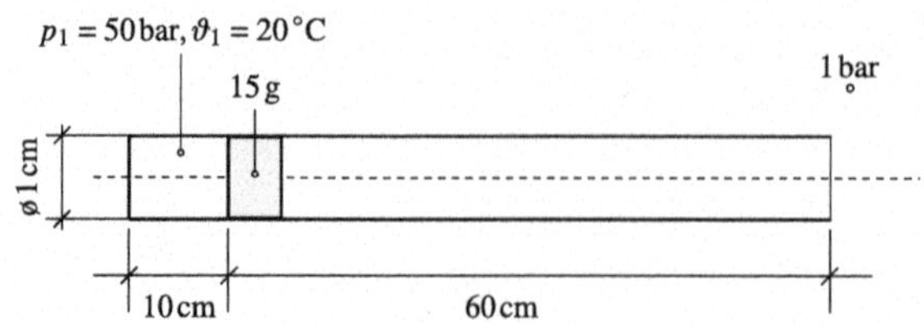

Fig. 3.2 Acceleration

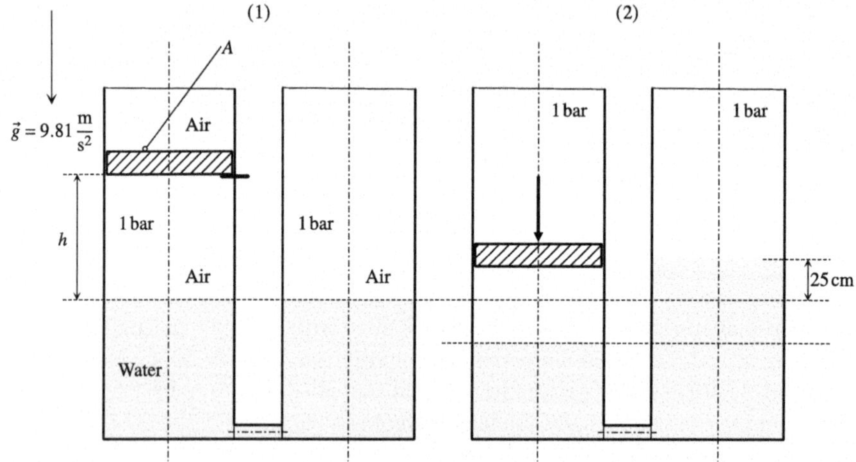

Fig. 3.3 Water columns

3.13 In an experiment, a test object with a mass of 15 g is to be accelerated by adiabatic, reversible expansion of compressed air (ideal gas, $\kappa = 1.4$, $R = 287\,\frac{J}{kgK}$). The mass is initially fixed, see Fig. 3.2. Now, the fixation is released. What velocity is achieved when the test object has just fully left the horizontal tube? The kinetic energy of the gas can be ignored. The piston shall have ambient temperature and moves frictionless.

3.14 Two cylindrical vessels of the same cross-section ($A = 200\,cm^2$) open at the top are in connection with each other and are partly filled with water ($\rho = 1000\,\frac{kg}{m^3}$), see Fig. 3.3. In the left vessel, $h = 60$ cm above the water surface, there is a tightly sealed piston which is to be very slowly pressed in so far that the water on the right rises by 25 cm. How much heat has to be exchanged with the environment so that the process is isothermal? Air is to be treated as an ideal gas; water vapour is to be neglected. Changes of the potential energy of the air can be ignored. Ambient pressure shall be 1 bar.

Fig. 3.4 Piston

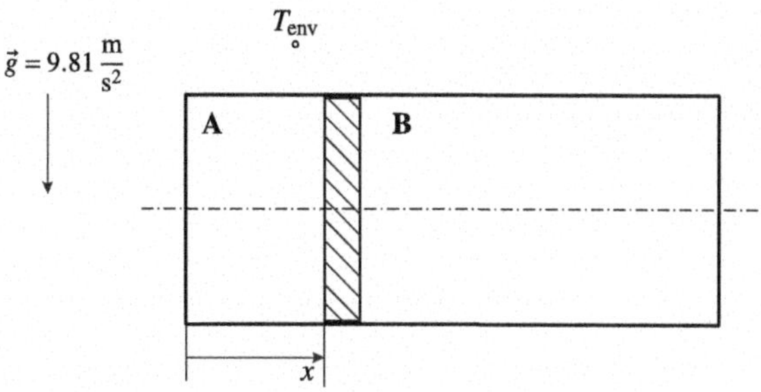

3.15 In an non-adiabatic, horizontal cylinder with a frictionless, freely movable, gas-tight piston, there is air (ideal, $c_p = 1.004\,\frac{\text{kJ}}{\text{kgK}}$, $\kappa = 1.4$) in Chambers A and B, see Fig. 3.4. At the beginning, the piston is fixed and the following parameters are given:

- Chamber A: $V_{\text{A},0} = 2\,\text{dm}^3$, $p_{\text{A},0} = 8\,\text{bar}$, $\vartheta_{\text{A},0} = \vartheta_0 = 20\,°\text{C}$
- Chamber B: $V_{\text{B},0} = 10\,\text{dm}^3$, $p_{\text{B},0} = 1\,\text{bar}$, $\vartheta_{\text{B},0} = \vartheta_0 = 20\,°\text{C}$
- Piston: $m_{\text{P}} = 1\,\text{kg}$, $c_{\text{P}} = 800\,\frac{\text{J}}{\text{kgK}}$, $\vartheta_{\text{P},0} = 80\,°\text{C}$
- Environment: $\vartheta_{\text{env}} = 20\,°\text{C}$

The fixation is now released. After some oscillations the piston finally comes to rest. What is the pressure in Chamber A once thermodynamic equilibrium is reached?

3.16 Two air flows are mixed isobarically and adiabatically in a steady state process. Air is to be regarded as an ideal gas with $c_p = 1004\,\frac{\text{J}}{\text{kgK}}$. The following parameters are given:

- Flow 1: $\dot{m}_1 = 12\,\frac{\text{kg}}{\text{s}}$, $T_1 = 608\,\text{K}$
- Flow 2: $\dot{m}_2 = 18\,\frac{\text{kg}}{\text{s}}$, $T_2 = 298\,\text{K}$

What is the generated flux of entropy $\dot{S}_{\text{i}}$, in case changes of potential/kinetic energies are ignored?

3.17 In an experiment, an adiabatic, frictionless test object with a mass of $m_{\text{K}} = 5\,\text{kg}$ is to be accelerated by adiabatic, reversible expansion of compressed air (ideal gas, $\kappa = 1.4$, $R = 287\,\frac{\text{J}}{\text{kgK}}$). The mass is initially fixed, see Fig. 3.5. Now, the fixation is released. What initial pressure p_1 is required to accelerate the object to a velocity of $2\,\frac{\text{m}}{\text{s}}$ when it just fully leaves the vertical tube?

3.18 Two cylindrical vessels of the same cross-section ($A = 200\,\text{cm}^2$) are in connection with each other and are partly filled with water ($\rho = 1000\,\frac{\text{kg}}{\text{m}^3}$, incompressible). Both vessels are closed at the interface to the air (ideal gas, $\kappa = 1.4$) with freely movable and frictionless pistons ($m_{\text{P}} = 2.3\,\text{kg}$). In the right vessel, $h_1 = 60\,\text{cm}$ above

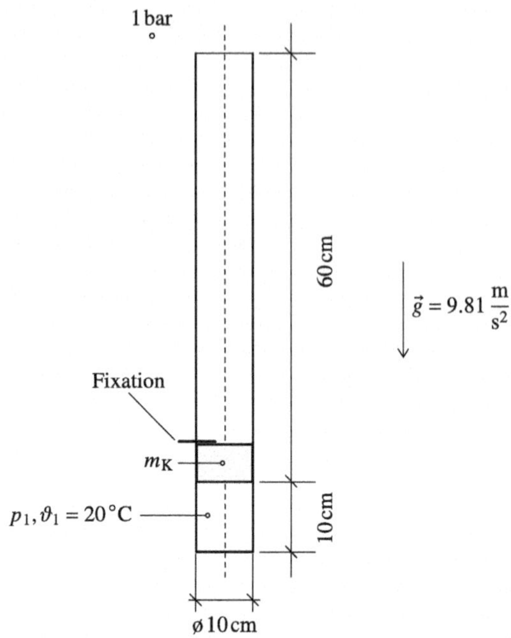

Fig. 3.5 Acceleration

the movable piston, there is a fixed piston, see Fig. 3.6. Initially the air pressure is 1 bar and the level of water is identical in both vessels. How much work W_{12} is required to increase the pressure in the right vessel to $p_2 = 2$ bar by pushing the left piston down very slowly? Changes of the potential energy of the air can be ignored. Ambient pressure shall be 1 bar. All pistons as well as vessels shall be adiabatic.

3.19 In a horizontal adiabatic cylinder there is air of $T_1 = 300$ K and $p_1 = 1$ bar at a volume of $V_1 = 2\,\text{dm}^3$. By a quasi-static and frictionless change of state the air is compressed to a volume of $V_2 = 1\,\text{dm}^3$. Determine the change in internal energy for the compression. Air can be regarded as an ideal gas with constant specific heat capacity ($R = 0.287\,\frac{\text{kJ}}{\text{kgK}}$, $\kappa = 1.4$).

3.20 In a tank open to the environment ($p_{\text{env}} = 1$ bar) there is $5\,\text{m}^3$ of a viscous liquid. Now a stirrer is immersed in the liquid, which rotates for 90 min at a speed of $100\,\frac{1}{\text{min}}$ and a torque of $M = 120$ N m. The volume of the liquid has increased by 10 % after the 90 min. Calculate the sum of all work exchanged across the system boundary if the liquid is considered a thermodynamic system. The change in the centre of gravity of the fluid is to be disregarded. What is the dissipation in the liquid?

3.21 A weight of 100 kg is placed on a piston (mass of the piston is negligible) which perfectly seals a cylinder ($D = 10$ cm), cf. Fig. 3.7. Initially, the entire system is at

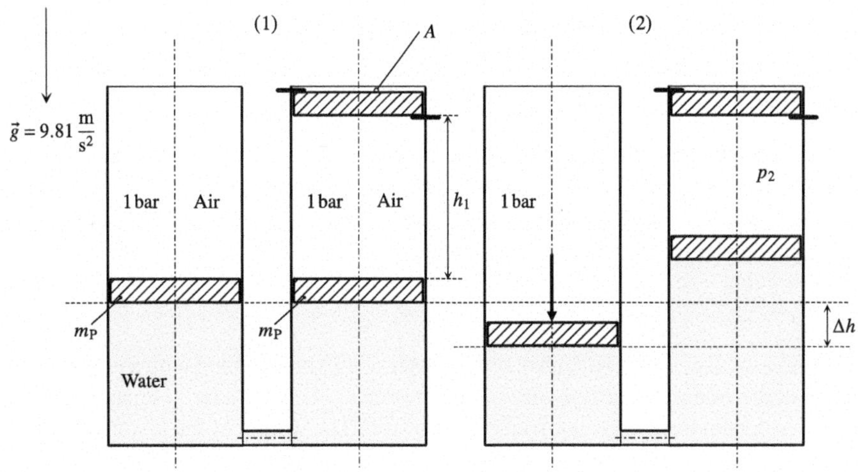

Fig. 3.6 Water columns

Fig. 3.7 Oscillation

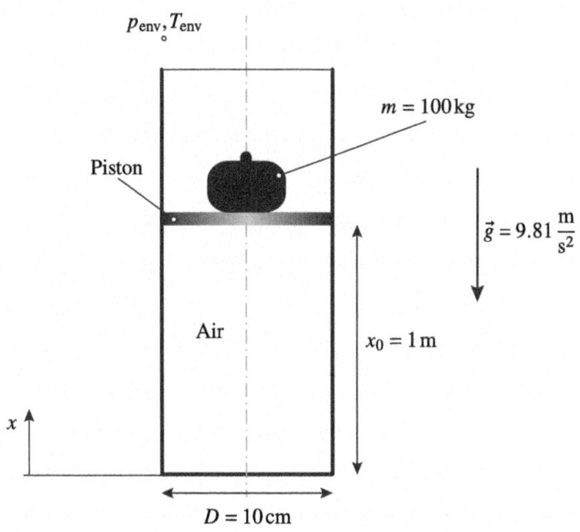

ambient thermal state ($T_0 = T_{env} = 300\,K$, $p_0 = p_{env} = 1\,bar$) with $x_0 = 1\,m$. Due to the weight, the piston starts to move downwards. The cylinder is filled with air (ideal gas, $R = 287.4\,\frac{J}{kgK}$, $\kappa = 1.401$). Changes in the temperature of the cylinder walls, the piston and the mass can generally be neglected. Likewise, the potential energy of the gas may be neglected. Two cases are considered:

- Case 1: The change of state of the air is adiabatic and frictionless. There is no friction between the piston and the cylinder wall.

- Case 2: Friction occurs between the piston and the cylinder wall. In addition, the system is perfectly thermally coupled with the environment. There shall be no dissipation within the gas.

(a) Sketch path-time diagrams of the mass for both cases.
(b) At what position x does the mass come to rest in Case 2?
(c) How much heat is released from the system in Case 2 until equilibrium is reached? How much energy is dissipated?
(d) Formulate the equations needed to determine the lowest position of the mass after release in Case 1!

3.22 A vertical cylinder with an inner diameter of 250 mm is filled with a gas mixture that is in thermal equilibrium with the environment. The cylinder is separated from the environment by an airtight, frictionless piston. The weight of the piston results in a pressure of 4 bar in the cylinder. Now the mixture is ignited and combustion takes place, as a result of which the piston moves upwards and heat is exchanged with the environment. The entire process takes place so slowly that the mixture is under a constant pressure of 4 bar. After this process, thermal equilibrium is reached. By what absolute value does the internal energy of the gas change when 10 kJ of heat is released to the environment and the piston comes to rest in a 75 mm higher position? The ambient pressure shall be $p_{\mathrm{env}} = 1$ bar. The potential energy of the gas can be disregarded. Derive the energy balances for the following system boundaries:

(A) Cylinder content on its own
(B) Cylinder content and piston

3.23 A vertical cylinder with a cross-section of $A = 0.8\,\mathrm{m}^2$ contains air which is considered an ideal gas ($c_v = 0.717\,\frac{\mathrm{kJ}}{\mathrm{kgK}}$, $\kappa = 1.4$). The air is enclosed by a frictionless, adiabatic piston with a mass of $m_{\mathrm{P}} = 2500\,\mathrm{kg}$. Ambient pressure $p_{\mathrm{env}} = 1$ bar is present above the piston. A heat quantity of 120 kJ is released very slowly from the gas. Determine the vertical displacement of the piston! The potential energy of the gas may be neglected. The gravitational acceleration can be assumed $g = 9.81\,\frac{\mathrm{m}}{\mathrm{s}^2}$.

3.24 An adiabatic turbine is operated with helium as the working fluid (ideal gas, $c_p = 5.18\,\frac{\mathrm{kJ}}{\mathrm{kgK}}$, $\kappa = 1.67$). The fluid enters the turbine with $p_1 = 70$ bar and $T_1 = 1130\,\mathrm{K}$. A fraction of 90 % of the supplied mass flux is expanded to $p_3 =$ 24 bar, reaching the temperature $T_3 = 770\,\mathrm{K}$. The remaining mass flux is taken from the turbine when pressure is $p_2 = 40$ bar, see Fig. 3.8.

- Ambient temperature is $\vartheta_{\mathrm{env}} = 19.85\,°\mathrm{C}$.
- The polytropic exponent n shall be constant in the turbine for simplicity.
- Changes of kinetic and potential energies are not to be considered.

(a) What is the polytropic exponent of the change of state in the turbine?
(b) Sketch the changes of state in a T, s-diagram and clearly mark states (1), (2) and (3).

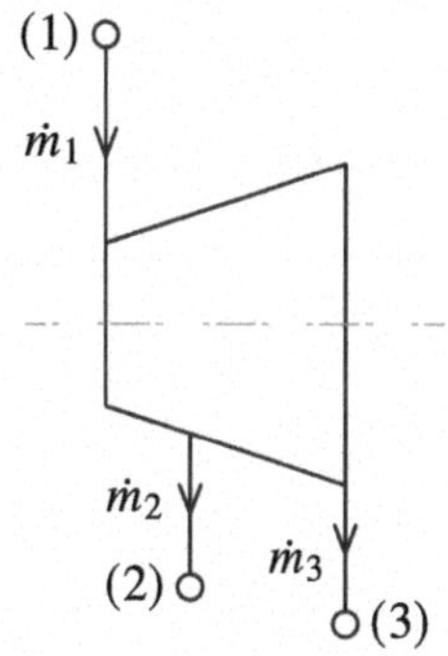

Fig. 3.8 Turbine

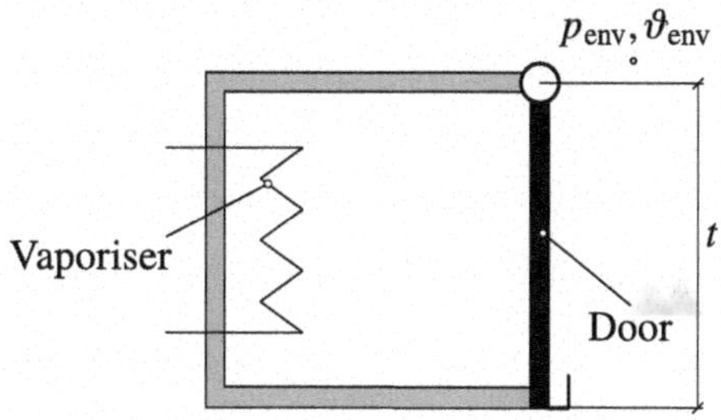

Fig. 3.9 Freezer

(c) What is the temperature of the fluid in state (2)?
(d) The turbine has a technical power of $P_t = -80\,\text{kW}$. What are the mass fluxes $\dot{m}_1$, $\dot{m}_2$ and $\dot{m}_3$?
(e) What is the power resulting from the pressure change? What is the dissipated power?
(f) Determine the entropy flow generated $\dot{S}_i$.
(g) What is the thermodynamic mean temperature in the turbine?

3.25 An ideally insulated freezer, see Fig. 3.9, with a volume of $0.4\,\text{m}^3$ is opened. A complete exchange of air takes place. Therefore, after the door is closed (state 1), there is ambient air in the freezer. Subsequently, the air is cooled down to $\vartheta_2 = -12\,°\text{C}$ by heat transfer to the evaporator (state 2).

- Air can be considered an ideal gas with $R = 0.287\,\frac{\text{kJ}}{\text{kgK}}$ and $c_p = 1.004\,\frac{\text{kJ}}{\text{kgK}}$.
- The door closes tightly (no air exchange with the environment possible after closing).
- Ambient state: $\vartheta_{env} = 25\,°\text{C}$, $p_{env} = 1\,\text{bar}$.

(a) Determine the mass of air in the freezer after closing the door.
(b) What heat must be transferred to the vaporiser to cool the air to $\vartheta_2 = -12\,°\text{C}$?
(c) What is the pressure of the air in the freezer in state (2)?
(d) What force is now needed to open the door if it has a height of $h = 0.6\,\text{m}$ and the distance of the handle from the hinge is $t = 0.5\,\text{m}$?

(e) What is the thermodynamic mean temperature $T_{m,12}$ during the change of state (1) → (2)?
(f) What is the entropy $S_{i,12}$ generated inside the freezer, i.e. for the system *gas*?

3.26 In a compressor, an air flow of 12 $\frac{kg}{h}$ is compressed from ambient pressure ($p_{env} = 1$ bar) and ambient temperature ($\vartheta_{env} = 20\,°C$) to 3.3 bar. Additionally, the compressor releases a heat flux of 350 W. The outlet temperature is finally 160 °C. Changes in kinetic and potential energy are negligible. Air may be regarded as an ideal gas ($R = 287\,\frac{J}{kgK}$, $c_p = 1004\,\frac{J}{kgK}$).

(a) What is the compressor's power?
(b) What is the polytropic exponent n?
(c) What is the exergy flux of the released heat?

3.2 Solutions

3.1 **(b)**
Two mechanisms are responsible for the change in the state value entropy, i.e.

$$\mathrm{d}S = \underbrace{\delta S_i}_{\geq 0} + \underbrace{\delta S_a}_{\gtreqless 0} = \frac{\delta \Psi_{12}}{T} + \frac{\delta Q_{12}}{T}. \tag{3.1}$$

- Entropy can be generated inside the system, e.g. by dissipation. In this case, the generated entropy is $\delta S_i > 0$. In the best case, the process is reversible, i.e. $\delta S_i = 0$. However, δS_i can never become negative.
- Entropy is transported with heat across the system boundary and has the same sign as heat. This means that in the case of a heated system, entropy is transported into the system and thus causes an increase in the state value entropy in the system. In the case of cooling, entropy is released by the system via the system boundary.

Referring to the task, this means that the entropy of a closed system can decrease by sufficient cooling. Answer (b) is therefore correct. Work done on the system and supplied from outside is composed of volume work, dissipation and mechanical work according to the partial energy equation, i.e.

$$W_{12} = W_{V,12} + W_{mech,12} + \Psi_{12}. \tag{3.2}$$

With the supplied work, dissipation could therefore be brought into the system, which leads to the generation of entropy. However, this effect is always responsible for an increase in entropy. Therefore, answer (c) does not apply.

3.2 **(a)**
The change of state proceeds very slowly, the mass of the piston and the ambient pressure remain constant, so that the change of state (1) → (2) is isobaric,[1] cf. [1] and Problem 1.11, i.e.

$$p = p_{\mathrm{env}} + \frac{m_{\mathrm{P}} g}{A} = \text{const.} \tag{3.3}$$

Furthermore, the change of state is isothermal, so that according to

$$s_2 - s_1 = c_p \ln \frac{T_2}{T_1} - R \ln \frac{p_2}{p_1} = 0 \tag{3.4}$$

the specific entropy is also constant. The following therefore applies to the change in extensive entropy

$$\Delta S = m_2 s_2 - m_1 s_1 = s_1 (m_2 - m_1) = -0.2 m_1 s_1 = -0.2 S_1 . \tag{3.5}$$

The entropy therefore decreases by 20 %, so answer (a) is correct.

3.3 **(a)**
Under the given boundary conditions, the first law of thermodynamics reads as follows

$$W_{12} + Q_{12} = U_2 - U_1 . \tag{3.6}$$

Since the change of state is supposed to be isothermal, the following further applies

$$W_{12} + Q_{12} = 0 \tag{3.7}$$

respectively due to the cooling

$$W_{12} = -Q_{12} > 0. \tag{3.8}$$

Thus, work is supplied to the system. Now the partial energy equation is applied for the given case, i.e.

$$W_{12} = \underbrace{W_{\mathrm{V},12}}_{=0} + \underbrace{W_{\mathrm{mech},12}}_{=0} + \Psi_{12} > 0 \tag{3.9}$$

$W_{\mathrm{V},12} = 0$ due to the constant volume of the system and $W_{\mathrm{mech},12} = 0$ since potential resp. kinetic energies of the system do not vary. Consequently, all the work supplied from outside is dissipated., i.e.

$$W_{12} = \Psi_{12} > 0. \tag{3.10}$$

Hence, answer (a) is correct. Since energy is dissipated, the change of state cannot be reversible. Therefore, answer (b) is wrong. A perpetual mobile of the 2nd kind

---

[1] It is assumed that the mass of the gas is small compared to the mass of the piston.

Fig. 3.10 Change of state of an adiabatic turbine in a T, s-diagram

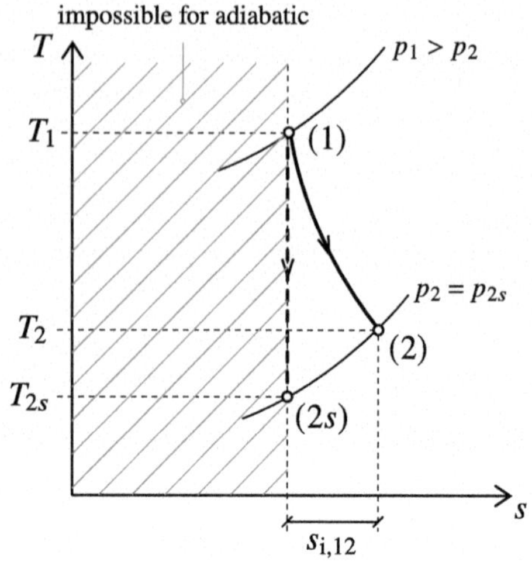

is a fictitious[2] machine that violates the second law of thermodynamics, i.e. would destroy entropy. In the problem described, the second law of thermodynamics is not violated, so answer (c) is wrong.

3.4 **(d)**
The change of state of an adiabatic turbine in a T, s-diagram is shown in Fig. 3.10. In a turbine, the gas is expanded, i.e. the pressure decreases during the change of state. Therefore, the two isobars between which the turbine operates are first sketched in a T, s-diagram. Answer (a) is therefore false. The first law of thermodynamics obeys

$$w_{t,12} + \underbrace{q_{12}}_{=0} = h_2 - h_1 + \underbrace{\Delta e_{a,12}}_{=0} \tag{3.11}$$

so that[3]

$$w_{t,12} = h_2 - h_1 = c_p \left(T_2 - T_1\right) < 0. \tag{3.12}$$

Due to the work release, the temperature must decrease between the inlet and outlet. The second law of thermodynamics reads as[4]

$$s_1 + s_{i,12} + \underbrace{s_{a,12}}_{=0} = s_2 \tag{3.13}$$

respectively

$$s_{i,12} = s_2 - s_1 \geq 0. \tag{3.14}$$

---

[2] Such a machine does not exist.

[3] Note that a turbine releases work.

[4] Since the turbine is adiabatic.

Hence, two cases are possible:

- Adiabatic, irreversible change of state (1) → (2): Due to dissipation, the entropy rises.
- Adiabatic, reversible, i.e. isentropic, change of state (1) → (2s): The entropy remains constant.

However, in both cases the temperature decreases, see Fig. 3.10. Answer (b) is wrong. Since the system is adiabatic, there is no entropy transportation by heat across the system boundary, i.e. answer (c) is false as well.

3.5 **(b)**
First, the partial energy equation is formulated, i.e.

$$w_t = 0 = \int_1^2 v\,dp + \psi_{12} + \underbrace{\Delta e_{a,12}}_{=\frac{1}{2}\left(c_2^2 - c_1^2\right)} \,. \tag{3.15}$$

For the external energies, the potential energies are not relevant because of the horizontal position of the pipe. The specific technical work $w_t = 0$, since the flow shall be work-insulated. Due to the incompressible liquid it is $v = \text{const.}$, so that

$$0 = v\left(p_2 - p_1\right) + \psi_{12} + \frac{1}{2}\left(c_2^2 - c_1^2\right) \tag{3.16}$$

respectively

$$p_2 - p_1 = \underbrace{-\rho\psi_{12}}_{\leq 0} - \frac{\rho}{2}\left(c_2^2 - c_1^2\right) . \tag{3.17}$$

Three cases are possible:

- Pressure remains constant, i.e.

$$\psi_{12} = \underbrace{-\frac{1}{2}\left(c_2^2 - c_1^2\right)}_{\geq 0} \tag{3.18}$$

  In a reversible flow, i.e. $\psi_{12} = 0$, the velocity at the inlet must be equal to the velocity at the outlet. For an irreversible flow, i.e. $\psi_{12} > 0$, the velocity at the inlet must be greater than the velocity at the outlet.
- Pressure decreases, i.e.

$$\psi_{12} > -\frac{1}{2}\left(c_2^2 - c_1^2\right) \tag{3.19}$$

- Pressure rises, i.e.

$$\psi_{12} < \underbrace{-\frac{1}{2}\left(c_2^2 - c_1^2\right)}_{>0} \tag{3.20}$$

That means that the flow must be decelerated. Hence, answer (b) is correct.

Secondly, in an incompressible flow, the velocity depends on the cross-section, as the continuity equation proves:

$$\dot{m}_1 = \rho c_1 A_1 = \rho c_2 A_2 = \dot{m}_2 \tag{3.21}$$

so that

$$\frac{c_1}{c_2} = \frac{A_2}{A_1}. \tag{3.22}$$

Finally, the first law of thermodynamics is not violated because

$$w_{t,12} + q_{12} = 0 = h_2 - h_1 + \frac{1}{2}\left(c_2^2 - c_1^2\right) \tag{3.23}$$

respectively

$$h_2 - h_1 = -\frac{1}{2}\left(c_2^2 - c_1^2\right) = v\left(p_2 - p_1\right) + \psi_{12}. \tag{3.24}$$

If the specific heat capacity c is constant, one obtains

$$c\left(T_2 - T_1\right) + v\left(p_2 - p_1\right) = v\left(p_2 - p_1\right) + \psi_{12} \tag{3.25}$$

so that

$$c\left(T_2 - T_1\right) = \psi_{12} \geq 0 \tag{3.26}$$

respectively

$$T_2 - T_1 = \frac{\psi_{12}}{c} \geq 0. \tag{3.27}$$

3.6 **(c)**
For a closed system, the first law of thermodynamics obeys[5]

$$W_{12} + Q_{12} = U_2 - U_1 = m c_v \left(T_2 - T_1\right). \tag{3.28}$$

Since this is an isothermal change of state, it follows that

$$W_{12} = -Q_{12}. \tag{3.29}$$

---

[5] The kinetic energy is neglected because the system is at rest at the beginning and at the end. The vertical position of the centre of gravity does not change either.

The partial energy equation reads as[6]

$$W_{12} = \underbrace{W_{V,12}}_{=0} + \underbrace{W_{mech,12}}_{=0} + \Psi_{12} = \Psi_{12}. \tag{3.30}$$

This means that all the work supplied is dissipated. Thus, answer (b) is incorrect. The entropy balance follows

$$S_2 = S_1 + S_i + S_a \tag{3.31}$$

respectively

$$S_i + S_a = S_2 - S_1 = m\left[c_v \ln\frac{T_2}{T_1} + R\ln\frac{v_2}{v_1}\right] = 0. \tag{3.32}$$

The entropy inside the system remains constant. Consequently, answer (a) is wrong. Hence, it is

$$S_i = -S_a = -\frac{Q_{12}}{T} = \frac{W_{12}}{T} = \frac{\Psi_{12}}{T}. \tag{3.33}$$

The loss of exergy then is

$$\Delta E_{x,V} = S_i T_{env} = \frac{\Psi_{12}}{T} T_{env} = \frac{\Psi_{12}}{1.2 \cdot T_{env}} T_{env} = 0.8333 \cdot \Psi_{12}. \tag{3.34}$$

Hence, answer (c) is correct.

3.7 **(c)**
Let's start with the first law of thermodynamics for closed systems, i.e.

$$W_{12} + Q_{12} = U_2 - U_1 \tag{3.35}$$

so that

$$U_2 - U_1 = 20\,\text{kJ} - 60\,\text{kJ} = -40\,\text{kJ} < 0 \tag{3.36}$$

respectively

$$T_2 = T_1 - \frac{40\,\text{kJ}}{mc_v}. \tag{3.37}$$

The temperature therefore decreases, answer (a) is therefore wrong. The partial energy equation is

$$W_{12} = W_{V,12} + W_{mech,12} + \Psi_{12}. \tag{3.38}$$

Since the volume does not change and the centre of gravity also remains unchanged, the equation simplifies to

$$W_{12} = \Psi_{12} = 20\,\text{kJ}. \tag{3.39}$$

[6] Note that the volume remains constant and the centre of gravity remains unchanged.

Hence, it is

$$S_{i,12} = \int_1^2 \frac{\delta\Psi}{T} = \frac{\Psi_{12}}{T_m} > 0. \tag{3.40}$$

The change of state is therefore irreversible and answer (b) is also wrong. However, the thermodynamic mean temperature obeys

$$T_m = \frac{q_{12} + \psi_{12}}{s_2 - s_1} = \frac{q_{12} + w_{12}}{s_2 - s_1} = \frac{u_2 - u_1}{s_2 - s_1}. \tag{3.41}$$

With the caloric equations of state for an isochoric change of state one gets

$$T_m = \frac{c_v\,(T_2 - T_1)}{c_v \ln \frac{T_2}{T_1}} = \frac{T_2 - T_1}{\ln \frac{T_2}{T_1}}. \tag{3.42}$$

Consequently, the loss of exergy follows

$$\Delta E_{x,V} = S_{i,12} T_{env} = T_{env} \cdot \frac{\Psi_{12}}{T_2 - T_1} \ln \frac{T_2}{T_1} > 0. \tag{3.43}$$

Thus, answer (c) is correct.

3.8 **(a)**
The caloric equation of state for incompressible fluids is as follows

$$h_2 - h_1 = c\,(T_2 - T_1) + v\,(p_2 - p_1) \tag{3.44}$$

so that

$$p_2 - p_1 = \rho\,[h_2 - h_1 - c\,(T_2 - T_1)] = -1 \times 10^5\ \text{Pa} = -1\ \text{bar}. \tag{3.45}$$

Answer (a) is correct, there is a pressure loss. Entropy can never be destroyed, answer (b) is wrong. According to the partial energy equation, it follows

$$w_t = \int_1^2 v\,\mathrm{d}p + \psi_{12} + \Delta e_{a,12}. \tag{3.46}$$

The first law of thermodynamics for an open system obeys

$$w_t + q_{12} = \underbrace{\int_1^2 v\,\mathrm{d}p}_{=v(p_2 - p_1)} + \psi_{12} + \Delta e_{a,12} + q_{12} = h_2 - h_1 + \Delta e_{a,12}. \tag{3.47}$$

respectively

$$v\,(p_2 - p_1) + \psi_{12} + q_{12} = c\,(T_2 - T_1) + v\,(p_2 - p_1) \tag{3.48}$$

so that

$$\psi_{12} + q_{12} = c\,(T_2 - T_1)\,. \tag{3.49}$$

Exchanged heat as well as dissipation are responsible for the temperature change of the fluid. With the thermodynamic mean temperature and the second law of thermodynamics one gets

$$\frac{\delta q}{T} + \frac{\delta\psi}{T} = \mathrm{d}s \tag{3.50}$$

respectively by integration[7]

$$\frac{q_{12}}{T_\mathrm{m}} + \frac{\psi_{12}}{T_\mathrm{m}} = s_2 - s_1 = c\ln\frac{T_2}{T_1} > 0. \tag{3.51}$$

Hence, it is

$$T_\mathrm{m} = \frac{q_{12} + \psi_{12}}{c\ln\frac{T_2}{T_1}} = \frac{T_2 - T_1}{\ln\frac{T_2}{T_1}}. \tag{3.52}$$

The first law of thermodynamics for open systems can be reformulated into the first law for closed systems in a few steps:

$$\boxed{\delta w_\mathrm{t} + \delta q = \mathrm{d}h + \mathrm{d}e_\mathrm{a}} \tag{3.53}$$

replacing the specific enthalpy by the specific internal energy leads to

$$\delta w_\mathrm{t} + \delta q = \mathrm{d}u + p\,\mathrm{d}v + v\,\mathrm{d}p + \mathrm{d}e_\mathrm{a}. \tag{3.54}$$

By applying the partial energy equation for open systems one gets

$$\underbrace{v\,\mathrm{d}p + \delta\psi + \mathrm{d}e_\mathrm{a}}_{=\delta w_\mathrm{t}} + \delta q = \mathrm{d}u + p\,\mathrm{d}v + v\,\mathrm{d}p + \mathrm{d}e_\mathrm{a} \tag{3.55}$$

respectively

$$\underbrace{-p\,\mathrm{d}v + \delta\psi + \mathrm{d}e_\mathrm{a}}_{=\delta w} + \delta q = \mathrm{d}u + \mathrm{d}e_\mathrm{a}. \tag{3.56}$$

This represents the first law of thermodynamics for closed systems, i.e.

$$\boxed{\delta w + \delta q = \mathrm{d}u + \mathrm{d}e_\mathrm{a}} \tag{3.57}$$

---

[7] Obviously, the entropy of the fluid increases, so answer (c) is wrong.

In general, the following applies to a polytropic, non-isothermal change of state

$$w_{t,12} + q_{12} = h_2 - h_1 + \Delta e_{a,12} = c_p\,(T_2 - T_1) + \Delta e_{a,12}. \tag{3.58}$$

With the partial energy equation for $w_{t,12}$ it becomes

$$y_{12} + \psi_{12} + \Delta e_{a,12} + q_{12} = c_p\,(T_2 - T_1) + \Delta e_{a,12}. \tag{3.59}$$

respectively

$$\psi_{12} + q_{12} = c_p\,(T_2 - T_1) - y_{12}. \tag{3.60}$$

The pressure work can be substituted according to Appendix E, so that

$$\psi_{12} + q_{12} = c_p\,(T_2 - T_1) - \frac{n}{n-1} R\,(T_2 - T_1) = (T_2 - T_1)\left[c_p - \frac{n}{n-1} R\right]. \tag{3.61}$$

The caloric equation of state for the specific entropy yields

$$s_2 - s_1 = c_p \ln \frac{T_2}{T_1} - R \ln \frac{p_2}{p_1} = \left[c_p - \frac{n}{n-1} R\right] \ln \frac{T_2}{T_1}. \tag{3.62}$$

The specific entropy follows

$$\frac{\delta q}{T} + \frac{\delta \psi}{T} = \mathrm{d}s \tag{3.63}$$

so that the integration leads to

$$\frac{q_{12}}{T_m} + \frac{\psi_{12}}{T_m} = s_2 - s_1. \tag{3.64}$$

Hence, the thermodynamic mean temperature is generally

$$\boxed{T_m = \frac{q_{12} + \psi_{12}}{s_2 - s_1} = \frac{(T_2 - T_1)\left[c_p - \frac{n}{n-1} R\right]}{\left[c_p - \frac{n}{n-1} R\right] \ln \frac{T_2}{T_1}} = \frac{T_2 - T_1}{\ln \frac{T_2}{T_1}}} \tag{3.65}$$

3.9 First, the specific heat capacity c_v is determined, it is

$$c_v = \frac{c_p}{\kappa} = 717.1429\,\frac{\mathrm{J}}{\mathrm{kgK}}. \tag{3.66}$$

The individual gas constant then is

$$R = c_p - c_v = 286.8571\,\frac{\mathrm{J}}{\mathrm{kgK}}. \tag{3.67}$$

The first law of thermodynamics for a closed system obeys

$$Q_{12} + W_{12} = U_2 - U_1. \tag{3.68}$$

Applying the caloric equation of state leads to

$$Q_{12} + W_{12} = mc_v \left(T_2 - T_1\right) \tag{3.69}$$

respectively with $W_{12} = P\,\Delta t$

$$T_2 = T_1 + \frac{P\,\Delta t}{mc_v} = 308.0902\,\mathrm{K}. \tag{3.70}$$

The pressures follow from the thermal equation of state

$$p_1 = \frac{mRT_1}{V} = 1.1773\,\mathrm{bar} \tag{3.71}$$

and

$$p_2 = \frac{mRT_2}{V} = 1.2373\,\mathrm{bar}. \tag{3.72}$$

The second law of thermodynamics for a closed system yields

$$S_2 = S_1 + S_{\mathrm{i},12} + \underbrace{S_{\mathrm{a},12}}_{=0}. \tag{3.73}$$

$S_{\mathrm{a},12}$ disappears, since the system is adiabatic, so that

$$S_{\mathrm{i},12} = S_2 - S_1 = m\left(s_2 - s_1\right). \tag{3.74}$$

By applying the caloric equation of state one gets

$$S_{\mathrm{i},12} = m\left[c_p \ln\frac{T_2}{T_1} - R\ln\frac{p_2}{p_1}\right] = 49.9071\,\frac{\mathrm{J}}{\mathrm{K}}. \tag{3.75}$$

The partial energy equation follows[8]

$$W_{12} = W_{\mathrm{V},12} + W_{\mathrm{mech},12} + \Psi_{12} = \Psi_{12} = 15\,000\,\mathrm{J}. \tag{3.76}$$

For the generation of entropy it is

$$\delta S_{\mathrm{i}} = \frac{\delta\Psi}{T}. \tag{3.77}$$

[8] The work W_{12} supplied via the stirrer is neither used in the system to change the position of the centre of gravity nor to increase or decrease the volume. It is therefore completely dissipated.

The integration results in

$$S_{\mathrm{i},12} = \int_1^2 \delta S_\mathrm{i} = \int_1^2 \frac{\delta \Psi}{T} = \frac{\Psi_{12}}{T_\mathrm{m}}. \tag{3.78}$$

Hence, the thermodynamic mean temperature is

$$T_\mathrm{m} = \frac{\Psi_{12}}{S_{\mathrm{i},12}} = \frac{P\,\Delta t}{S_{\mathrm{i},12}} = 300.56\,\mathrm{K}. \tag{3.79}$$

3.10 The isentropic exponent is

$$\kappa = \frac{c_p}{c_v} = 1.6796. \tag{3.80}$$

Since

$$\underbrace{\frac{\delta q}{T}}_{=0} + \underbrace{\frac{\delta \psi}{T}}_{=0} = \mathrm{d}s \tag{3.81}$$

the change of state is isentropic, so that

$$s_2 - s_1 = 0 = c_p \ln \frac{T_2}{T_1} - R \ln \frac{p_2}{p_1} \tag{3.82}$$

respectively

$$\ln \frac{T_2}{T_1} - \underbrace{\frac{R}{c_p}}_{=\frac{\kappa-1}{\kappa}} \ln \frac{p_2}{p_1} = 0. \tag{3.83}$$

Solving for T_2 results in

$$T_2 = T_1 \left(\frac{p_2}{p_1}\right)^{\frac{\kappa-1}{\kappa}} = 572.3061\,\mathrm{K}. \tag{3.84}$$

Finally, by applying the caloric equation of state, it follows

$$\Delta u = u_2 - u_1 = c_v\,(T_2 - T_1) = 8.4143 \times 10^4\,\frac{\mathrm{J}}{\mathrm{kg}}. \tag{3.85}$$

3.11 First, the specific heat capacity c_v is determined, it is

$$c_v = \frac{c_p}{\kappa} = 717.1429\,\frac{\mathrm{J}}{\mathrm{kgK}}. \tag{3.86}$$

The first law of thermodynamics for a closed system obeys

$$\underbrace{Q_{12}}_{=0} + W_{12} = U_2 - U_1 = mc_v\,(T_2 - T_1) \tag{3.87}$$

respectively with $W_{12} = P_{\mathrm{el}} \cdot \Delta t$

$$W_{12} = P_{\mathrm{el}} \cdot \Delta t = mc_v\,(T_2 - T_1)\,. \tag{3.88}$$

Solving for T_2 results in

$$T_2 = T_1 + \frac{P_{\mathrm{el}} \cdot \Delta t}{mc_v} = 308.0902\,\mathrm{K}. \tag{3.89}$$

3.12 Due to the initially different pressures in chambers A and B, opposing forces act on the piston, the amounts of which are different. Therefore, there is a resulting force which, according to Newton, causes an acceleration of the mass. Such an acceleration leads to an increase in velocity which continues until the forces acting on the piston balance out, i.e. the pressures in chambers A and B are identical. Thus, maximum speed is reached when $p_{\mathrm{A}} = p_{\mathrm{B}}$. According to the task, the change of state of the gas should be isentropic, i.e. adiabatic and frictionless. Therefore, $pv^{\kappa} = \mathrm{const.}$ applies. The pressures in chambers A and B can thus be formulated as a function of the initial pressures:

$$p_{\mathrm{A}} = p_{\mathrm{A},0}\left(\frac{V_{\mathrm{A},0}}{V_{\mathrm{A}}}\right)^{\kappa} = p_{\mathrm{A},0}\left(\frac{V_{\mathrm{A},0}}{x\frac{\pi}{4}d^2}\right)^{\kappa}. \tag{3.90}$$

With $L_{\mathrm{A},0} = \frac{V_{\mathrm{A},0}}{\frac{\pi}{4}d^2} = 0.1768\,\mathrm{m}$ it follows

$$p_{\mathrm{A}} = p_{\mathrm{A},0}\left(\frac{L_{\mathrm{A},0}\frac{\pi}{4}d^2}{x\frac{\pi}{4}d^2}\right)^{\kappa} = p_{\mathrm{A},0}\left(\frac{L_{\mathrm{A},0}}{x}\right)^{\kappa}. \tag{3.91}$$

For chamber B applies analogously

$$p_{\mathrm{B}} = p_{\mathrm{B},0}\left(\frac{L_{\mathrm{B},0}}{L_{\mathrm{A},0} + L_{\mathrm{B},0} - x}\right)^{\kappa} \tag{3.92}$$

with $L_{\mathrm{B},0} = \frac{V_{\mathrm{B},0}}{\frac{\pi}{4}d^2} = 0.8842\,\mathrm{m}$. As discussed at the beginning, the velocity is maximum when $p_{\mathrm{A}} = p_{\mathrm{B}}$, i.e.

$$p_{\mathrm{A},0}\left(\frac{L_{\mathrm{A},0}}{x_{@c_{\max}}}\right)^{\kappa} = p_{\mathrm{B},0}\left(\frac{L_{\mathrm{B},0}}{L_{\mathrm{A},0} + L_{\mathrm{B},0} - x_{@c_{\max}}}\right)^{\kappa}. \tag{3.93}$$

Fig. 3.11 Energy balance for the object

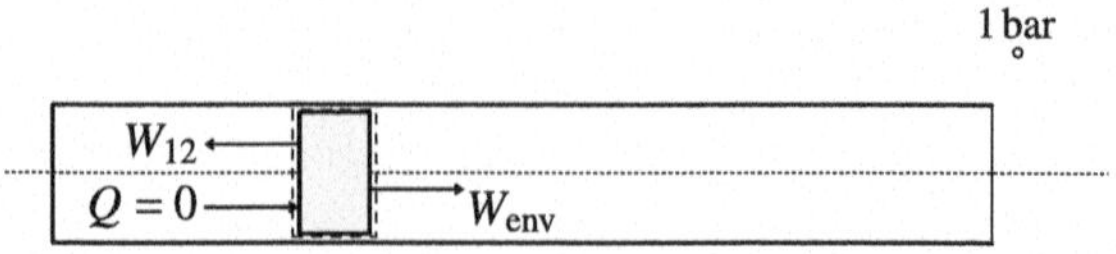

The position $x_{@c_{max}}$ at which the pressures are identical and thus maximum piston speed is reached therefore follows:

$$x_{@c_{max}} = \frac{L_{A,0}^2 + L_{A,0}L_{B,0}}{L_{A,0} + L_{B,0}\left(\frac{p_{B,0}}{p_{A,0}}\right)^{\frac{1}{\kappa}}} = 0.4976\,\text{m}. \tag{3.94}$$

3.13 The initial gas volume is

$$V_1 = \frac{\pi}{4}d^2\,L_1 = 7.854 \times 10^{-6}\,\text{m}^3 \quad \text{with} \quad L_1 = 10\,\text{cm} \tag{3.95}$$

and the gas volume when the object has just fully left the tube

$$V_2 = \frac{\pi}{4}d^2\,L_2 = 5.4978 \times 10^{-5}\,\text{m}^3 \quad \text{with} \quad L_2 = 70\,\text{cm}. \tag{3.96}$$

The work W_{12} follows from the partial energy equation for the compressed air[9]

$$W_{12} = W_{V,12} + \underbrace{W_{mech,12}}_{=0} + \underbrace{\Psi_{12}}_{=0} = \frac{p_1 V_1}{\kappa - 1}\left[\left(\frac{V_1}{V_2}\right)^{\kappa-1} - 1\right]. \tag{3.97}$$

The work that is required to compress[10] the environment equals

$$W_{env} = p_{env}\,(V_2 - V_1) > 0. \tag{3.98}$$

According to Fig. 3.11 the partial energy equation for the object is

$$W_{Obj} = -W_{12} - W_{env} = \underbrace{W_{V,Obj}}_{=0} + W_{mech,Obj} + \underbrace{\Psi_{Obj}}_{=0}. \tag{3.99}$$

[9] The change of state is isentropic.

[10] Work is supplied, i.e. positive work, to compress the environment.

Thus, it is

$$\underbrace{-\frac{p_1 V_1}{\kappa - 1}\left[\left(\frac{V_1}{V_2}\right)^{\kappa-1} - 1\right] - p_{\text{env}}\left(V_2 - V_1\right)}_{=48.3848\,\text{J}} = \frac{1}{2} m_{\text{Obj}} c_2^2 + \underbrace{\Psi_{\text{Obj}}}_{=0} \tag{3.100}$$

respectively the velocity of the object

$$c = \sqrt{\frac{2 \cdot \left\{-\frac{p_1 V_1}{\kappa-1}\left[\left(\frac{V_1}{V_2}\right)^{\kappa-1} - 1\right] - p_{\text{env}}\left(V_2 - V_1\right)\right\}}{m_{\text{Obj}}}} = 80.32\,\frac{\text{m}}{\text{s}}. \tag{3.101}$$

Generally speaking, the first law of thermodynamics for the object obeys

$$W_{\text{Obj}} + Q = \Delta U_{\text{Obj}} + \frac{1}{2} m_{\text{Obj}}\left(c_2^2 - c_1^2\right) \tag{3.102}$$

with

$$W_{\text{Obj}} = -W_{12} - W_{\text{env}} = \underbrace{W_{\text{V,Obj}}}_{=0} + W_{\text{mech,Obj}} + \Psi_{\text{Obj}} = \frac{1}{2} m_{\text{Obj}}\left(c_2^2 - c_1^2\right) + \Psi_{\text{Obj}}. \tag{3.103}$$

Hence, it is

$$\frac{1}{2} m_{\text{Obj}}\left(c_2^2 - c_1^2\right) + \Psi_{\text{Obj}} + Q = \Delta U_{\text{Obj}} + \frac{1}{2} m_{\text{Obj}}\left(c_2^2 - c_1^2\right) \tag{3.104}$$

respectively

$$\Psi_{\text{Obj}} + Q = \Delta U_{\text{Obj}}. \tag{3.105}$$

The internal energy, and thus the temperature, of the object can therefore only be changed by friction or heat transfer. In this case, the object moves without friction and no heat is transferred: the gas space is adiabatic and the object is at ambient temperature. The same result can also be obtained with the fundamental equation of thermodynamics:

$$T\,\mathrm{d}S = \mathrm{d}U_{\text{Obj}} + \underbrace{p\,\mathrm{d}V}_{=0} \tag{3.106}$$

with

$$T\,\mathrm{d}S = \delta Q + \delta\Psi_{\text{Obj}} \tag{3.107}$$

so that

$$\delta Q + \delta\Psi_{\text{Obj}} = \mathrm{d}U_{\text{Obj}}. \tag{3.108}$$

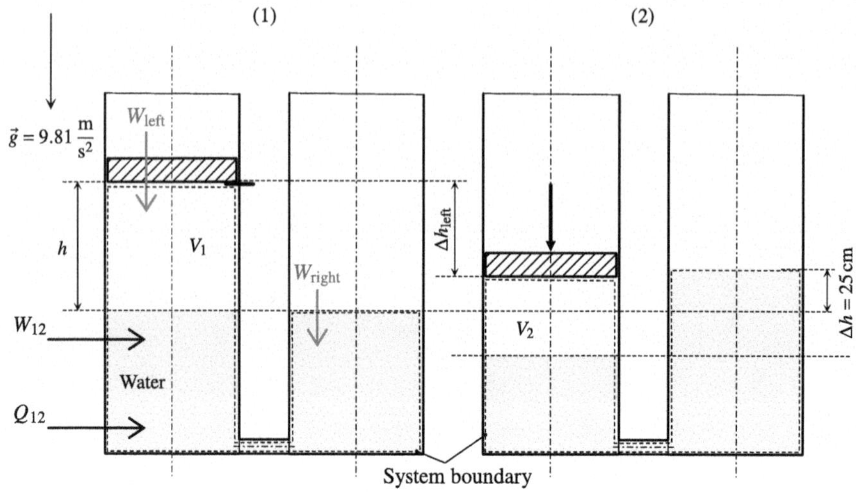

Fig. 3.12 Water columns

3.14 Air pressure in state (2) in the left vessel equals the pressure in the right vessel at the same vertical position, i.e.

$$p_2 = p_{env} + 2\,\Delta h\,\rho g = 1.049\,05\,\text{bar} \quad \text{with} \quad \Delta h = 25\,\text{cm}. \tag{3.109}$$

The total work for the selected system boundary, see Fig. 3.12 is composed of volume work, mechanical work and dissipation,[11] i.e.

$$W_{12} = W_{V,12} + W_{mech,12} + \underbrace{\Psi_{12}}_{=0} = -p_1 \underbrace{Ah}_{=V_1} \ln\frac{p_1}{p_2} + \underbrace{\rho\,\Delta h\,A}_{=\Delta m}\,g\,\Delta h = 69.7245\,\text{J}. \tag{3.110}$$

The first law of thermodynamics for the system boundary obeys

$$W_{12} + Q_{12} = \Delta U_{Air} + \Delta U_{Water} + \rho\,\Delta h\,Ag\,\Delta h \tag{3.111}$$

respectively with the caloric equation of state for an isothermal change of state

$$W_{12} + Q_{12} = m_{Air}c_v\underbrace{(T_2 - T_1)}_{=0} + m_{Water}c_W\underbrace{(T_2 - T_1)}_{=0} + \rho\,\Delta h\,Ag\,\Delta h. \tag{3.112}$$

[11] Since the change of state according to the problem discription is very slow, dissipation is neglected in this case.

Solving for the exchanged heat Q_{12} yields

$$Q_{12} = \rho\, \Delta h\, Ag\, \Delta h - W_{12} = -57.4620\,\text{J}. \tag{3.113}$$

The work W_{12} is supplied via the left tank and the right tank respectively, i.e.

$$W_{12} = W_{\text{left}} + W_{\text{right}}. \tag{3.114}$$

W_{right} is the volume work done on the environment on the right side by the rising water,[12] i.e.

$$W_{\text{right}} = W_{\text{V,env,right}} < 0. \tag{3.115}$$

The work on the left side W_{left} consists of volume change work on the environment,[13] the work supplied by the piston and the external work by slowly pushing down the piston, i.e.

$$W_{\text{left}} = \underbrace{W_{\text{V,env,left}}}_{>0} + \underbrace{W_{\text{P}}}_{=m_{\text{P}}g\,\Delta h_{\text{left}}} + W_{\text{ext}}. \tag{3.116}$$

Consequently, the external work required to conduct the change of state (1) → (2) is

$$\begin{aligned} W_{\text{ext}} &= W_{12} - W_{\text{V,env,left}} - W_{\text{V,env,right}} - m_{\text{P}}g\,\Delta h_{\text{left}} \\ &= W_{12} - p_{\text{env}} \underbrace{A\,(\Delta h_{\text{left}} - \Delta h)}_{=V_1 - V_2 > 0} - m_{\text{P}}g\,\Delta h_{\text{left}}. \end{aligned} \tag{3.117}$$

This means that the change in volume of the environment is only due to the compression of the air in the system, as the water is considered an incompressible liquid. The volume of the system becomes smaller, i.e. the environment expands and supports the required work from (1) → (2) - the external work is therefore smaller than W_{12}, i.e.

$$\begin{aligned} W_{\text{ext}} &= W_{12} - p_{\text{env}}\,(V_1 - V_2) - m_{\text{P}}g\,\Delta h_{\text{left}} \\ &= W_{12} - p_{\text{env}} \left(V_1 - \frac{p_1}{p_2} V_1 \right) - m_{\text{P}}g\,\Delta h_{\text{left}}. \end{aligned} \tag{3.118}$$

With

$$\begin{aligned} \Delta h_{\text{left}} &= h - (h_2 - \Delta h) = h - \left(\frac{V_2}{A} - \Delta h \right) \\ &= h - \left(\frac{p_1}{p_2 A} V_1 - \Delta h \right) = 0.2781\,\text{m} \end{aligned} \tag{3.119}$$

Hence, in case the mass of the piston is 3 kg the remaining external work equals

---

[12] This work is negative from a system point of view, as the volume of the environment is reduced.

[13] $W_{\text{V,env,left}}$ is positive since the environment supports pressing down the piston.

$$W_{\text{ext}} = W_{12} - p_{\text{env}} \left(V_1 - \frac{p_1}{p_2} V_1 \right) - m_{\text{P}} g \, \Delta h_{\text{left}} = 5.4335\,\text{J}. \tag{3.120}$$

3.15 After a very long time, a thermodynamic equilibrium is established, i.e. due to the thermal coupling to the environment, the very large environment will impose its temperature on the system.[14] It therefore applies

$$T_{\text{A}} = T_{\text{B}} = T_{\text{P}} = T_{\text{env}} = 293.15\,\text{K}. \tag{3.121}$$

In addition to the thermal equilibrium, a mechanical equilibrium is also found. This is as follows:

$$p_{\text{A}} = p_{\text{B}} = p. \tag{3.122}$$

Since the initial and final temperature are identical, the application of the thermal equation of state yields

$$p_{\text{A}} V_{\text{A}} = \underbrace{m_{\text{A}} RT}_{=\text{const.}} = p_{\text{A},0} V_{\text{A},0} \tag{3.123}$$

respectively

$$p_{\text{B}} V_{\text{B}} = \underbrace{m_{\text{B}} RT}_{=\text{const.}} = p_{\text{B},0} V_{\text{B},0}. \tag{3.124}$$

The entire volume of the cylinder remains constant so that

$$V_{\text{A}} + V_{\text{B}} = V_{\text{A},0} + V_{\text{B},0} \tag{3.125}$$

respectively

$$\frac{p_{\text{A},0} V_{\text{A},0}}{p_{\text{A}}} + \frac{p_{\text{B},0} V_{\text{B},0}}{p_{\text{B}}} = V_{\text{A},0} + V_{\text{B},0}. \tag{3.126}$$

Solving for $p = p_{\text{A}} = p_{\text{B}}$ results in

$$p = \frac{p_{\text{A},0} V_{\text{A},0} + p_{\text{B},0} V_{\text{B},0}}{V_{\text{A},0} + V_{\text{B},0}} = 2.1667\,\text{bar}. \tag{3.127}$$

Friction within the system is responsible for the equilibrium that is established. Without friction, the piston would not come to rest and would continuously perform an oscillating movement.

3.16 The first law of thermodynamics, see Fig. 3.13a, obeys

$$\underbrace{\dot{Q} + P}_{=0} + \dot{m}_1 h_1 + \dot{m}_2 h_2 = \dot{m}_3 h_3 = (\dot{m}_1 + \dot{m}_2)\, h_3 \tag{3.128}$$

[14] The initial temperature of the piston is irrelevant, as it will also be in thermodynamic equilibrium with the environment due to the thermal coupling.

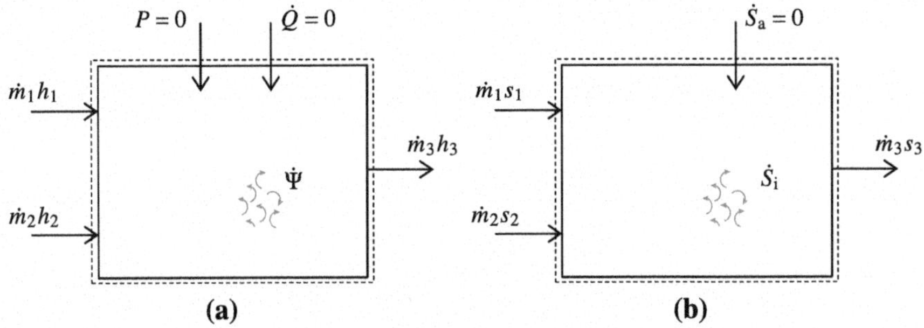

Fig. 3.13 Mixing: **a** First law of thermodynamics **b** Second law of thermodynamcis

respectively

$$\dot{m}_1 (h_3 - h_1) + \dot{m}_2 (h_3 - h_2) = 0. \tag{3.129}$$

Applying the caloric equation of state leads to

$$\dot{m}_1 c_p (T_3 - T_1) + \dot{m}_2 c_p (T_3 - T_2) = 0. \tag{3.130}$$

Finally, solving for T_3 yields

$$T_3 = \frac{\dot{m}_1 T_1 + \dot{m}_2 T_2}{\dot{m}_1 + \dot{m}_2} = 422.0\,\mathrm{K}. \tag{3.131}$$

The second law of thermodynamics, see Fig. 3.13b obeys

$$\dot{m}_1 s_1 + \dot{m}_2 s_2 + \underbrace{\dot{S}_a}_{=0} + \dot{S}_i = (\dot{m}_1 + \dot{m}_2) s_3 \tag{3.132}$$

respectively

$$\dot{S}_i = \dot{m}_1 (s_3 - s_1) + \dot{m}_2 (s_3 - s_2). \tag{3.133}$$

With the caloric equation of state for an isobaric change of state, one finally obtains

$$\dot{S}_i = \dot{m}_1 c_p \ln \frac{T_3}{T_1} + \dot{m}_2 c_p \ln \frac{T_3}{T_2} = 1.8879\,\frac{\mathrm{kW}}{\mathrm{K}}. \tag{3.134}$$

3.17 Let us start with the volumes occupied by the gas in state (1)

$$V_1 = \frac{\pi}{4} d^2 L_1 = 7.854 \times 10^{-4}\,\mathrm{m}^3 \tag{3.135}$$

and in state (2)

$$V_2 = \frac{\pi}{4} d^2 L_2 = 0.0055\,\mathrm{m}^3. \tag{3.136}$$

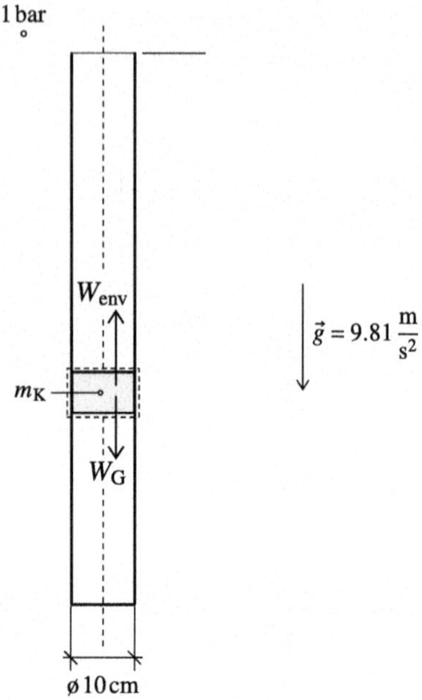

Fig. 3.14 Acceleration

According to Fig. 3.14, the entire work[15] acting on the test object is

$$W_{Obj} = -\underbrace{W_G}_{<0} - \underbrace{W_{env}}_{>0}. \tag{3.137}$$

The change of state of the gas is adiabatic and frictionless, i.e. isentropic, so that W_G only consists of the volume work of the trapped gas, i.e.

$$W_G = W_{V,G} + \underbrace{\Psi_G}_{=0} = \frac{p_1 V_1}{\kappa - 1}\left[\left(\frac{V_1}{V_2}\right)^{\kappa-1} - 1\right] \tag{3.138}$$

respectively

$$W_{Obj} = -\underbrace{\frac{p_1 V_1}{\kappa - 1}\left[\left(\frac{V_1}{V_2}\right)^{\kappa-1} - 1\right]}_{=W_G} - \underbrace{p_{env}\left(V_2 - V_1\right)}_{=W_{env}}. \tag{3.139}$$

The partial energy equation from the point of view of the object is therefore

[15] This is the net work crossing the system boundary.

$$W_{\text{Obj}} = \underbrace{W_{\text{V,Obj}}}_{=0} + \underbrace{\Psi_{\text{Obj}}}_{=0} + W_{\text{mech,Obj}} = m_{\text{K}} \left(\frac{1}{2} c_2^2 + g\, \Delta h \right). \tag{3.140}$$

Thus, it is

$$\underbrace{-\frac{p_1 V_1}{\kappa - 1} \left[\left(\frac{V_1}{V_2} \right)^{\kappa - 1} - 1 \right] - p_{\text{env}} (V_2 - V_1)}_{=W_{\text{Obj}}} = m_{\text{K}} \left(\frac{1}{2} c_2^2 + g\, \Delta h \right). \tag{3.141}$$

Finally, solving for p_1 yields

$$p_1 = -\frac{\kappa - 1}{V_1 \left[\left(\frac{V_1}{V_2} \right)^{\kappa - 1} - 1 \right]} \left(\frac{1}{2} m_{\text{K}} c_2^2 + m_{\text{K}} g\, \Delta h + p_{\text{env}} (V_2 - V_1) \right) = 4.8088\,\text{bar}. \tag{3.142}$$

The first law of thermodynamics for the object reads as

$$W_{\text{Obj}} + Q = \Delta U_{\text{Obj}} + m_{\text{K}} \left(\frac{1}{2} c_2^2 + g\, \Delta h \right). \tag{3.143}$$

respectively by substitution of W_{Obj} according to Eq. 3.140

$$m_{\text{K}} \left(\frac{1}{2} c_2^2 + g\, \Delta h \right) + Q = \Delta U_{\text{Obj}} + m_{\text{K}} \left(\frac{1}{2} c_2^2 + g\, \Delta h \right). \tag{3.144}$$

Hence, it follows

$$Q = \Delta U_{\text{Obj}}. \tag{3.145}$$

For a frictionless movement of the object only heat can modify the temperature of the object. However, in this case the object is adiabatic, so that the temperature of the moving object remains unchanged.

3.18 For the isentropic[16] change of the enclosed gas, see Fig. 3.15, it is

$$p_1 V_1^{\kappa} = p_2 V_2^{\kappa} \quad \text{with} \quad V_1 = A h_1. \tag{3.146}$$

Solving for V_2 leads to

$$V_2 = A h_1 \left(\frac{p_1}{p_2} \right)^{\frac{1}{\kappa}} = 0.0073\,\text{m}^3. \tag{3.147}$$

---

[16] The change of state shall be adiabatic and very slowly, i.e. free of friction.

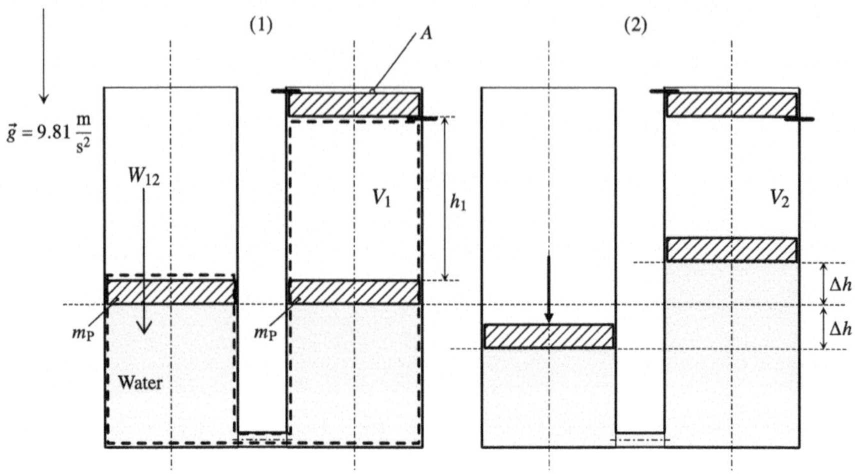

Fig. 3.15 Water columns

The distance by which the left piston must be pressed down is therefore

$$\Delta h = h_1 - \frac{V_2}{A} = 0.2343\,\text{m}. \tag{3.148}$$

The partial energy equation for the system is

$$W_{12} = W_{V,12} + W_{\text{mech},12} + \underbrace{\Psi_{12}}_{=0} \tag{3.149}$$

with

$$W_{V,12} = \frac{p_1 V_1}{\kappa - 1}\left[\left(\frac{V_1}{V_2}\right)^{\kappa-1} - 1\right] = 657.0410\,\text{J} \tag{3.150}$$

and

$$W_{\text{mech},12} = -m_P g\,\Delta h + m_P g\,\Delta h + \rho\,\Delta h\,A g\,\Delta h = 10.7703\,\text{J}. \tag{3.151}$$

Finally, the total work W_{12} obeys

$$W_{12} = W_{V,12} + W_{\text{mech},12} = 667.8113\,\text{J}. \tag{3.152}$$

3.19 The mass of the enclosed gas in the cylinder is according to the thermal equation of state

$$m = \frac{p_1 V_1}{R T_1} = 0.0023\,\text{kg}. \tag{3.153}$$

The specific heat capacity at constant volume c_v equals

$$c_v = \frac{R}{\kappa - 1} = 717.5\,\frac{\text{J}}{\text{kgK}}. \tag{3.154}$$

Due to the adiabatic and frictionless change of state, the isentropic correlation

$$p_1 V_1^\kappa = p_2 V_2^\kappa \tag{3.155}$$

can be applied. Solving for p_2 brings

$$p_2 = p_1 \left(\frac{V_1}{V_2}\right)^\kappa = 2.639\,\text{bar}. \tag{3.156}$$

The temperature in state (2) then follows

$$T_2 = T_1 \left(\frac{p_2}{p_1}\right)^{\frac{\kappa-1}{\kappa}} = 395.8524\,\text{K}. \tag{3.157}$$

The change of the internal energy obeys the caloric equation of state, i.e.

$$\Delta U = mc_v\,(T_2 - T_1) = 159.7540\,\text{J}. \tag{3.158}$$

3.20 The work exchanged with the environment across the system boundary follows according to Fig. 3.16. Thus, from an environmental's point of view it applies:

$$W_{12} = \underbrace{W_{\text{Stirrer}}}_{\text{IN}} - \underbrace{W_{\text{V,env,12}}}_{\text{OUT}}. \tag{3.159}$$

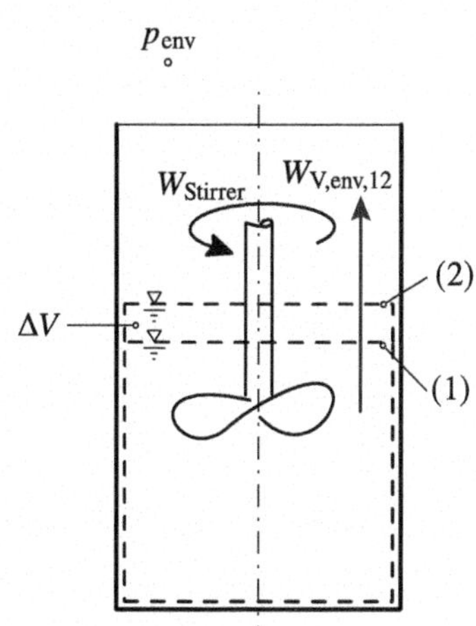

Fig. 3.16 Viscous liquid

W_{Stirrer} is supplied to the system by the stirrer, i.e. it has a positive impact, whereas $W_{\text{V,env},12}$ is the work that is required to compress the environment. This part of the work is taken from the expansion of the fluid. The total work W_{12} is thus reduced by the work to compress the environment, so that a negative sign appears in Eq. 3.159. W_{12} is therefore the total work that is supplied to the system from outside. It further is

$$W_{\text{Stirrer}} = M\omega\,\Delta t. \tag{3.160}$$

As mentioned, $W_{\text{V,env},12}$ is the work required to compress the environment due to the expansion of the fluid, i.e. the environment is supplied with volume work. Therefore, from the environment's point of work it must be positive:

$$W_{\text{V,env},12} = p_{\text{env}}\,\Delta V > 0. \tag{3.161}$$

The total work follows

$$W_{12} = M\omega\,\Delta t - p_{\text{env}}\,\Delta V = M2\pi n\,\Delta t - p_{\text{env}}\,(1.1V_0 - V_0) = 6.7358 \times 10^6\,\text{J}. \tag{3.162}$$

From the point of view of the system, the partial energy equation can be applied:

$$W_{12} = \underbrace{W_{\text{V},12}}_{<0} + W_{\text{mech},12} + \Psi_{12}. \tag{3.163}$$

$W_{\text{V},12}$ represents the volume work of the viscous fluid and should not be confused with $W_{\text{V,env},12}$. The mechanical work can be neglected since the position of the centre of gravity shall be disregarded, so that

$$W_{12} = -\int_1^2 p\,\text{d}V + \Psi_{12}. \tag{3.164}$$

Since the system is open at the top, the ambient pressure is imposed on the fluid, i.e.

$$W_{12} = -p_{\text{env}}\,\Delta V + \Psi_{12}. \tag{3.165}$$

From Eqs. 3.162 and 3.165 it follows

$$\Psi_{12} = M2\pi n\,\Delta t = 6.7858 \times 10^6\,\text{J}. \tag{3.166}$$

Let us assume that the change of state is adiabatic. The first law of thermodynamics then obeys

$$W_{12} + \underbrace{Q_{12}}_{=0} = \Delta U. \tag{3.167}$$

However, since the fluid is not ideal, i.e. neither an ideal gas nor an incompressible liquid, the change of state cannot be calculated easily. Under these conditions the second law of thermodynamics reads as

$$\int_1^2 \frac{\delta \Psi}{T} + \underbrace{S_a}_{=0} = S_2 - S_1 . \tag{3.168}$$

Here, too, the caloric state variable S cannot be calculated so easily because a fluid model is required. A balance of exergy yields

$$E_{x,2} = E_{x,1} + \underbrace{E_{x,Q}}_{=0} + W_{eff} - \Delta E_{x,V} \tag{3.169}$$

with

$$W_{eff} = W_{12} + p_{env}\, \Delta V = M\omega\, \Delta t . \tag{3.170}$$

Hence, the loss of exergy is

$$\Delta E_{x,V} = E_{x,1} - E_{x,2} + \underbrace{M\omega\, \Delta t}_{=\Psi_{12}} . \tag{3.171}$$

3.21

(a) Figure 3.17 shows the path-time diagrams for both cases. Although not explicitly asked, the oscillation equation is derived below. The key is the formulation of

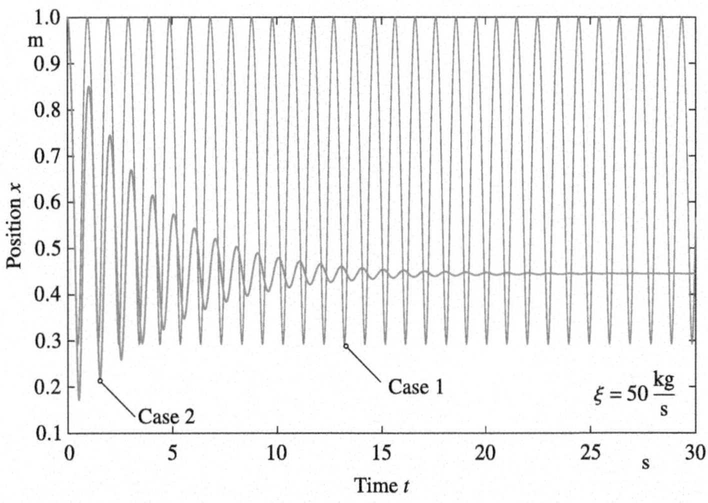

Fig. 3.17 Path-time diagram

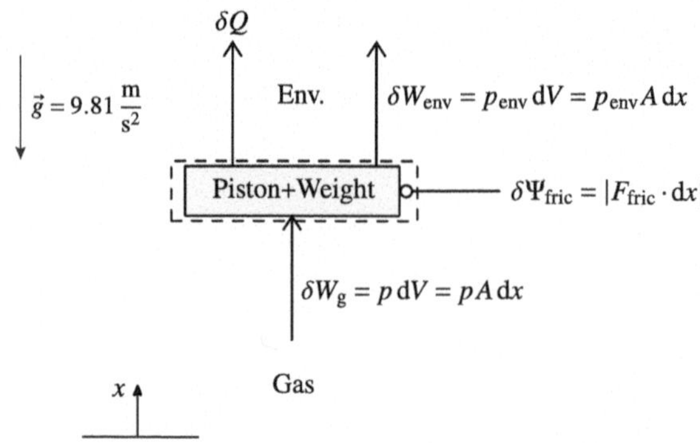

Fig. 3.18 First law of thermodynamics: Piston

the first law of thermodynamics for the piston. For this purpose, the frictional Case 2 is examined first, from which Case 1 can be derived later by simplification. According to Fig. 3.18, the 1st law of thermodynamics is in differential formulation:

$$\delta Q + \delta W_{\mathrm{g}} - \delta W_{\mathrm{env}} = \mathrm{d}U_{\mathrm{P}} + \mathrm{d}E_{\mathrm{a,P}} \tag{3.172}$$

δW_{g} is the work from the system Gas,[17] i.e.

$$\delta W_{\mathrm{g}} = -\delta W_{\mathrm{gas}} = -\left(-p\,\mathrm{d}V + \underbrace{\delta W_{\mathrm{gas,\,mech}}}_{=0} + \underbrace{\delta \Psi_{\mathrm{gas}}}_{=0}\right) = p\,\mathrm{d}V. \tag{3.173}$$

δW_{env} is the work at the uniform environment that is compressed, respectively expands, i.e.

$$\delta W_{\mathrm{env}} = p_{\mathrm{env}}\,\mathrm{d}V. \tag{3.174}$$

Following the fundamental equation of thermodynamics

$$T\,\mathrm{d}S = \mathrm{d}U_{\mathrm{P}} + p\,\mathrm{d}V \tag{3.175}$$

it applies for the incompressible piston[18]

$$T\,\mathrm{d}S = \delta Q + \delta \Psi_{\mathrm{fric}} = \mathrm{d}U_{\mathrm{P}} = 0. \tag{3.176}$$

The change of mechanical energies in differential notation obeys[19]

$$\mathrm{d}E_{\mathrm{a,P}} = mg\,\mathrm{d}x + mc\,\mathrm{d}c. \tag{3.177}$$

[17] The minus sign is due to the arrow that is pointing out of the system Gas.

[18] According to the problem description the temperature of the piston shall not change.

[19] By integration it can easily be shown that $\Delta E_{\mathrm{a,12}} = mg\,\Delta x + \frac{1}{2}m\left(c_2^2 - c_1^2\right)$.

Therefore, the 1st law of thermodynamics is as follows[20]:

$$-\delta\Psi_{\text{fric}} + p\,\mathrm{d}V - p_{\text{env}}\,\mathrm{d}V = mg\,\mathrm{d}x + mc\,\mathrm{d}c. \tag{3.178}$$

Reformulation leads to

$$-\delta\Psi_{\text{fric}} + p\,\mathrm{d}V - p_{\text{env}}\,\mathrm{d}V = mg\,\mathrm{d}x + m\frac{\mathrm{d}x}{\mathrm{d}t}\,\mathrm{d}c \tag{3.179}$$

respectively

$$-\delta\Psi_{\text{fric}} + p\,\mathrm{d}V - p_{\text{env}}\,\mathrm{d}V = mg\,\mathrm{d}x + m\frac{\mathrm{d}c}{\mathrm{d}t}\,\mathrm{d}x \tag{3.180}$$

and finally

$$-\delta\Psi_{\text{fric}} + pA\,\mathrm{d}x - p_{\text{env}}A\,\mathrm{d}x = mg\,\mathrm{d}x + m\frac{\mathrm{d}^2x}{\mathrm{d}t^2}\,\mathrm{d}x. \tag{3.181}$$

A model must be provided for the friction. We assume that friction is proportional to the velocity of the piston and the distance moved,[21] i.e.

$$\delta\Psi_{\text{fric}} \propto \frac{\mathrm{d}x}{\mathrm{d}t}\mathrm{d}x \tag{3.182}$$

respectively

$$\delta\Psi_{\text{fric}} = \xi\frac{\mathrm{d}x}{\mathrm{d}t}\mathrm{d}x = |F_{\text{fric}}\cdot\mathrm{d}x| \tag{3.183}$$

Substituting Eq. 3.183 into Eq. 3.181 yields

$$\frac{\mathrm{d}^2x}{\mathrm{d}t^2} = \frac{pA}{m} - \frac{p_{\text{env}}A}{m} - g - \frac{\xi}{m}\frac{\mathrm{d}x}{\mathrm{d}t} \tag{3.184}$$

with

$$A = \frac{\pi}{4}D^2. \tag{3.185}$$

Now the pressure is determined depending on the position of the piston. For the most general case, a polytropic change of state is assumed, i.e.

$$p = p_0\left(\frac{V_0}{V}\right)^n = p_0\left(\frac{x_0}{x}\right)^n. \tag{3.186}$$

---

[20] One can also apply the partial energy equation for the piston. The total work that occurs at the piston is $\delta W = \delta W_g - \delta W_{\text{env}}$. This work is composed of mechanical work that equals $\mathrm{d}E_{\text{a,P}}$ and dissipation while there is no volume work, i.e. $\delta W = \delta W_g - \delta W_{\text{env}} = \mathrm{d}E_{\text{a,P}} + \delta\Psi$.

[21] This can ensure that $\delta\Psi \geq 0$.

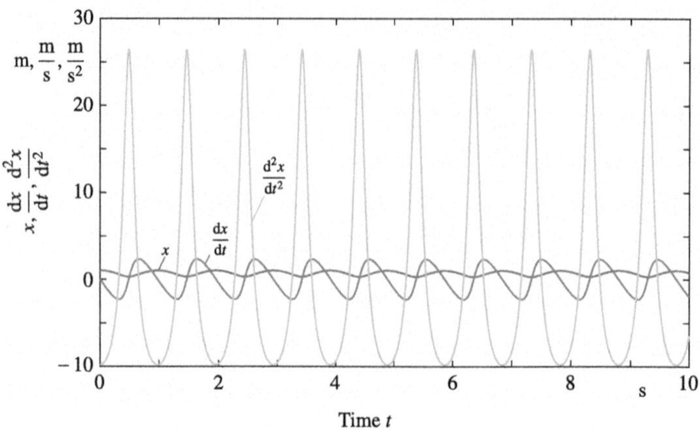

Fig. 3.19 Case 1: Position, velocity, acceleration

The equation of motion is therefore

$$\boxed{\frac{d^2x}{dt^2} = \frac{p_0 x_0^n A}{m x^n} - \frac{p_{env} A}{m} - g - \frac{\xi}{m}\frac{dx}{dt}} \tag{3.187}$$

This equation of motion can now be applied to both cases:

- Case 1:
  In this case, the change of state for the gas is isentropic. This means that

$$n = \kappa. \tag{3.188}$$

  Furthermore, no friction occurs between the piston and the wall, so that

$$\xi = 0. \tag{3.189}$$

  The differential equation for this case is hence

$$\boxed{\frac{d^2x}{dt^2} = \frac{p_0 x_0^\kappa A}{m x^\kappa} - \frac{p_{env} A}{m} - g} \tag{3.190}$$

  This differential equation can be solved numerically. Figure 3.19 shows position, velocity and acceleration of the piston for Case 1.
- Case 2:
  Due to the perfect thermal coupling to the environment, the gas in the cylinder always has the imposed ambient temperature. The change of state is therefore isothermal, so that

$$n = 1. \tag{3.191}$$

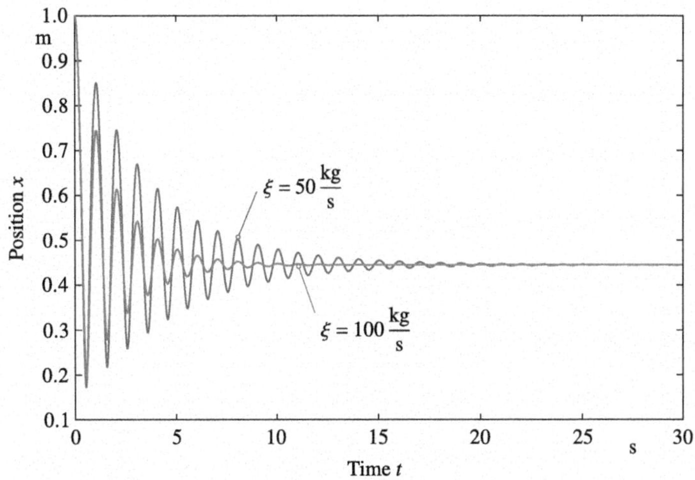

Fig. 3.20 Case 2: Influence of friction parameter ξ

Here, friction occurs between the piston and the wall, so that

$$\xi > 0. \tag{3.192}$$

The differential equation for this case is hence

$$\boxed{\frac{d^2x}{dt^2} = \frac{p_0 x_0 A}{mx} - \frac{p_{env} A}{m} - g - \frac{\xi}{m}\frac{dx}{dt}} \tag{3.193}$$

Figure 3.17 depicts a comparison of the position x for Cases 1 and 2. $\xi = 50\,\frac{kg}{s}$ has been assumed for Case 2. However, the resulting state of equilibrium is independent of ξ. Only the time needed to reach this state varies with ξ, cf. Fig. 3.20.

(b) In Case 2, a state of equilibrium is reached after a long period of time due to friction, resulting in the following equilibrium of forces:

$$p_\infty A = p_{env} A + mg. \tag{3.194}$$

The resulting pressure is therefore

$$p_\infty = p_{env} + \frac{mg}{A} = 2.249\,\text{bar}. \tag{3.195}$$

In part (a) it has been shown that the change of state is isothermal because of the perfect thermal coupling to the environment, i.e.

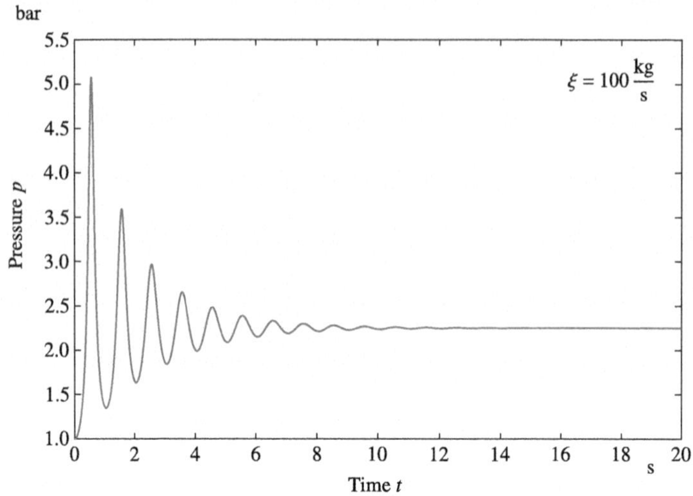

Fig. 3.21 Case 2: Pressure p as function of time t

$$T_\infty = 300\,\text{K}. \tag{3.196}$$

In order to infer the position of the piston, the volume of air at equilibrium is first determined. For an isothermal change of state, the following therefore applies, cf. also Fig. 3.21,

$$p_0 V_0 = p_\infty V_\infty. \tag{3.197}$$

Thus it is

$$V_\infty = V_0 \frac{p_0}{p_\infty} = \frac{\pi}{4} D^2 x_0 \frac{p_0}{p_\infty}. \tag{3.198}$$

Dividing by $A = \frac{\pi}{4} D^2$ yields the position in equilibrium, i.e.

$$x_\infty = x_0 \frac{p_0}{p_\infty} = 0.4446\,\text{m}. \tag{3.199}$$

This result can also be taken from the numerical solution in Fig. 3.20.

(c) To determine the amount of heat, the 1st law of thermodynamics is formulated from (0) to (∞) for the entire system, see Fig. 3.22. The work that acts on the environment $W_{0\to\text{env}}$ and the heat exchanged with the environment $Q_{0\to\text{env}}$ is exchanged across the system boundary. Friction Ψ_{fric} is part of the system and is not exchanged at the system boundary. The 1st law of thermodynamics therefore obeys

$$W_{0\to\text{env}} + Q_{0\to\text{env}} = \Delta U_{\text{P}} + \Delta U_{\text{W}} + \Delta U_{\text{g}} + mg\left(x_\infty - x_0\right). \tag{3.200}$$

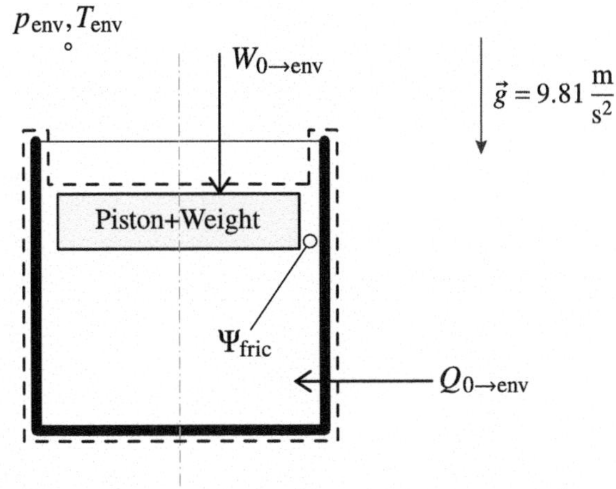

Fig. 3.22 Case 2: First law of thermodynamics

The overall system consisting of piston, wall and gas is isothermal, so that the 1st law of thermodynamics is further simplified, i.e.

$$W_{0\to env} + Q_{0\to env} = mg\,(x_\infty - x_0)\,. \tag{3.201}$$

The work on the environment is volume work and is calculated as follows:

$$W_{0\to env} = -p_{env} A\,(x_\infty - x_0) = 436.1846\,\text{J} > 0. \tag{3.202}$$

The work must be positive because the environment expands as the piston moves downwards. Thus, work is supplied to the system. The heat therefore results in

$$Q_{0\to env} = mg\,(x_\infty - x_0) + p_{env} A\,(x_\infty - x_0) = -981\,\text{J}. \tag{3.203}$$

The partial energy equation from the point of view of the system, see Fig. 3.22, is:

$$W_{0\to env} = W_V + \Psi_{fric} + mg\,(x_\infty - x_0)\,. \tag{3.204}$$

Since the change of state is supposed to be isothermal, the following applies for the volume work of the gas

$$W_V = -p_0 V_0 \ln\frac{V_\infty}{V_0} = -p_0 A x_0 \ln\frac{x_\infty}{x_0}. \tag{3.205}$$

The dissipated energy thus is

$$\Psi_{fric} = W_{0\to env} + p_0 A x_0 \ln\frac{x_\infty}{x_0} - mg\,(x_\infty - x_0) = 344.4293\,\text{J}. \tag{3.206}$$

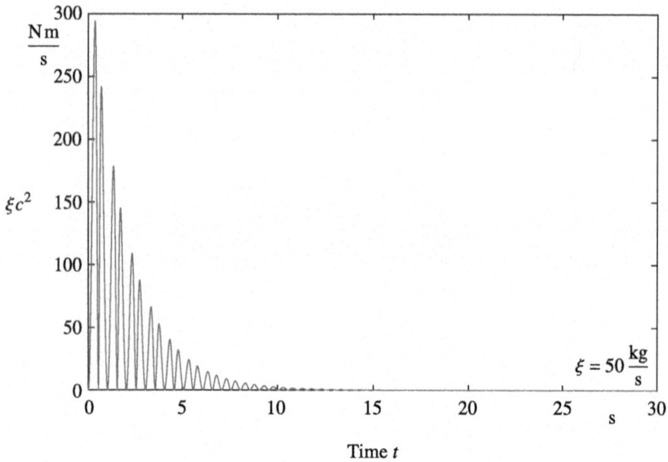

Fig. 3.23 Friction

Alternatively, one can also consider the introduced model for the friction, see Eq. 3.183

$$\delta\Psi_{\text{fric}} = \xi\frac{\mathrm{d}x}{\mathrm{d}t}\mathrm{d}x = \xi c\,\mathrm{d}x = \xi c^2\,\mathrm{d}t \quad \text{since} \quad \mathrm{d}x = c\,\mathrm{d}t. \tag{3.207}$$

For the total dissipated energy one obtains

$$\Psi_{\text{fric}} = \int_0^\infty \delta\Psi_{\text{fric}} = \int_0^\infty \xi c^2\,\mathrm{d}t. \tag{3.208}$$

The course of $f(t) = \xi c^2$ is shown in Fig. 3.23 and follows from the numerical solution of the differential equation. The integration of this function results in

$$\Psi_{\text{fric}} = \int_0^\infty \xi c^2\,\mathrm{d}t = 344.4293\,\mathrm{J}. \tag{3.209}$$

The result of Ψ_{fric} is independent of the friction parameter ξ. This only influences how long it takes until a state of equilibrium is reached.

(d) For Case 1, the 1st law of thermodynamics is formulated for the piston from state (0) to bottom dead centre (BDC), see Fig. 3.24. The piston does not exchange heat with the environment, as it is always at ambient temperature according to

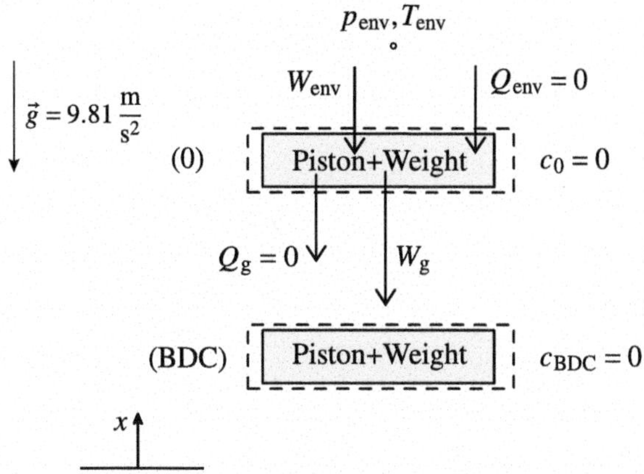

Fig. 3.24 Case 1: 1st law of thermodynamics from (0) to (BDC)

the task, i.e. $Q_{env} = 0$. The change of state of the gas shall also be adiabatic and frictionless, so that $Q_g = 0$. The first law of thermodynamics is therefore[22]

$$W_{env} - W_g = \underbrace{\Delta U_P}_{=0} + mg\,(x_{BDC} - x_0) + \underbrace{\frac{1}{2}m\left(c_{BDC}^2 - c_0^2\right)}_{=0}. \tag{3.210}$$

The work on the gas is calculated as

$$W_g = -\int_0^{BDC} p\,dV + \underbrace{\Psi_g}_{=0} > 0. \tag{3.211}$$

The change of state is isentropic, i.e.

$$W_g = -\int_0^{BDC} p\,dV = \frac{p_0 V_0}{\kappa - 1}\left[\left(\frac{p_{BDC}}{p_0}\right)^{\frac{\kappa-1}{\kappa}} - 1\right]. \tag{3.212}$$

With

$$\frac{p_{BDC}}{p_0} = \left(\frac{V_0}{V_{BDC}}\right)^{\kappa} = \left(\frac{x_0}{x_{BDC}}\right)^{\kappa} \tag{3.213}$$

it yields

$$W_g = \frac{p_0 V_0}{\kappa - 1}\left[\left(\frac{x_0}{x_{BDC}}\right)^{\kappa-1} - 1\right]. \tag{3.214}$$

[22] The fundamental equation for a solid is $T\,dS = dU$. According to the second law of thermodynamics it further applies $T\,dS = \delta Q + \delta\Psi$. Here it is $\delta\Psi = 0$, so that $\delta Q = dU$. I.e. if U does not change, the sum of the heats is also 0.

The work on the environment is

$$W_{\text{env}} = p_{\text{env}} A \left(x_0 - x_{\text{BDC}}\right) > 0. \tag{3.215}$$

The overall equation therefore follows

$$\boxed{p_{\text{env}} A \left(x_0 - x_{\text{BDC}}\right) - \frac{p_0 V_0}{\kappa - 1}\left[\left(\frac{x_0}{x_{\text{BDC}}}\right)^{\kappa-1} - 1\right] = mg\left(x_{\text{BDC}} - x_0\right)} \tag{3.216}$$

This equation can be solved numerically for x_{BDC} and one obtains

$$x_{\text{BDC}} = 0.2918\,\text{m}. \tag{3.217}$$

The pressure then follows

$$p_{\text{BDC}} = p_0 \left(\frac{x_0}{x_{\text{BDC}}}\right)^{\kappa} = 5.6158\,\text{bar} \tag{3.218}$$

and the temperature

$$T_{\text{BDC}} = T_0 \left(\frac{p_{\text{BDC}}}{p_0}\right)^{\frac{\kappa-1}{\kappa}} = 491.6105\,\text{K}. \tag{3.219}$$

The graph of pressure and temperature over the oscillation can also be taken from Fig. 3.25.

3.22 The sketch for solving the task for both options A and B is shown in Fig. 3.26. According to the task, the piston moves very slowly, i.e. the dissipation in the piston is neglected in the following. It therefore applies

$$\Psi_{12,\text{G}} = 0. \tag{3.220}$$

The change of state is isothermal and isobaric, i.e. the temperature of the gas does not change. Despite the constant temperature of the gas, its internal energy will change because the thermal properties of the gas have changed from state (1) to state (2) due to the chemical reaction, i.e.

$$\Delta U_{\text{G}} \neq 0. \tag{3.221}$$

The internal energy of the piston, on the other hand, does not change because the temperature in both states is ambient temperature and the chemical composition of the piston has not changed, so that

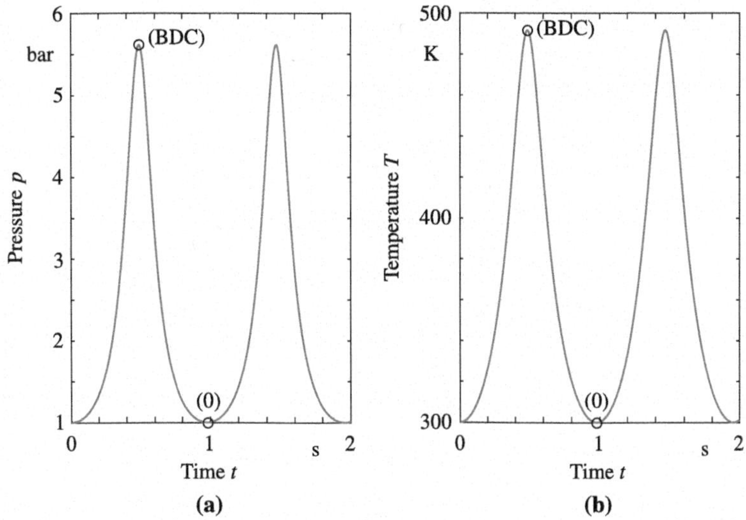

Fig. 3.25 Case 1: **a** Pressure p and **b** Temperature T as function of time t

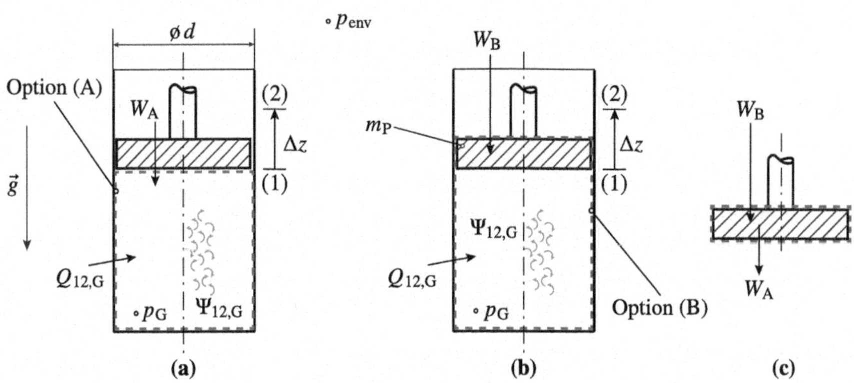

Fig. 3.26 **a** Energy balance according to Option (A) **b** Energy balance according to Option (B) **c** Partial energy equation for the piston

$$\Delta U_P = 0. \tag{3.222}$$

(A) Let us start with the balancing according to the system boundary shown in Fig. 3.26a. The first law of thermodynamics obeys

$$Q_{12,G} + W_A = \Delta U_G + \underbrace{\Delta E_{pot,G}}_{=0}. \tag{3.223}$$

In this equation, $Q_{12,G}$ represents the heat that was released by the chemical process inside the cylinder and is now transferred to the environment. This can happen via the cylinder wall and via the piston, i.e. across the selected system boundary, and totals $Q_{12,\mathrm{G}} = -10\,\mathrm{kJ}$. The partial energy equation clarifies what work is required on the system to achieve the change of state. This work is exchanged at the upper system edge and is composed as follows:

$$W_{\mathrm{A}} = W_{\mathrm{V,G}} + \underbrace{W_{\mathrm{mech,G}}}_{=0} + \underbrace{\Psi_{12,\mathrm{G}}}_{=0} = -\int_1^2 p_{\mathrm{G}}\,\mathrm{d}V. \tag{3.224}$$

Since the pressure inside the cylinder shall be constant, it yields

$$W_{\mathrm{A}} = -p_{\mathrm{G}} \int_1^2 \mathrm{d}V = -p_{\mathrm{G}}\,\Delta V = -p_{\mathrm{G}} A\,\Delta z \tag{3.225}$$

with

$$A = \frac{\pi}{4} d^2. \tag{3.226}$$

The energy balance is therefore

$$\boxed{\Delta U_{\mathrm{G}} = Q_{12,\mathrm{G}} - p_{\mathrm{G}} A\,\Delta z = -11.4726\,\mathrm{kJ}} \tag{3.227}$$

(B) Now the energy balance is carried out according to Fig. 3.26b. The gas inside the cylinder also releases the amount of heat $Q_{12,\mathrm{G}}$, which is partly released via the cylinder wall and partly via the piston. This means that the piston may initially receive part of the heat from the gas, but at the upper end it is completely released into the environment, as the piston has the same temperature in state (1) and (2). An incompressible solid can only change its temperature due to friction or heat:

$$\delta q + \delta w = \mathrm{d}u + \mathrm{d}e_{\mathrm{a}} \tag{3.228}$$

with

$$\delta w = \underbrace{\delta w_{\mathrm{V}}}_{=0} + \underbrace{\delta w_{\mathrm{mech}}}_{=\mathrm{d}e_{\mathrm{a}}} + \delta\psi_{\mathrm{fric}}. \tag{3.229}$$

So that

$$\mathrm{d}u = c\,\mathrm{d}T = \delta\psi_{\mathrm{fric}} + \delta q. \tag{3.230}$$

There is no friction for the piston, and in order to keep its temperature constant, the net heat must be zero, i.e. the piston must release the heat absorbed from the gas. So in total, the entire heat crossing the system boundary in option (B) is $Q_{12,\mathrm{G}} = -10\,\mathrm{kJ}$ as it is in option (A). The first law of thermodynamics reads as

$$Q_{12,\mathrm{G}} + W_{\mathrm{B}} = \Delta U_{\mathrm{G}} + \underbrace{\Delta U_{\mathrm{P}}}_{=0} + \underbrace{\Delta E_{\mathrm{pot,G}}}_{=0} + \Delta E_{\mathrm{pot,P}} = \Delta U_{\mathrm{G}} + m_{\mathrm{P}} g\, \Delta z. \tag{3.231}$$

The partial energy equation describes how much work W_{B} is required for the change of state of the system:

$$W_{\mathrm{B}} = W_{\mathrm{V,G}} + \underbrace{W_{\mathrm{mech,G}}}_{=0} + \underbrace{W_{\mathrm{mech,P}}}_{=m_{\mathrm{P}} g\, \Delta z} + \underbrace{\Psi_{12,\mathrm{G}}}_{=0} + \underbrace{\Psi_{\mathrm{fric,P}}}_{=0} = -\int_1^2 p_{\mathrm{G}}\, \mathrm{d}V + m_{\mathrm{P}} g\, \Delta z. \tag{3.232}$$

Since the pressure in the cylinder is constant, one obtains

$$W_{\mathrm{B}} = -p_{\mathrm{G}} A\, \Delta z + m_{\mathrm{P}} g\, \Delta z. \tag{3.233}$$

Combining the first law of thermodynamics and the partial energy equation, it yields

$$\boxed{\Delta U_{\mathrm{G}} = Q_{12,\mathrm{G}} - p_{\mathrm{G}} A\, \Delta z = -11.4726\,\mathrm{kJ}} \tag{3.234}$$

Furthermore, the correlation between W_{A} and W_{B} results from the first law of thermodynamics for the piston, see Fig. 3.26c. It has already been shown that the net heat is zero, so that only the work has to be considered

$$W_{\mathrm{B}} - W_{\mathrm{A}} = \underbrace{\Delta U_{\mathrm{P}}}_{=0} + \underbrace{\Delta E_{\mathrm{pot,P}}}_{=m_{\mathrm{P}} g\, \Delta z}. \tag{3.235}$$

Hence it is

$$\boxed{W_{\mathrm{A}} = W_{\mathrm{B}} - m_{\mathrm{P}} g\, \Delta z} \tag{3.236}$$

W_{B} vividly depicts the work done to compress the environment, i.e.

$$W_{\mathrm{B}} = p_{\mathrm{env}} \underbrace{\Delta V_{\mathrm{env}}}_{<0} = -p_{\mathrm{env}} A\, \Delta z < 0. \tag{3.237}$$

Thus

$$W_{\mathrm{A}} = -p_{\mathrm{env}} A\, \Delta z - m_{\mathrm{P}} g\, \Delta z \tag{3.238}$$

With Eq. 3.225 it is

$$-p_{\mathrm{G}} A\, \Delta z = -p_{\mathrm{env}} A\, \Delta z - m_{\mathrm{P}} g\, \Delta z \tag{3.239}$$

respectively

$$\boxed{p_\mathrm{G} = p_\mathrm{env} + \frac{m_\mathrm{P} g}{A}} \tag{3.240}$$

3.23 This is an isobaric change of state, as the external forces due to ambient pressure and the weight of the piston on the gas remain constant. The change of state also takes place very slowly, i.e. dissipation is neglected, and without acceleration. The balance of forces on the piston therefore obeys

$$p_\mathrm{env} A + m_\mathrm{P} g = pA. \tag{3.241}$$

The gas pressure is therefore

$$p = p_\mathrm{env} + \frac{m_\mathrm{P} g}{A} = 1.3066 \times 10^5\,\mathrm{Pa} = \mathrm{const.} \tag{3.242}$$

The first law of thermodynamics for the gas yields[23]

$$Q_{12} + W_{12} = U_2 - U_1 \quad \text{with} \quad Q_{12} = -120\,\mathrm{kJ}. \tag{3.243}$$

With the partial energy equation

$$W_{12} = W_{\mathrm{V},12} + \underbrace{W_{\mathrm{mech},12} + \Psi_{12}}_{=0} = -\int_1^2 p\,\mathrm{d}V = -pA\,\Delta z \tag{3.244}$$

the first law of thermodynamics simplifies to

$$Q_{12} - pA\,\Delta z = U_2 - U_1 = m\,(u_2 - u_1)\,. \tag{3.245}$$

Applying the caloric equation of state results in

$$Q_{12} - pA\,\Delta z = mc_v\,(T_2 - T_1) = c_v\,(mT_2 - mT_1)\,. \tag{3.246}$$

According to the thermal equation of state it is

$$mT_1 = \frac{pV_1}{R} \quad \text{and} \quad mT_2 = \frac{pV_2}{R}. \tag{3.247}$$

Hence, it follows

$$Q_{12} - pA\,\Delta z = \frac{c_v}{R}\,(pV_2 - pV_1) = p\frac{c_v}{R}A\,\Delta z \tag{3.248}$$

---

[23] The potential energy of the gas is to be neglected. Likewise, the gas is at rest in state (1) and state (2).

respectively

$$Q_{12} = A\,\Delta z\,p\left(\frac{c_v}{R}+1\right) \quad \text{with} \quad R = c_v\,(\kappa - 1) = 0.2868\,\frac{\text{kJ}}{\text{kgK}}. \tag{3.249}$$

Finally, the vertical displacement obeys

$$\Delta z = \frac{Q_{12}}{Ap\left(\frac{c_v}{R}+1\right)} = -0.3280\,\text{m}. \tag{3.250}$$

3.24

(a) The change of state is to proceed with a constant polytropic exponent n, i.e. for (1)–(3) applies

$$\frac{T_3}{T_1} = \left(\frac{p_3}{p_1}\right)^{\frac{n-1}{n}}. \tag{3.251}$$

The states (1) and (3) are each determined by pressure and temperature, so that the equation can be solved for n, i.e.

$$n = \frac{\ln\frac{p_1}{p_3}}{\ln\frac{T_3 p_1}{T_1 p_3}} = 1.5585. \tag{3.252}$$

(b) p, v- and T, s-diagram are shown in Fig. 3.27. The function $p = f(v)$ follows

$$pv^n = p_1 v_1^n \tag{3.253}$$

so that

$$\boxed{p(v) = p_1 v_1^n \frac{1}{v^n}} \tag{3.254}$$

For the function $T = f(s)$ it is

$$\begin{aligned} s - s_1 &= c_p \ln\frac{T}{T_1} - R\ln\frac{p}{p_1} \\ &= c_p \ln\frac{T}{T_1} - R\ln\left(\frac{T}{T_1}\right)^{\frac{n}{n-1}} \\ &= c_p \ln\frac{T}{T_1} - R\frac{n}{n-1}\ln\frac{T}{T_1} \end{aligned} \tag{3.255}$$

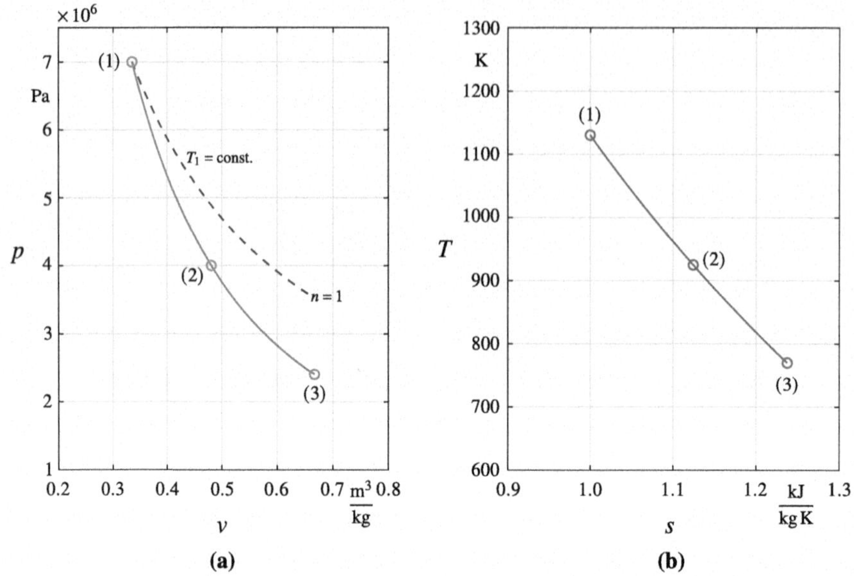

Fig. 3.27 **a** p, v- and **b** T, s-diagram

so it yields[24]

$$\boxed{T(s) = T_1 \cdot \mathrm{e}^{\frac{s-s_1}{c_p - \frac{n}{n-1}R}}} \tag{3.256}$$

with

$$R = c_p - \frac{c_p}{\kappa} = 2.0782 \times 10^3 \, \frac{\mathrm{J}}{\mathrm{kgK}}. \tag{3.257}$$

(c) Since the polytropic exponent n is now known, the temperature in state (2) can be determined, i.e.

$$T_2 = T_1 \left(\frac{p_2}{p_1}\right)^{\frac{n-1}{n}} = 924.6728\,\mathrm{K}. \tag{3.258}$$

(d) The mass balance in steady state obeys

$$\dot{m}_1 = \dot{m}_2 + \dot{m}_3. \tag{3.259}$$

It further is

$$\dot{m}_3 = 0.9 \cdot \dot{m}_1 \tag{3.260}$$

so that

$$\dot{m}_2 = 0.1 \cdot \dot{m}_1. \tag{3.261}$$

[24] For state (1), the specific entropy in Fig. 3.27b was arbitrarily set to $s_1 = 1\,\frac{\mathrm{kJ}}{\mathrm{kgK}}$.

The energy balance in steady state yields, see Fig. 3.28a

$$\dot{m}_1 h_1 + \dot{Q} + P_t = \dot{m}_2 h_2 + \dot{m}_3 h_3. \tag{3.262}$$

respectively with Eq. 3.259

$$P_t = \dot{m}_2 h_2 + \dot{m}_3 h_3 - \dot{m}_2 h_1 - \dot{m}_3 h_1 = \dot{m}_2 [h_2 - h_1] + \dot{m}_3 [h_3 - h_1]. \tag{3.263}$$

With Eqs. 3.260 and 3.261 it follows

$$P_t = 0.1 \cdot \dot{m}_1 [h_2 - h_1] + 0.9 \cdot \dot{m}_1 [h_3 - h_1]. \tag{3.264}$$

The mass flow then is

$$\dot{m}_1 = \frac{P_t}{0.1 \cdot [h_2 - h_1] + 0.9 \cdot [h_3 - h_1]} \tag{3.265}$$

respectively with the caloric equation of state

$$\dot{m}_1 = \frac{P_t}{0.1 \cdot c_p [T_2 - T_1] + 0.9 \cdot c_p [T_3 - T_1]} = 0.0448 \,\frac{\text{kg}}{\text{s}}. \tag{3.266}$$

The mass flow rates in state (2) and state (3) follow accordingly

$$\dot{m}_2 = 0.1 \cdot \dot{m}_1 = 0.0045 \,\frac{\text{kg}}{\text{s}} \tag{3.267}$$

and

$$\dot{m}_3 = 0.9 \cdot \dot{m}_1 = 0.0403 \,\frac{\text{kg}}{\text{s}}. \tag{3.268}$$

(e) The partial energy equation obeys

$$P_t = \dot{m}_2 \left[\int_1^2 v \, \mathrm{d}p + \psi_{12} \right] + \dot{m}_3 \left[\int_1^3 v \, \mathrm{d}p + \psi_{13} \right] = \dot{Y} + \dot{\Psi} \tag{3.269}$$

with

$$\int_1^2 v \, \mathrm{d}p = y_{12} = \frac{nR}{n-1} (T_2 - T_1) \tag{3.270}$$

and

$$\int_1^3 v \, \mathrm{d}p = y_{13} = \frac{nR}{n-1} (T_3 - T_1). \tag{3.271}$$

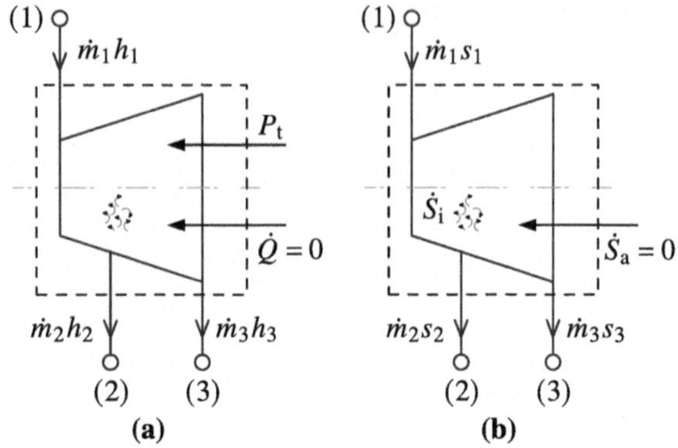

Fig. 3.28 **a** Energy and **b** entropy balance

Hence, the dissipated power is

$$\dot{\Psi} = \dot{m}_2 \psi_{12} + \dot{m}_3 \psi_{13} = P_t - \dot{m}_2 y_{12} - \dot{m}_3 y_{13} = 9.5679 \times 10^3\,\text{W} \tag{3.272}$$

and the power for the pressure change

$$\dot{Y} = \dot{m}_2 y_{12} + \dot{m}_3 y_{13} = -8.9568 \times 10^4\,\text{W}. \tag{3.273}$$

(f) The entropy balance in steady state yields, see Fig. 3.28b

$$\dot{m}_1 s_1 + \dot{S}_i + \underbrace{\dot{S}_a}_{=0} = \dot{m}_2 s_2 + \dot{m}_3 s_3. \tag{3.274}$$

respectively with Eq. 3.259

$$\dot{S}_i = \dot{m}_2 \left[s_2 - s_1 \right] + \dot{m}_3 \left[s_3 - s_1 \right]. \tag{3.275}$$

With the caloric equation of state it is

$$\dot{S}_i = \dot{m}_2 \left[c_p \ln \frac{T_2}{T_1} - R \ln \frac{p_2}{p_1} \right] + \dot{m}_3 \left[c_p \ln \frac{T_3}{T_1} - R \ln \frac{p_3}{p_1} \right] = 10.1440\,\frac{\text{W}}{\text{K}}. \tag{3.276}$$

(g) It applies

$$\delta \dot{S}_i = \frac{\delta \dot{\Psi}}{T} \tag{3.277}$$

so that

$$\dot{S}_i = \int_1^{2,3} \frac{\delta \dot{\Psi}}{T} = \frac{1}{T_m} \dot{\Psi}. \tag{3.278}$$

Finally, the thermodynamic mean temperature yields

$$T_{\mathrm{m}} = \frac{\dot{\Psi}}{\dot{S}_{\mathrm{i}}} = 943.2083\,\mathrm{K}. \tag{3.279}$$

3.25

(a) The mass of the air in the freezer follows from the ideal gas equation

$$p_{\mathrm{env}} V = m_{\mathrm{a}} R T_{\mathrm{env}}. \tag{3.280}$$

It follows

$$m_{\mathrm{a}} = \frac{p_{\mathrm{env}} V}{R T_{\mathrm{env}}} = 0.4675\,\mathrm{kg}. \tag{3.281}$$

(b) The first law of thermodynamics obeys

$$Q_{12} + W_{12} = U_2 - U_1 + \underbrace{\Delta E_{\mathrm{a},12}}_{=0}. \tag{3.282}$$

No work is exchanged across the system boundary, so $W_{12} = 0$. The specific heat capacity c_v follow

$$c_v = c_p - R = 717\,\frac{\mathrm{J}}{\mathrm{kgK}}. \tag{3.283}$$

With the caloric equation of state one gets

$$Q_{12} = U_2 - U_1 = m_{\mathrm{a}} c_v \left(T_2 - T_1\right) = -1.2401 \times 10^4\,\mathrm{J}. \tag{3.284}$$

(c) The pressure in state (2) yields[25]

$$p_2 = \frac{m_{\mathrm{a}} R T_2}{V} = 8.7590 \times 10^4\,\mathrm{Pa}. \tag{3.285}$$

(d) The balance of torques at the door is shown in Fig. 3.29. Clockwise torque through the effective force $(p_{\mathrm{env}} - p_2)\, h\, \mathrm{d}x$ obeys

$$M_1 = \int_0^t x \left(p_{\mathrm{env}} - p_2\right) h\, \mathrm{d}x = \left(p_{\mathrm{env}} - p_2\right) h \frac{1}{2} t^2. \tag{3.286}$$

[25] Both the gas mass and the volume are constant.

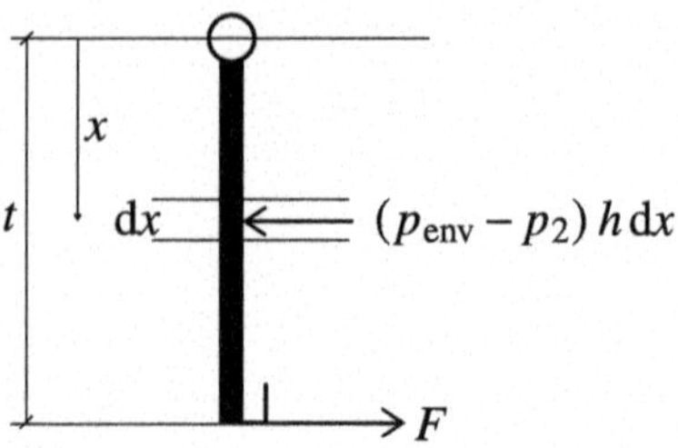

Fig. 3.29 Balance of torques

The counterclockwise torque due to the force at the handle is

$$M_2 = Ft. \tag{3.287}$$

To open the door, both torques must be balanced, i.e.

$$Ft = (p_{env} - p_2)\,h\frac{1}{2}t^2 \tag{3.288}$$

respectively

$$F = (p_{env} - p_2)\,h\frac{1}{2}t = 1.8615 \times 10^3\,\text{N}. \tag{3.289}$$

(e) It applies

$$\delta S_a = \frac{\delta Q}{T} \tag{3.290}$$

so that

$$S_{a,12} = \int_1^2 \frac{\delta Q}{T} = \frac{1}{T_m} Q_{12}. \tag{3.291}$$

The thermodynamic mean temperature then follows

$$T_m = \frac{Q_{12}}{\int_1^2 \frac{\delta Q}{T}}. \tag{3.292}$$

According to the first law of thermodynamics it applies

$$Q_{12} = U_2 - U_1 = m_a c_v\,(T_2 - T_1) \tag{3.293}$$

respectively in differential notation

$$\delta Q = m_a c_v\,dT. \tag{3.294}$$

Thus, it yields

$$\int_1^2 \frac{\delta Q}{T} = m_a c_v \int_1^2 \frac{dT}{T} = m_a c_v \ln \frac{T_2}{T_1}. \tag{3.295}$$

Finally, Eq. 3.292 simplifies to

$$T_m = \frac{Q_{12}}{\int_1^2 \frac{\delta Q}{T}} = \frac{Q_{12}}{m_a c_v \ln \frac{T_2}{T_1}} = \frac{T_2 - T_1}{\ln \frac{T_2}{T_1}} = 279.2416\,\text{K}. \tag{3.296}$$

(f) To determine the generated entropy two alternatives are possible:

- Alternative 1

$$W_{12} = 0 = \underbrace{W_{V,12}}_{=0} + \underbrace{W_{mech,12}}_{=0} + \Psi_{12} \tag{3.297}$$

so that

$$\Psi_{12} = 0 \tag{3.298}$$

Since no dissipation occurs in the gas, no entropy is generated,[26] i.e. $S_i = 0$.

- Alternative 2

$$S_2 - S_1 = S_{a,12} + S_{i,12} \tag{3.299}$$

so that

$$S_{i,12} = S_2 - S_1 - S_{a,12}. \tag{3.300}$$

With[27]

$$S_2 - S_1 = m_a \left[c_v \ln \frac{T_2}{T_1} + R \ln \frac{v_2}{v_1} \right] = m_a c_v \ln \frac{T_2}{T_1} \tag{3.301}$$

and

$$S_{a,12} = \frac{1}{T_m} Q_{12} = \frac{Q_{12}}{\frac{T_2 - T_1}{\ln \frac{T_2}{T_1}}} = m_a c_v \ln \frac{T_2}{T_1}. \tag{3.302}$$

Finally, it applies

$$S_{i,12} = S_2 - S_1 - S_{a,12} = m_a c_v \ln \frac{T_2}{T_1} - m_a c_v \ln \frac{T_2}{T_1} = 0. \tag{3.303}$$

[26] Furthermore, only the system *gas* is examined, i.e. the heat transfer from the environment through the wall is not part of the consideration.

[27] Note that the change of state runs isochorically.

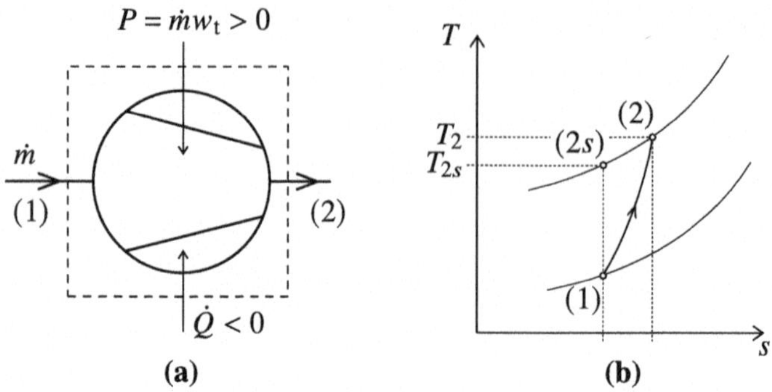

Fig. 3.30 **a** Energy balance and **b** T, s-diagram

3.26 Figure 3.30 shows the schematic change of state in a compressor as well as the illustration in a T, s-diagram. In Fig. 3.30b it is assumed that the curve is non-isentropic, since the fictitious isentropic temperature

$$T_{2s} = T_1 \left(\frac{p_2}{p_1} \right)^{\frac{\kappa-1}{\kappa}} = 412.3923\,\text{K} \tag{3.304}$$

is smaller than $T_2 = 433.15$ K.

(a) To calculate the required driving power of the compressor, the first law of thermodynamics is applied, i.e.

$$P + \dot{Q} + \dot{m}h_1 = \dot{m}h_2 \quad \text{with} \quad \dot{Q} = -350\,\text{W}. \tag{3.305}$$

Mind that in steady state, the energy flux into the system boundary is balanced by the energy flux out. By applying the caloric equation of state, P is solved for. One obtains

$$P = \dot{m}\,(h_2 - h_1) - \dot{Q} = \dot{m}c_p\,(T_2 - T_1) - \dot{Q} = 818.5333\,\text{W}. \tag{3.306}$$

(b) We assume a polytropic change of state that obeys the following relationship:

$$\frac{T_2}{T_1} = \left(\frac{p_2}{p_1} \right)^{\frac{n-1}{n}}. \tag{3.307}$$

Solving for the polytropic exponent n yields

$$\ln \frac{T_2}{T_1} = \frac{n-1}{n} \ln \frac{p_2}{p_1} \tag{3.308}$$

respectively

$$n = \frac{\ln \frac{p_2}{p_1}}{\ln \frac{p_2}{p_1} - \ln \frac{T_2}{T_1}} = 1.4859. \tag{3.309}$$

(c) The exergy of the released heat obeys

$$\dot{E}_{x,Q} = \dot{Q}\left(1 - \frac{T_{env}}{T_m}\right) \tag{3.310}$$

i.e. the thermodynamic mean temperature T_m is required. First, the partial energy equation is applied to determine the specific dissipation in the compressor, i.e.

$$w_{t,12} = \frac{P}{\dot{m}} = \underbrace{\int_1^2 v \, dp}_{=y_{12}} + \psi_{12}. \tag{3.311}$$

Thus, the specific dissipation follows

$$\psi_{12} = \frac{P}{\dot{m}} - y_{12} = \frac{P}{\dot{m}} - \frac{n}{n-1} R\,(T_2 - T_1) = 1.2268 \times 10^5 \,\frac{\mathrm{J}}{\mathrm{kg}}. \tag{3.312}$$

The dissipated energy $\dot{\Psi} = \dot{m}\psi_{12} = 408.9333\,\mathrm{W}$ is greater than the released heat of $\dot{Q} = -350\,\mathrm{W}$. This is the cause of the increase in entropy from state (1) to state (2).

Consequently, the thermodynamic mean temperature reads as

$$T_m = \frac{q_{12} + \psi_{12}}{s_2 - s_1} = \frac{\frac{\dot{Q}}{\dot{m}} + \psi_{12}}{c_p \ln \frac{T_2}{T_1} - R \ln \frac{p_2}{p_1}} = 358.6069\,\mathrm{K}. \tag{3.313}$$

Alternatively, the thermodynamic mean temperature is, cf. Problem 3.8,

$$T_m = \frac{T_2 - T_1}{\ln \frac{T_2}{T_1}} = 358.6069\,\mathrm{K}. \tag{3.314}$$

Finally, the exergy of the released heat obeys

$$\dot{E}_{x,Q} = \dot{Q}\left(1 - \frac{T_{env}}{T_m}\right) = -63.8858\,\mathrm{W}. \tag{3.315}$$

Chapter 4
Properties of Ideal Fluids and Changes of State

In order to describe states and their changes, thermodynamic properties of fluids are required. In addition to the thermal state values that can be measured directly and instantly, there are caloric state values to describe the energetic state that cannot be measured directly and are therefore calculated from other, accessible values. The focus in this chapter is on such caloric equations of state of ideal gases and incompressible liquids. In doing so, the findings from the previous chapters are deepened. The concept of entropy is further sharpened in order to evaluate thermodynamic processes and to show the limits of energetic conversion. Technical components such as turbines and compressors are explained by means of tasks. Fluids are subject to several successive changes of state in many applications. This is also deepened with the tasks presented.

4.1 Problems

4.1 What applies for the state value specific entropy of incompressible liquids?

(a) It increases, when the temperature rises.
(b) Entropy can increase only.
(c) It depends on temperature as well as pressure.
(d) None of the above statements is true.

4.2 How is a polytropic change of state of an ideal gas with $n = 1$ called?

(a) Isobaric
(b) Isochoric
(c) Adiabatic
(d) Isothermal

A. Schmidt, *Technical Thermodynamics Workbook for Engineers*,
https://doi.org/10.1007/978-3-031-50172-2_4

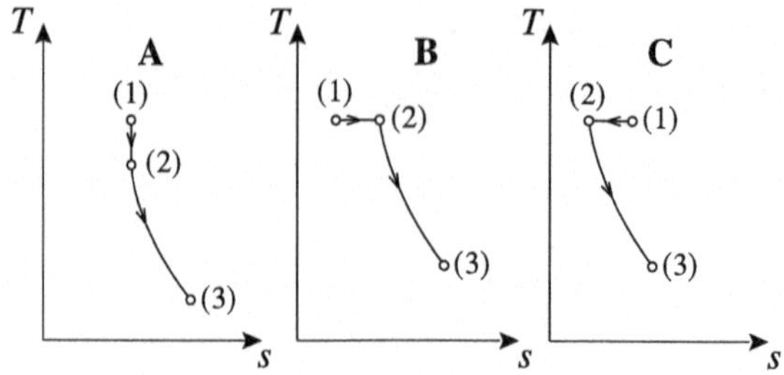

Fig. 4.1 Changes of state

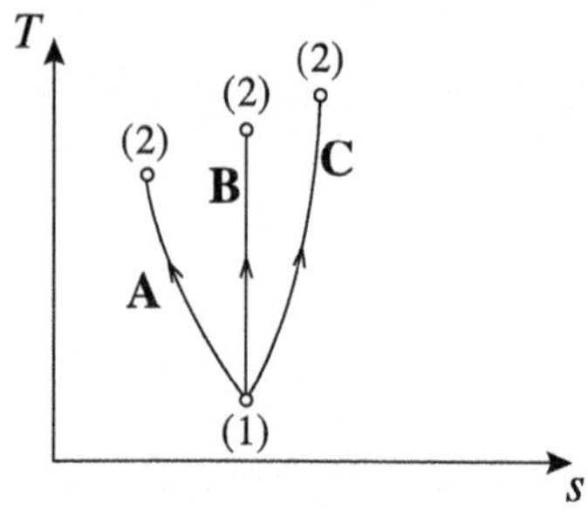

Fig. 4.2 Adiabatic compressor

4.3 An ideal gas underlies the following changes of state in a steady state flow process:

- (1) → (2): Adiabatic throttle
- (2) → (3): Irreversible expansion in an adiabatic turbine

Kinetic/potential energies can be ignored. Which graph, cf. Fig. 4.1, shows the changes of state (1) → (2) → (3) in a T, s-diagram?

(a) A
(b) B
(c) C
(d) None of the graphs shown.

4.4 Which of the given curves, cf. Fig. 4.2, shows a change of state of an ideal gas in an adiabatic compressor qualitatively? The isentropic efficiency is 85 %. Changes of potential/kinetic energies shall be ignored.

(a) A
(b) B
(c) C
(d) None of the graphs shown.

4.5 What applies for the specific enthalpy?

(a) It can be measured directly.
(b) It only exists in open systems.

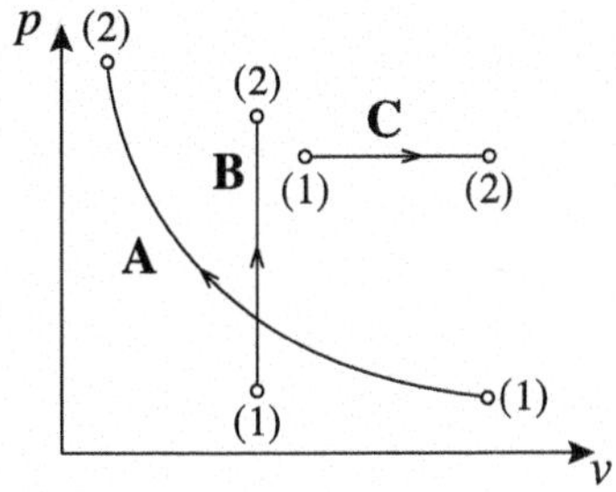

Fig. 4.3 Change of state

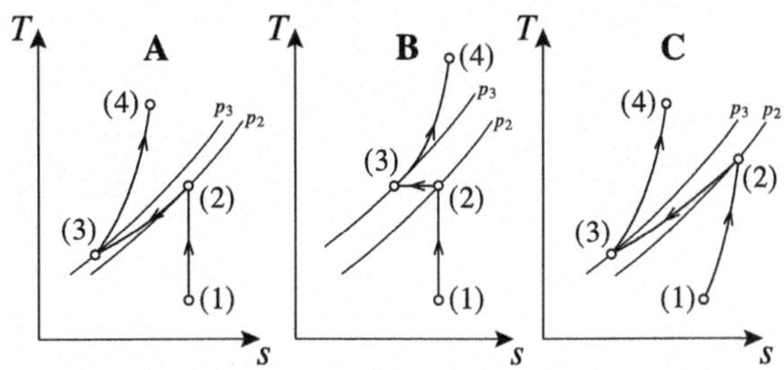

Fig. 4.4 Change of state

(c) It is a state value.
(d) None of the above statements is true.

4.6 Which of the shown curves, cf. Fig. 4.3, represents an isothermal change of state of an incompressible liquid?

(a) A
(b) B
(c) C
(d) None of the graphs shown.

4.7 An ideal gas underlies the following changes of state in a steady state flow process (kinetic, potential energies to be ignored):

- (1) → (2): isentropic compression
- (2) → (3): cooling down in a heat exchanger
- (3) → (4): polytropic, adiabatic compression

Which graph, cf. Fig. 4.4, shows the changes of state (1) → (2) → (3) → (4) in a T, s-diagram qualitatively?

(a) A
(b) B
(c) C
(d) None of the graphs shown.

4.8 An adiabatic compressor (isentropic efficiency: 85 %) increases the pressure of air (ideal gas, $c_p = 1004\frac{\text{J}}{\text{kgK}}$, $c_v = 717\frac{\text{J}}{\text{kgK}}$) from 1 bar to 5 bar. The initial temperature is 293K. Changes of potential/kinetic energies shall be ignored. What is the temperature after the compression?

(a) 304.37 K
(b) 438.49 K
(c) 464.17 K
(d) 494.37 K

4.9 Which of the listed changes of state of an ideal gas represents a polytropic change of state with $n = \frac{c_p}{c_v}$?

(a) Isobaric
(b) Isochoric
(c) Adiabatic, reversible
(d) Isothermal

4.10 By what amount does the entropy of 1kg air (ideal gas, $M = 28.96\frac{\text{kg}}{\text{kmol}}$) change when it expands isothermally to twice its volume?

(a) $-75.2\,\text{J}\,\text{K}^{-1}$
(b) $187.8\,\text{J}\,\text{K}^{-1}$
(c) $199.0\,\text{J}\,\text{K}^{-1}$
(d) $213.1\,\text{J}\,\text{K}^{-1}$

4.11 An ideal gas is heated in a stationary flow process. Which statement is true?

(a) The change of state can never be isentropic.
(b) The change of state can be isentropic if mechanical power is released from the system.
(c) The change of state is adiabatic.
(d) None of the above statements applies.

4.12 What happens during adiabatic throttling of an ideal gas when the external energies are neglected?

(a) The anergy of the gas increases.
(b) The gas cools down.
(c) The exergy of the gas increases.
(d) None of the above statements applies.

4.13 Which of the following statements applies to an isochoric change of state of an ideal gas?

(a) Isochore through a thermodynamic state is steeper than the isobar in the T, s-diagram.
(b) The slope of an isochore in the T, s-diagram is constant.
(c) The area under an isochore in the T, s-diagram corresponds to the specific exergy loss.

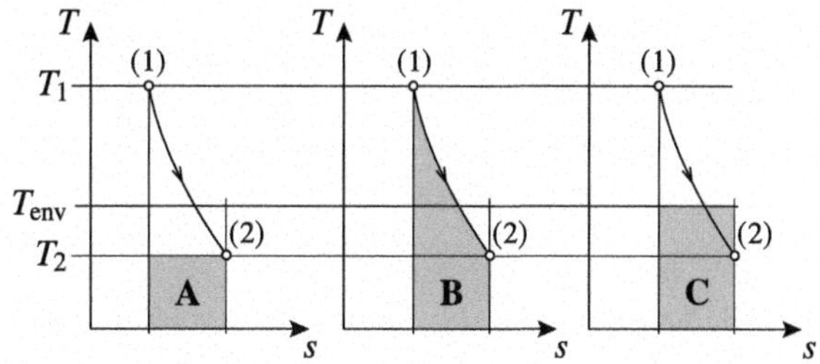

Fig. 4.5 Cooled turbine

(d) None of the above statements applies.

4.14 What is true about the state value specific entropy for incompressible fluids?

(a) It increases when the temperature rises.
(b) It is both pressure and temperature dependent.
(c) The specific entropy can only increase.
(d) None of the above statements applies.

4.15 Which of the marked areas, cf. Fig. 4.5, shows the specific dissipation of a cooled turbine? Changes in kinetic/potential energy are to be neglected.

(a) A
(b) B
(c) C
(d) None of the graphs shown.

4.16 An ideal gas underlies the following changes of state:

- (1) → (2): Adiabatic, irreversible compression: $p_1 = p_{env} = 1\,\text{bar}$, $p_2 = 4\,\text{bar}$, $T_1 = T_{env}$
- (2) → (3): Cooling down: $p_3 = p_2$
- (3) → (4): Cooled, frictionless compression: $p_4 = 10\,\text{bar}$

Which diagram, see Fig. 4.6, shows the changes of state in a T, s diagram qualitatively?

(a) A
(b) B
(c) C
(d) None of the graphs shown.

4.17 Air ($c_p = 1.004\,\frac{\text{kJ}}{\text{kgK}}$, $R = 0.287\,\frac{\text{kJ}}{\text{kgK}}$) at 0 °C is adiabatically, reversibly compressed in two successive stages, with the pressure doubling each time. What final temperature does the air reach?

(a) −89.37 °C
(b) 59.86 °C

Fig. 4.6 Changes of state

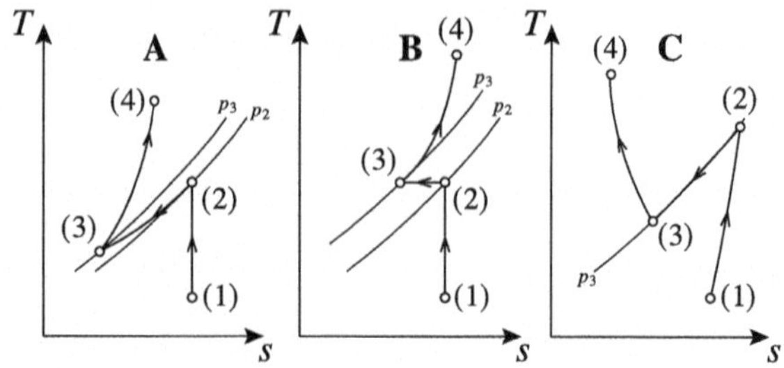

Fig. 4.7 Adiabatic throttle

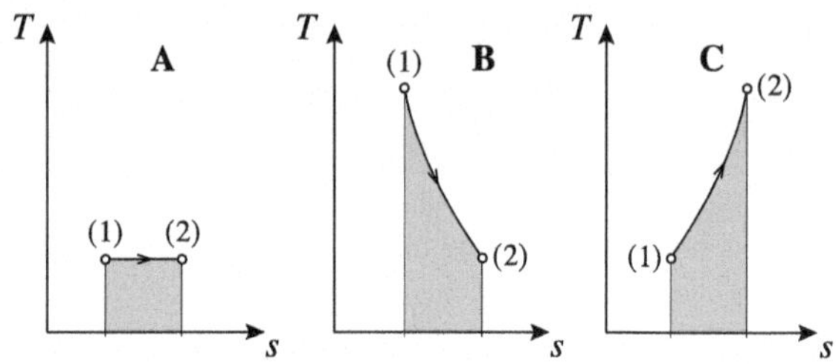

(c) 132.83 °C
(d) 305.98 °C

4.18 Which of the following statements applies to an isochoric change of state of an ideal gas?

(a) It equals a polytropic change of state with $n \to \infty$.
(b) The slope of an isochore in the T, s-diagram is constant.
(c) Isochoric changes of state are associated with an increase in entropy.
(d) None of the above statements applies.

4.19 An ideal gas is trapped in a closed system. Let the change of state be isochoric. What is true for the specific enthalpy?

(a) $\mathrm{d}h = c_p \, \mathrm{d}T$
(b) $\mathrm{d}h = c_v \, \mathrm{d}T$
(c) Since the system is closed, no specific enthalpy exists.
(d) None of the above statements applies.

4.20 Which of the marked areas, cf. Fig. 4.7, shows the specific dissipation of an adiabatic throttle? Changes in kinetic/potential energy are to be neglected. The throttle is operated with an ideal gas.

(a) A
(b) B
(c) C
(d) None of the graphs shown.

4.21 In a parallel-flow heat exchanger, $1\,\frac{\text{kg}}{\text{s}}$ of water ($c_W = 4.19\,\frac{\text{kJ}}{\text{kgK}}$) is to be heated isobarically from 70 °C to 90 °C. For this purpose, $1.5\,\frac{\text{kg}}{\text{s}}$ of air ($c_p = 1.004\,\frac{\text{kJ}}{\text{kgK}}$) of 130 °C is available. Which statement applies?

(a) Heat cannot be transferred from air to water because the specific heat capacity of water is greater than that of air.
(b) A parallel connection does not work. The apparatus must be operated in a counterflow.
(c) Yes, the heat exchanger works!
(d) None of the above statements applies.

4.22 In an adiabatic turbine, air (ideal gas, $\kappa = 1.4$) expands from 10 bar and 500 K polytropically ($n = 1.35$) to 2 bar. What is the isentropic efficiency of the turbine?

(a) 0.8333
(b) 0.9255
(c) 0.9512
(d) 1

4.23 The change of state in an open system is isentropic. Which statement applies?

(a) The pressure can never rise.
(b) Dissipated energy is completely released to the environment by heat transfer.
(c) Such a flow is not possible because it violates the second law of thermodynamics.
(d) None of the above statements applies.

4.24 Which of the following statements applies to an isobaric change of state of an ideal gas?

(a) Temperature must rise.
(b) The isobar in a p, v,-diagram shows a hyperbola.
(c) The ratio of specific volume to temperature remains constant.
(d) None of the above statements applies.

4.25 An ideal gas is subject to the following changes of state:

- Irreversible, adiabatic compression from p_1 to $p_2 > p_1$
- Isobaric heating to $T_3 > T_2$
- Irreversible, adiabatic expansion to $p_4 = p_1$

Which of the following sketches, cf. Fig. 4.8, shows the changes of state qualitatively?

(a) A
(b) B
(c) C
(d) None of the graphs shown.

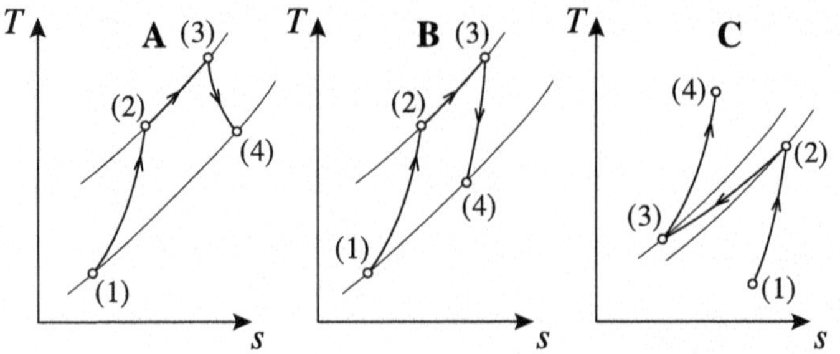

Fig. 4.8 Changes of state

4.26 Two air flows are mixed in a steady state process. 50 kW of heat are additionally supplied from the environment. Air is to be regarded as an ideal gas with $c_p = 1004 \frac{\text{J}}{\text{kgK}}$. The following parameters are given:

- Flow 1: $\dot{m}_1 = 6 \frac{\text{kg}}{\text{s}}$, $T_1 = 512\,\text{K}$
- Flow 2: $\dot{m}_2 = 15 \frac{\text{kg}}{\text{s}}$, $T_2 = 290\,\text{K}$

What is the temperature after the mixing, in case changes of potential/kinetic energies are ignored?

4.27 Helium passes a horizontal, heated channel in a steady state flow. The change of state (1) → (2) is polytropic with $n = 0.61$. Helium is to be regarded as an ideal gas with $c_p = 5193 \frac{\text{J}}{\text{kgK}}$ and $R = 2077 \frac{\text{J}}{\text{kgK}}$. The following information is given:

- $T_1 = 300\,\text{K}$, $p_1 = 1\,\text{bar}$, $c_1 = 1.2\,\text{m}\,\text{s}^{-1}$
- $p_2 = 0.95\,\text{bar}$, $c_2 = c_1 = 1.2\,\text{m}\,\text{s}^{-1}$

What is the specific dissipation ψ_{12}?

4.28 An ideal gas underlies an adiabatic, reversible compression in a horizontal cylinder. The following is know of the equilibrium states (1) and (2):

- State (1): $T_1 = 300\,\text{K}$, $p_1 = 1.013\,25\,\text{bar}$
- State (2): $p_2 = 5\,\text{bar}$

What is the temperature in state (2) when $c_p = 0.519 \frac{\text{kJ}}{\text{kgK}}$ and $c_v = 0.309 \frac{\text{J}}{\text{kgK}}$?

4.29 Two air flows are mixed adiabatically in a steady state process. Air is to be regarded as an ideal gas with $c_p = 1004 \frac{\text{J}}{\text{kgK}}$. The following parameters are given:

- Flow 1: $\dot{m}_1 = 12 \frac{\text{kg}}{\text{s}}$, $T_1 = 608\,\text{K}$
- Flow 2: $\dot{m}_2 = 18 \frac{\text{kg}}{\text{s}}$, $T_2 = 298\,\text{K}$

What is the temperature after the mixing, in case changes of potential/kinetic energies are ignored?

4.30 Air, to be treated as an ideal gas with $R = 0.287\,\frac{\text{kJ}}{\text{kgK}}$ and $c_p = 1.0045\,\frac{\text{kJ}}{\text{kgK}}$, enters a heated, horizontal channel with $T_1 = T_\text{env} = 303.15\,\text{K}$, $p_1 = 2.9\,\text{bar}$ and $c_1 = 8.0\,\text{m}\,\text{s}^{-1}$. A specific heat of $q_{12} = 85\,\frac{\text{kJ}}{\text{kg}}$ is supplied; the air expands polytropically ($n = 0.8457$) to $p_2 = 1.0\,\text{bar}$ at the end of the channel. Calculate the thermodynamic mean temperature. What is the specific loss of exergy? Determine the ratio of outlet cross section A_2 to inlet cross section A_1.

4.31 An adiabatic turbine consumes an air flow of $\dot{V}_\text{L} = 40\,\frac{\text{m}^3}{\text{h}}$ (ideal gas, $c_p = 1.004\,\frac{\text{kJ}}{\text{kgK}}$, $R = 0,287\,\frac{\text{kJ}}{\text{kgK}}$) at a pressure of $p_1 = 6\,\text{bar}$ and a temperature of $\vartheta_1 = 30\,°\text{C}$. Let the inlet velocity of the air be equal to the outlet velocity. The change in potential energy from (1) to (2) is negligible. The isentropic efficiency of the turbine is $\eta_\text{sT} = 0.93$. What is the technical power of the turbine when the air is expanded to 1 bar?

4.32 A gas flows isobarically through an adiabatic horizontal channel. Its velocity decreases from $c_1 = 100\,\text{m}\,\text{s}^{-1}$ to $c_2 = 20\,\text{m}\,\text{s}^{-1}$. The specific heat capacity shall be $c_p = 1.004\,\frac{\text{kJ}}{\text{kgK}}$. Its inlet temperature is $T_1 = 400\,\text{K}$.

a) Calculate is the specific entropy generation s_i.
b) What is the specific dissipation ψ_{12}?
c) Determine the ratio of outlet to inlet cross section.
d) Sketch the change of state in a p, v- as well as in a T, s-diagram.

4.33 To generate a high velocity in a wind tunnel, a tank (3) filled with air is partially evacuated to the state 3_I ($p_{3_\text{I}} = 0.2\,\text{bar}$, $T_{3_\text{I}} = 300\,\text{K}$). By opening a valve, air from the environment ($p_1 = p_\text{env} = 1.0\,\text{bar}$, $T_1 = T_\text{env} = 300\,\text{K}$) is then sucked in frictionlessly for a time interval of 10 seconds and expanded to such an extent that the velocity c_2 at the outlet of the nozzle is 80 % of the local velocity of sound. By regulating the valve, the pressure p_2 at the nozzle outlet remains constant for the duration of the experiment, while the pressure in the container increases in the process up to the final pressure p_{3_II}, see Fig. 4.9.

- Nozzle outlet cross-section: $A_2 = 0.4\,\text{m}^2$
- $R = 287\,\frac{\text{J}}{\text{kgK}}$, $c_p = 1004\,\frac{\text{J}}{\text{kgK}}$
- The plant is adiabatic.
- Air can be regarded as an ideal gas with constant c_p and c_v. Changes in potential energies are negligible.
- The velocity of sound a in state (2) depends on the local temperature: $a = \sqrt{\kappa RT}$

a) Determine the state values T_2, p_2, c_2 at the outlet of the nozzle and the mass Δm_in supplied to the tank during the given time period.
b) Calculate the temperature T_{3_II} and the pressure p_{3_II} in the vessel at the end of the experiment. The vessel's volume is $V_3 = 6000\,\text{m}^3$. The volumes of the feed lines are small compared to V_3.
c) What is the loss of exergy?

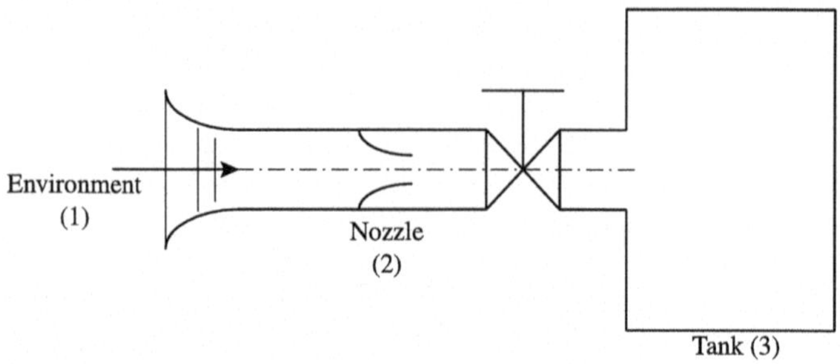

Fig. 4.9 Wind tunnel

Fig. 4.10 Changes of state

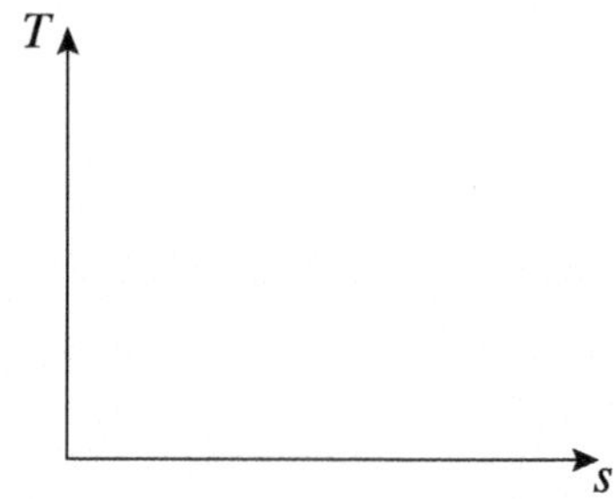

4.34 An ideal gas undergoes the following changes of state:

- Expansion in a first adiabatic turbine ($\eta_{s,T} < 1$) from $T_1,\ p_1$ to (2)
- Isobaric heating to (3) with $T_3 = T_1$
- Expansion in a second adiabatic turbine ($\eta_{s,T} < 1$) to (4)

Changes of kinetic/potential energies can be disregarded. Sketch the given process in a $T,\ s$-diagram, see Fig. 4.10!

4.35 What is true for a $T,\ s$-diagram of an ideal gas?

(a) Isotherms are vertical lines.
(b) $\frac{dT}{ds}\Big|_{p=\text{const.}} < \frac{dT}{ds}\Big|_{v=\text{const.}}$
(c) The specific work corresponds to the area beneath the change of state.
(d) None of the above statements applies.

4.2 Solutions

4.1 **(a)**
It applies that

$$ds = \frac{du}{T} + \frac{p\,dv}{T}. \tag{4.1}$$

For an incompressible liquid with $\mathrm{d}v = 0$ it simplifies to

$$\mathrm{d}s = \frac{\mathrm{d}u}{T} = c\frac{\mathrm{d}T}{T}. \tag{4.2}$$

In other words, the specific entropy does not depend on the pressure,[1] answer (c) is not true. However, according to Eq. 4.2, the entropy can rise, it can decrease or it even may be constant depending on the change in temperature $\mathrm{d}T$. Answer (b) therefore is wrong. With rising temperature, i.e. $\mathrm{d}T > 0$, the specific entropy rises as well. Answer (a) is correct.

4.2 **(d)**
The definition of a polytropic change of state is

$$pv^n = \text{const.} \tag{4.3}$$

In the case that $n = 1$, it results

$$pv = \text{const.} \tag{4.4}$$

With the thermal equation of state for ideal gases follows

$$pv = RT = \text{const.} \tag{4.5}$$

so that

$$T = \text{const.} \tag{4.6}$$

Answer (d) is correct. An adiabatic change of state for ideal gases results with $n = \kappa$. An isobaric change of state for ideal gases results with $n = 0$. For $n \to \infty$ the change of state is isochoric since

$$\lim_{n\to\infty} pv^n = \text{const.} \tag{4.7}$$

respectively

$$\lim_{n\to\infty} p^{\frac{1}{n}} v = \text{const.} \tag{4.8}$$

Finally, one gets

$$v \cdot \lim_{n\to\infty} p^{\frac{1}{n}} = v \cdot 1 = v = \text{const.} \tag{4.9}$$

4.3 **(b)**
For the adiabatic throttling of an ideal gas with external energies neglected, the following applies

$$\underbrace{P_\mathrm{t} + \dot{Q}}_{=0} = \dot{m}\,(h_2 - h_1) = \dot{m} c_p\,(T_2 - T_1)\,. \tag{4.10}$$

[1] The specific heat capacity c does not depend on pressure, but only on temperature.

Hence, the temperature remains constant, i.e. $T_2 = T_1$. Answer (a) must be wrong. Furthermore, dissipation must occur in an adiabatic throttle, since the pressure should decrease, i.e.

$$w_\mathrm{t} = 0 = \int_1^2 v\,\mathrm{d}p + \psi_{12} + \underbrace{\Delta e_{\mathrm{a},12}}_{=0} \tag{4.11}$$

respectively

$$\psi_{12} = -\underbrace{\int_1^2 v\,\mathrm{d}p}_{<0} > 0. \tag{4.12}$$

Due to dissipation, according to the second law of thermodynamics, there is an increase in entropy from state (1) to state (2), i.e.

$$\mathrm{d}s = \underbrace{\frac{\delta q}{T}}_{=0} + \frac{\delta \psi}{T} = \frac{\delta \psi}{T} > 0. \tag{4.13}$$

Answer (c) is therefore not possible. Consequently, answer (b) is correct, since an irreversible expansion from (2) → (3) means that entropy rises and the pressure decreases.

4.4 **(c)**
The entropy balance for an adiabatic compressor yields

$$\dot{m}s_1 + \dot{S}_\mathrm{i} + \underbrace{\dot{S}_\mathrm{a}}_{=0} = \dot{m}s_2 \tag{4.14}$$

and in specific notation

$$s_2 - s_1 = s_\mathrm{i} = \int_1^2 \frac{\delta \psi}{T} = \frac{\psi_{12}}{T_\mathrm{m}}. \tag{4.15}$$

Since the isentropic efficiency is less than 100 %, energy is dissipated in the compressor, i.e. entropy is generated, so that $s_\mathrm{i} > 0$. Thus, the specific entropy in state (2) is greater than in state (1). Answers (a) and (b) thus are incorrect. Answer (c), on the other hand, shows correct compression to a higher pressure level, i.e. $p_2 > p_1$.

4.5 **(c)**
The specific enthalpy is a (caloric) state variable and cannot be measured directly. Answer (a) is therefore not applicable, while answer (c) is correct. State values are fluid properties and independent of the system in which the fluid is located. Although the specific enthalpy has a descriptive meaning especially for open systems, since it

represents the sum of specific internal energy and shift work. Nevertheless, it also occurs in closed systems, as the defining equation

$$h = u + pv \tag{4.16}$$

can also be applied to closed systems, since it is only formed from state values: Consequently, answer (b) is wrong.

4.6 **(b)**
An incompressible fluid is characterised by the fact that the specific volume changes neither under the influence of pressure nor temperature, i.e.

$$\mathrm{d}v = 0. \tag{4.17}$$

Answer (a) depicts the compression of an ideal gas whose specific volume decreases with an increase in pressure. Answer (c) shows the isobaric change of state of a compressible fluid as the specific volume increases. Answer (b) shows the pressure increase of an incompressible fluid, because with pressure increase the specific volume remains constant. If this happens, for example, in an open system when passing a pump, the process can be isothermal by releasing heat to the environment[2]:
The first law of thermodynamics with the caloric equation of state for incompressible liquids obeys

$$q + w_t = h_2 - h_1 = c\underbrace{(T_2 - T_1)}_{=0} + v\,(p_2 - p_1) \tag{4.18}$$

respectively with the partial energy equation

$$q + \overbrace{\underbrace{\int_1^2 v\,\mathrm{d}p}_{=v(p_2-p_1)}}^{=w_t} + \psi_{12} = c\underbrace{(T_2 - T_1)}_{=0} + v\,(p_2 - p_1) \tag{4.19}$$

so that

$$q = -\psi_{12}. \tag{4.20}$$

Graph B thus shows a potentially isothermal increase in pressure for an incompressible fluid.

4.7 **(d)**
The change of state (1) after (2) is an isentropic compression, i.e. the specific entropy remains constant. The change of state must therefore run vertically in a T, s-diagram. Since the pressure increases, the change of state must run vertically upwards. This is fulfilled in Graphs A and B. Graph C is therefore ruled out.

---

[2] Outer energies are neglected.

In the heat exchanger the temperature should decrease, but this is only fulfilled in Graph A.[3] Nevertheless, Graph A is not correct: According to the partial energy equation

$$w_t = 0 = \int_1^2 v\,dp + \psi_{12} + \underbrace{\Delta e_{a,12}}_{=0} \tag{4.21}$$

respectively

$$\int_1^2 v\,dp = -\psi_{12} \le 0 \tag{4.22}$$

Therefore, the pressure can remain constant if the change of state is frictionless, or it must decrease. In Graph A, however, the pressure increases, so that none of the changes of state shown apply. So answer (d) is correct.

4.8 **(d)**
The isentropic exponent equals

$$\kappa = \frac{c_p}{c_v} = 1.4003. \tag{4.23}$$

For an adiabatic, isentropic compression,[4] the resulting temperature is therefore

$$T_{2s} = T_1 \left(\frac{p_2}{p_1}\right)^{\frac{\kappa-1}{\kappa}} = 464.1654\,\text{K}. \tag{4.24}$$

Applying the definition of isentropic efficiency of a turbine

$$\eta_{sT} = \frac{h_2 - h_1}{h_{2s} - h_1} = \frac{T_2 - T_1}{T_{2s} - T_1} \tag{4.25}$$

leads to

$$T_2 = T_1 + \eta_{sT}\,(T_{2s} - T_1) = 438.4906\,\text{K}. \tag{4.26}$$

4.9 **(c)**
Since

$$n = \frac{c_p}{c_v} = \kappa \tag{4.27}$$

this is an isentropic change of state. Two possible cases are conceivable for an isentropic change of state:

[3] Graph C has already been excluded.

[4] Hence, the compression is reversible.

1. The dissipated energy is released by cooling, i.e

$$ds = \frac{\delta q}{T} + \frac{\delta \psi}{T} = 0 \tag{4.28}$$

respectively

$$q = -\psi < 0. \tag{4.29}$$

2. The change of state is adiabatic and reversible, so that

$$ds = \underbrace{\frac{\delta q}{T}}_{=0} + \underbrace{\frac{\delta \psi}{T}}_{=0} = 0 \tag{4.30}$$

Therefore, answer (c) is correct in this case. For more information on polytropic changes of state, see Problem 4.2.

4.10 **(c)**
The caloric equation of state for an ideal gas obeys

$$S_2 - S_1 = m\left[c_v \ln\frac{T_2}{T_1} + R\ln\frac{v_2}{v_1}\right]. \tag{4.31}$$

Since the change of state is isothermal, it simplifies[5] to

$$S_2 - S_1 = m\frac{R_M}{M}\ln\frac{V_2}{V_1} = 0.1990\,\text{kJ}\,\text{K}^{-1}. \tag{4.32}$$

4.11 **(a)**
According to Problem 4.9, there are two cases for an isentropic change of state:

$$ds = \frac{\delta q}{T} + \frac{\delta \psi}{T} = 0. \tag{4.33}$$

Since dissipation is always positive, cooling is required in the friction case. Conversely, a heated system would require the destruction of entropy. This is not possible. Therefore, a heated system can never be isentropic, answer (a) is correct.

4.12 **(a)**
The first law of thermodynamics for an adiabatic throttle obeys

$$q + w_t = 0 = h_2 - h_1 + \underbrace{\Delta e_{a,12}}_{=0}. \tag{4.34}$$

[5] Note that $\frac{v_2}{v_1} = \frac{V_2}{V_1}$ and $R = \frac{R_M}{M}$.

Consequently, for an ideal gas it simplifies to

$$0 = c_p\,(T_2 - T_1) \tag{4.35}$$

so that

$$T_2 = T_1. \tag{4.36}$$

Thus, answer (b) must be wrong. Let us now have a closer look at the partial energy equation

$$w_\mathrm{t} = 0 = \int_1^2 v\,\mathrm{d}p + \psi_{12} + \underbrace{\Delta e_{\mathrm{a},12}}_{=0} \tag{4.37}$$

respectively

$$\int_1^2 v\,\mathrm{d}p = -\psi_{12} \le 0. \tag{4.38}$$

Dissipation is therefore required to reduce the pressure in a throttle, i.e. $\psi_{12} > 0$. Thus, due to dissipation, entropy is generated,[6] i.e.

$$s_{\mathrm{i},12} = \int_1^2 \frac{\delta\psi}{T} = \frac{\psi_{12}}{T_\mathrm{m}} > 0. \tag{4.39}$$

The entropy generated is responsible for a loss of working capability, i.e. an exergy loss

$$e_{\mathrm{x,V}} = T_\mathrm{env} s_{\mathrm{i},12} > 0. \tag{4.40}$$

Since energy is constant and composed of exergy and anergy,[7] if the exergy e_x decreases, the anergy b_x must increase:

$$e = e_\mathrm{x} + b_\mathrm{x} = \mathrm{const.} \tag{4.41}$$

Answer (c) is incorrect, while answer (a) applies.

4.13 **(a)**
The first law of thermodynamics for a closed system reads as

$$\delta q + \delta w = \mathrm{d}u + \mathrm{d}e_\mathrm{a} \tag{4.42}$$

and the partial equation obeys

---

[6] In this case, due to the isothermal change of state, it applies $T_\mathrm{m} = T_1 = T_2$.

[7] The part of the energy that cannot be converted further.

$$\delta w = -p\,\mathrm{d}v + \delta\psi + \mathrm{d}e_{\mathrm{a}}. \tag{4.43}$$

Hence, the first law of thermodynamics simplifies to

$$\delta q + \psi = \mathrm{d}u + p\,\mathrm{d}v \tag{4.44}$$

respectively with the second law of thermodynamics

$$T\,\mathrm{d}s = \mathrm{d}u + p\,\mathrm{d}v. \tag{4.45}$$

This equation is denoted fundamental equation of thermodynamics. With

$$\mathrm{d}h = \mathrm{d}u + p\,\mathrm{d}v + v\,\mathrm{d}p \tag{4.46}$$

the fundamental equation of thermodynamics also obeys

$$T\,\mathrm{d}s = \mathrm{d}h - v\,\mathrm{d}p. \tag{4.47}$$

For an isochoric change of state, i.e. $\mathrm{d}v = 0$, Eq. 4.45 simplifies for an ideal gas to

$$T\,\mathrm{d}s = \mathrm{d}u = c_v\,\mathrm{d}T. \tag{4.48}$$

The slope of an isochore in a $T,\ s$-diagram is therefore

$$\boxed{\left(\frac{\mathrm{d}T}{\mathrm{d}s}\right)_v = \frac{T}{c_v}} \tag{4.49}$$

As the temperature increases, the slope of the isochore increases, so answer (b) is wrong.
For an isobaric change of state, i.e. $\mathrm{d}p = 0$, Eq. 4.47 simplifies for an ideal gas to

$$T\,\mathrm{d}s = \mathrm{d}h = c_p\,\mathrm{d}T. \tag{4.50}$$

The slope of an isobar in a $T,\ s$-diagram is therefore

$$\boxed{\left(\frac{\mathrm{d}T}{\mathrm{d}s}\right)_p = \frac{T}{c_p}} \tag{4.51}$$

Since $c_p > c_v$, it applies that

$$\boxed{\left(\frac{\mathrm{d}T}{\mathrm{d}s}\right)_v > \left(\frac{\mathrm{d}T}{\mathrm{d}s}\right)_p} \tag{4.52}$$

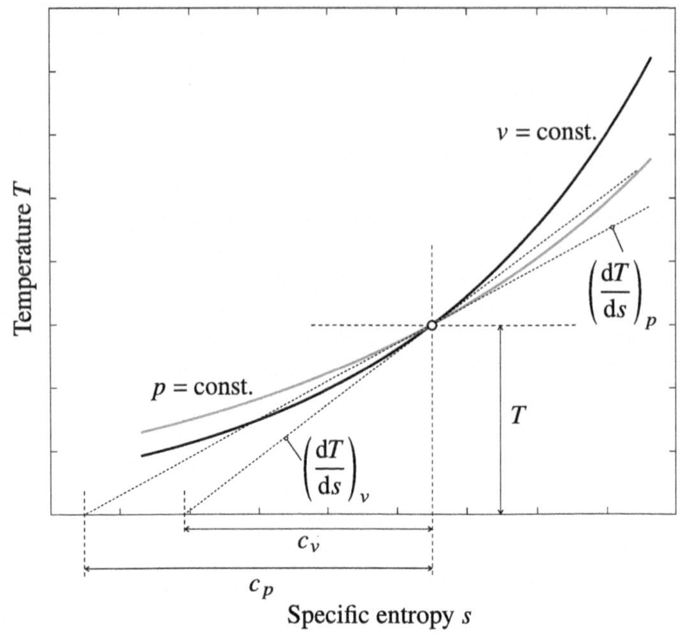

Fig. 4.11 Isochore versus isobar

The slope of an isochore in the T, s-diagram for ideal gases is greater than the slope of an isobar see Fig. 4.11.

Answer (a) is correct. The area under a change of state in the T, s-diagram corresponds to $q + \psi$. Answer (c) is therefore wrong.

4.14 **(a)**
Specific entropy as a state value is subject to two mechanisms, i.e.

$$ds = \frac{\delta q}{T} + \frac{\delta \psi}{T}. \tag{4.53}$$

While the expression $\frac{\delta \psi}{T}$ can only be positive or equal to zero, the term $\frac{\delta q}{T}$ can be both positive and negative. The specific entropy can therefore decrease, remain constant or increase. Answer (c) is thus wrong. According the caloric equation of state

$$s_2 - s_1 = c \ln \frac{T_2}{T_1} \tag{4.54}$$

yields that the specific entropy for an incompressible liquid depends merely on temperature. Answer (b) is wrong. So in case $T_2 > T_1$ it is $s_2 > s_1$, answer (a) therefore is correct.

4.15 **(d)**
The area under a change of state in the T, s-diagram shows the sum of specific heat q and specific dissipation ψ. Only in the case that the change of state is adiabatic, the specific dissipation can be read directly as the area under the change of state. Since in this case the turbine is cooled, no direct information regarding the specific dissipation can be taken from the diagram. Therefore, answer (d) is correct.

4.16 **(c)**
Due to

$$ds = \underbrace{\frac{\delta q}{T}}_{=0} + \frac{\delta \psi}{T} > 0 \tag{4.55}$$

the specific entropy from (1) to (2) must rise as shown in Graph C. Thus, Graphs A and B are wrong. The isobaric cooling from (2) to (3) is also correctly shown in Graph C. Due to the irreversibility from (3) to (4), the specific entropy would increase, but additional heat, and thus also entropy, is removed from the system. If more heat is released by the system than dissipation occurs internally, a process as in Graph C is possible. Therefore, answer (c) is correct.

4.17 **(c)**
Let us start with the obvious:

$$c_v = c_p - R = 717 \frac{\mathrm{J}}{\mathrm{kgK}} \tag{4.56}$$

so that

$$\kappa = \frac{c_p}{c_v} = 1.4003. \tag{4.57}$$

For the first isentropic compression[8] applies

$$T_1 = T_0 \pi_1^{\frac{\kappa-1}{\kappa}} \quad \text{with} \quad \pi_1 = \frac{p_1}{p_0} = 2 \tag{4.58}$$

and the second compression accordingly

$$T_2 = T_1 \pi_2^{\frac{\kappa-1}{\kappa}} \quad \text{with} \quad \pi_2 = \frac{p_2}{p_1} = 2 = \pi_1 = \pi. \tag{4.59}$$

Combining these two equations, one gets:

$$T_2 = T_0 \pi^{\frac{2(\kappa-1)}{\kappa}} = 405.9794\,\mathrm{K}. \tag{4.60}$$

Consequently, the temperature in the Celsius scale corresponds to $\vartheta_2 = 132.8294\,°\mathrm{C}$. Option (c) is correct.

[8] With π, not the circle number but the pressure ratio is meant.

4.18 **(a)**
A detailed answer can be found in Problems 4.2 and 4.13. For isochoric changes of state of an ideal gas, the entropy does not necessarily have to remain constant. The caloric equation of state

$$s_2 - s_1 = c_v \ln \frac{T_2}{T_1} + \underbrace{R \ln \frac{v_2}{v_1}}_{=0} \tag{4.61}$$

shows that the change in entropy in this case depends only on the temperature change. If the temperature rises, the specific entropy increases; if the temperature falls, the specific entropy decreases.

4.19 **(a)**
The specific enthalpy is a state value, i.e. it has nothing to do with the system in which the fluid is located, but describes the thermodynamic state of the fluid. Answer (c) is therefore incorrect: the specific enthalpy can also be determined in a closed system: For an ideal gas, it is composed of measurable state values according to

$$h = u + pv. \tag{4.62}$$

The caloric equation of state for an ideal gas obeys

$$\mathrm{d}h = c_p \, \mathrm{d}T. \tag{4.63}$$

The equation is regardless of which change of state the ideal gas undergoes. Answer (a) is correct.
Students often erroneously assume that the specific enthalpy only occurs in open systems, since it has a descriptive meaning there when a fluid particle enters an open system. In this case, the specific enthalpy represents the internal energy and the specific work to push the fluid particle in. However, even in a closed system the specific enthalpy has a physical motivation: Consider a horizontal cylinder closed by a freely moving, gas-tight piston so that the enclosed ideal gas undergoes an isobaric, reversible change of state with $p = p_{\mathrm{env}}$. The first law of thermodynamics obeys

$$Q_{12} + W_{12} = U_2 - U_1 \tag{4.64}$$

with the partial energy equation

$$W_{12} = W_{\mathrm{V},12} + \underbrace{W_{\mathrm{mech},12} + \Psi_{12}}_{=0} = -\int_1^2 p \, \mathrm{d}V = -p \left(V_2 - V_1\right). \tag{4.65}$$

Hence, it is

$$Q_{12} = U_2 - U_1 + p\,(V_2 - V_1) = H_2 - H_1 = m\,(h_2 - h_1)\,. \tag{4.66}$$

4.20 **(a)**
The area under a change of state in the T, s-diagram corresponds to the sum of specific heat q and specific dissipation ψ. Since the described change of state is supposed to be adiabatic, all three areas shown correspond to specific dissipation. It must now be clarified which of the three changes of state depicts the adiabatic throttling of an ideal gas. The first law of thermodynamics follows

$$q_{12} + w_{12} = h_2 - h_1 + \underbrace{\Delta e_{a,12}}_{=0}\,. \tag{4.67}$$

The throttle is adiabatic, i.e. $q_{12} = 0$, and it is work-isolated, i.e. there is no work exchange with the environment, so that the first law of thermodynamics simplifies to

$$0 = h_2 - h_1\,. \tag{4.68}$$

Applying the caloric equation of state, one gets

$$T_1 = T_2. \tag{4.69}$$

In other words, the change of state is isothermal under the given premises. Consequently, answer (a) applies. The specific entropy increases within an adiabatic throttle caused by the generation of entropy due to friction.

4.21 **(b)**
To warm the water isobarically, according to the 1st law of thermodynamics, a heat flux of

$$\dot{Q} = \dot{m}_{\mathrm{w}} c_{\mathrm{w}}\,\Delta T_{\mathrm{w}} = 83.8\,\mathrm{kW} \tag{4.70}$$

is required. This heat flux is taken from the air, which cools down by ΔT_{air} according to the first law of thermodynamics i.e.

$$\dot{Q} = -\dot{m}_{\mathrm{air}} c_p\,\Delta T_{\mathrm{air}} \tag{4.71}$$

respectively

$$\Delta T_{\mathrm{air}} = -\frac{\dot{Q}}{\dot{m}_{\mathrm{air}} c_p} = -55.6441\,\mathrm{K} \tag{4.72}$$

Since

$$\Delta T_{\mathrm{air}} = \vartheta_{\mathrm{air,out}} - \vartheta_{\mathrm{air,in}} \tag{4.73}$$

the air outlet temperature is

$$\vartheta_{\mathrm{air,out}} = \vartheta_{\mathrm{air,in}} + \Delta T_{\mathrm{air}} = 74.3559\,^{\circ}\mathrm{C}. \tag{4.74}$$

In principle, it is therefore possible to heat water with the help of an air flow. Answer (a) is therefore not applicable. However, it must now be clarified whether the heat exchanger can be operated in parallel flow. Since the specific enthalpy of the water and the air is linearly dependent on the temperature due to [9]

$$dh_\mathrm{W} = c_\mathrm{W}\,\mathrm{d}T_\mathrm{W} \quad \text{resp.} \quad dh_\mathrm{air} = c_p\,\mathrm{d}T_\mathrm{air}, \tag{4.75}$$

the $\vartheta = f(h_\mathrm{air})$ curve is now plotted, see Fig. 4.12. First, the temperature profile of the air is plotted against the specific enthalpy of the air. The specific enthalpy is maximum for the air inlet temperature and decreases linearly up to the air outlet temperature. In Fig. 4.12a, the inlet and outlet temperatures of the two fluids correspond because of the parallel flow. The temperature curve of the water must also be linear in this plot. Shortly after the inlet of the fluids, the heat exchanger works very well because the maximum driving temperature potential is present and heat is transferred from the air to the water. As the flow continues, the driving potential becomes smaller and smaller and the fluid temperatures become more and more equal. Finally, a thermal equilibrium of air and water is reached at point (Eq). From this point on, therefore, no further heat flux is transferred. The outlet temperatures of air and water are now identical and the desired water temperature of 90 °C is not reached. The heat exchanger therefore does not provide the desired performance in parallel flow. Thus, answer (c) is wrong.

However, Fig. 4.12b shows the heat exchanger in counter flow. The representation of the temperature curve of the air remains unchanged in this representation, only the inlet and outlet of the water flow are swapped. Compared to the direct current, it can be seen that the driving temperature potential is now present in the entire apparatus, i.e. the temperature curves of air and water do not intersect. The desired water outlet temperature is reached. The heat exchanger achieves the desired performance. Answer (b) therefore is correct.

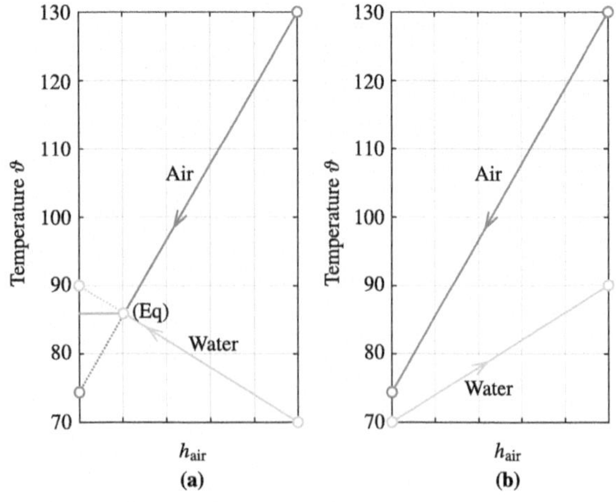

Fig. 4.12 Heat exchanger: **a** Parallel flow and **b** Counter flow

[9] The specific heat capacities shall be constant.

4.22 **(b)**
First, the outlet temperature of the polytropic, adiabatic turbine is calculated:

$$T_2 = T_1 \left(\frac{p_2}{p_1}\right)^{\frac{n-1}{n}} = 329.4242\,\text{K}. \tag{4.76}$$

In order to determine the isentropic efficiency, T_{2s} of a reversible, adiabatic turbine, which serves as a benchmark, must be determined in a next step, i.e.

$$T_{2s} = T_1 \left(\frac{p_{2s}}{p_1}\right)^{\frac{\kappa-1}{\kappa}} = 315.6925\,\text{K} \quad \text{with} \quad p_{2s} = p_2. \tag{4.77}$$

The isentropic efficiency of a turbine finally results in

$$\eta_{s\text{T}} = \frac{h_2 - h_1}{h_{2s} - h_1} = \frac{T_2 - T_1}{T_{2s} - T_1} = 0.9255. \tag{4.78}$$

4.23 **(b)**
Answer (a) is not applicable: For an ideal gas, for example, the caloric equation of state

$$s_2 - s_1 = c_p \ln \frac{T_2}{T_1} - R \ln \frac{p_2}{p_1} \tag{4.79}$$

can be applied to determine the entropy change during a change of state. It can be seen that when the pressure increases, the specific entropy remains constant if the temperature also increases. This correlation is also illustrated in a $T,\ s$-diagram, see Fig. 4.13.

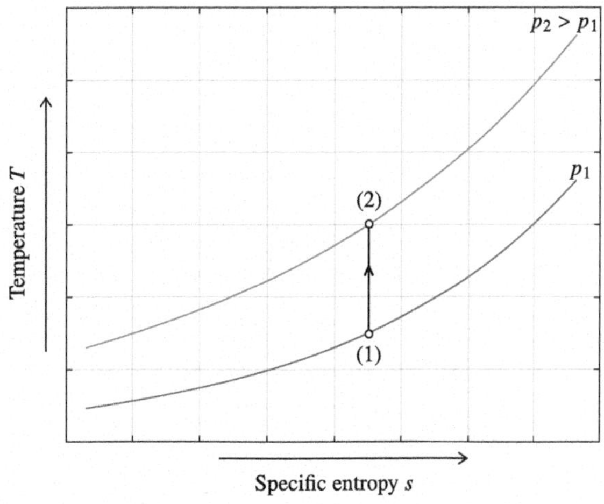

Fig. 4.13 Isentropic change of state of an ideal gas

The change of entropy underlies two mechanisms:

$$s_2 - s_1 = \underbrace{\frac{\psi_{12}}{T_m}}_{=s_i} + \underbrace{\frac{q_{12}}{T_m}}_{=s_a} . \tag{4.80}$$

Dissipation inside the system is responsible for generation, i.e. an increase, of entropy. Heat exchange on the other hand also has an influence on the state value entropy, e.g. entropy decreases when the system is cooled. According to Eq. 4.80 the change of state is isentropic, i.e. $s_2 = s_1$, in case the dissipated energy is fully released by heat transfer to the environment. Consequently, answer (b) is correct. The second law of thermodynamics is not violated because an isentropic change of state is also possible for $s_i > 0$ if the system is cooled as just described. A heated, isentropic system, on the other hand, is not possible and would violate the 2nd law of thermodynamics, since in this case entropy would have to be destroyed, i.e. $s_i < 0$.

4.24 **(c)**
The thermal equation of state of an ideal gas follows

$$pv = RT. \tag{4.81}$$

Since the change of state is supposed to be isobaric, it is

$$\frac{v}{T} = \frac{R}{p} = \text{const.} \tag{4.82}$$

Thus, answer (c) is correct. The isobar in a p, v-diagram is a horizontal line - but the isotherm of an ideal gas follows a hyperbola. Therefore, answer (b) does not apply. According to Eq. 4.82 it is

$$\frac{v_1}{T_1} = \frac{v_2}{T_2}. \tag{4.83}$$

The temperature can increase or it may rise. Answer (a) does not make sense.

4.25 **(a)**
In case of an irreversible, adiabatic compression, the pressure and the specific entropy must increase due to the entropy generation by the irreversibility. This is guaranteed in all three Graphs. The change of state from (2) to (3) must follow an isobar such that the temperature increases. This is not the case in Graph C because the temperature decreases. In the final adiabatic, irreversible expansion, the pressure must decrease and the specific entropy according to

$$s_4 - s_3 = \underbrace{\frac{\psi_{34}}{T_m}}_{=s_i>0} + \underbrace{\frac{q_{34}}{T_m}}_{=s_a=0} \tag{4.84}$$

needs to increase. This is only the case in Graph A. Therefore, answer (a) is applicable.

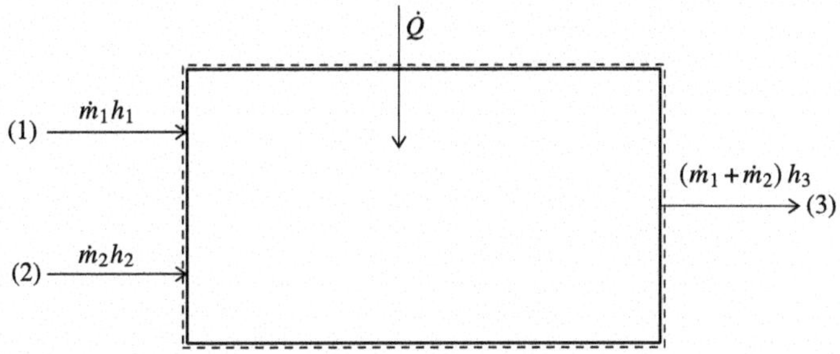

Fig. 4.14 Mixing of two air flows

4.26 A sketch of the Problem is shown in Fig. 4.14. The first law of thermodynamics for the diabatic mixing process is therefore as follows

$$\dot{m}_1 h_1 + \dot{m}_2 h_2 + \dot{Q} = (\dot{m}_1 + \dot{m}_2)\, h_3. \tag{4.85}$$

Rearranging leads to

$$\dot{Q} = \dot{m}_1 \left(h_3 - h_1\right) + \dot{m}_2 \left(h_3 - h_2\right) \tag{4.86}$$

respectively with the caloric equation of state

$$\dot{Q} = \dot{m}_1 c_p \left(T_3 - T_1\right) + \dot{m}_2 c_p \left(T_3 - T_2\right). \tag{4.87}$$

Finally, solving for T_3 yields

$$T_3 = \frac{\frac{\dot{Q}}{c_p} + \dot{m}_1 T_1 + \dot{m}_2 T_2}{\dot{m}_1 + \dot{m}_2} = 355.80\,\mathrm{K}. \tag{4.88}$$

4.27 Since the change of state is polytropic with known polytropic exponents n, the outlet temperature can be calculated, i.e.

$$T_2 = T_1 \left(\frac{p_2}{p_1}\right)^{\frac{n-1}{n}} = 310.0013\,\mathrm{K}. \tag{4.89}$$

To calculate the specific dissipation, one applies the partial energy equation for the open system. This is[10]:

[10] Outer energies do not play a role because inlet and outlet velocities are identical and the channel is horizontal.

$$w_\mathrm{t} = \underbrace{\int_1^2 v\,\mathrm{d}p}_{=y_{12}} + \psi_{12} = 0 \tag{4.90}$$

so that the specific dissipation follows

$$\psi_{12} = -y_{12}. \tag{4.91}$$

For a polytropic change of state the specific pressure work y_{12} can be calculated according to Appendix E, i.e.

$$\psi_{12} = -n\frac{RT_1}{n-1}\left(\frac{T_2}{T_1} - 1\right) = 3.2491 \times 10^4 \frac{\mathrm{J}}{\mathrm{kg}}. \tag{4.92}$$

Although not explicitly asked, the specific heat can be calculated using the first law of thermodynamics, i.e.

$$q_{12} + \underbrace{w_\mathrm{t}}_{=0} = h_2 - h_1 \tag{4.93}$$

respectively

$$q_{12} = c_p\,(T_2 - T_1) = 5.1937 \times 10^4 \frac{\mathrm{J}}{\mathrm{kg}}. \tag{4.94}$$

The thermodynamic mean temperature then follows

$$s_2 - s_1 = c_p \ln\frac{T_2}{T_1} - R\ln\frac{p_2}{p_1} = \frac{q_{12}}{T_\mathrm{m}} + \frac{\psi_{12}}{T_\mathrm{m}} \tag{4.95}$$

respectively

$$T_\mathrm{m} = \frac{q_{12} + \psi_{12}}{c_p \ln\frac{T_2}{T_1} - R\ln\frac{p_2}{p_1}} = 304.9733\,\mathrm{K} = \frac{T_2 - T_1}{\ln\frac{T_2}{T_1}}. \tag{4.96}$$

4.28 The change of state is adiabatic and reversible, i.e. isentropic, with an isentropic exponent of

$$\kappa = \frac{c_p}{c_v} = 1.6796. \tag{4.97}$$

The temperature T_2 therefore follows

$$T_2 = T_1\left(\frac{p_2}{p_1}\right)^{\frac{\kappa-1}{\kappa}} = 572.3061\,\mathrm{K}. \tag{4.98}$$

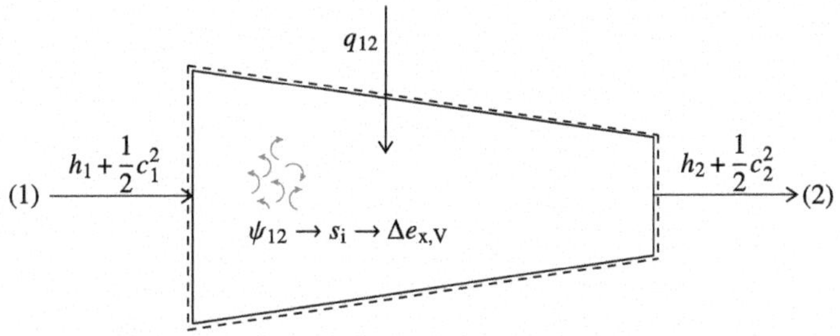

Fig. 4.15 Heated, horizontal flow channel

4.29 The solution essentially corresponds to Problem 4.26, whereby the heat flux is set to $\dot{Q} = 0$. This results in a mixing temperature of

$$T_3 = \frac{\dot{m}_1 T_1 + \dot{m}_2 T_2}{\dot{m}_1 + \dot{m}_2} = 422.0\,\text{K}. \tag{4.99}$$

4.30 For the polytropic change of state, the polytropic exponent n is known, so that the temperature T_2 is

$$T_2 = T_1 \left(\frac{p_2}{p_1}\right)^{\frac{n-1}{n}} = 368.1486\,\text{K}. \tag{4.100}$$

The outlet velocity follows from the 1st law of thermodynamics, cf. Fig. 4.15,

$$q_{12} + \underbrace{w_t}_{=0} = h_2 - h_1 + \frac{1}{2}\left(c_2^2 - c_1^2\right) = c_p\left(T_2 - T_1\right) + \frac{1}{2}\left(c_2^2 - c_1^2\right) \tag{4.101}$$

so that

$$c_2 = \sqrt{2\left[c_p\left(T_1 - T_2\right) + \frac{1}{2}c_1^2 + q_{12}\right]} = 198.7001\,\text{m}\,\text{s}^{-1}. \tag{4.102}$$

The partial energy equation obeys

$$w_t = 0 = \underbrace{\int_1^2 v\,\text{d}p}_{=y_{12}} + \psi_{12} + \frac{1}{2}\left(c_2^2 - c_1^2\right). \tag{4.103}$$

The solution for ψ_{12} applying the specific pressure work according to Appendix E gives

$$\psi_{12} = -n\frac{RT_1}{n-1}\left(\frac{T_2}{T_1} - 1\right) - \frac{1}{2}\left(c_2^2 - c_1^2\right) = 8.2535 \times 10^4 \frac{\mathrm{J}}{\mathrm{kg}}. \quad (4.104)$$

Finally, the thermodynamic mean temperature is

$$T_\mathrm{m} = \frac{\psi_{12} + q_{12}}{s_2 - s_1} = \frac{\psi_{12} + q_{12}}{c_p \ln\frac{T_2}{T_1} - R\ln\frac{p_2}{p_1}} = 334.5978\,\mathrm{K}. \quad (4.105)$$

Alternatively, according to Problem 3.8, it also yields, cf. Eq. 3.65,

$$T_\mathrm{m} = \frac{T_2 - T_1}{\ln\frac{T_2}{T_1}} = 334.5978\,\mathrm{K}. \quad (4.106)$$

The specific loss of exergy is

$$\Delta e_{\mathrm{x,V}} = T_\mathrm{env} s_\mathrm{i} = T_\mathrm{env}\frac{\psi_{12}}{T_\mathrm{m}} = 7.4778 \times 10^4 \frac{\mathrm{J}}{\mathrm{kg}}. \quad (4.107)$$

The ratio of outlet cross section A_2 to inlet cross section A_1 can be derived with the equation of continuity, i.e.

$$\dot{m}_1 = c_1 A_1 \rho_1 = \frac{c_1 A_1}{v_1} \quad \text{with} \quad v_1 = \frac{RT_1}{p_1} = 0.3\frac{\mathrm{m}^3}{\mathrm{kg}} \quad (4.108)$$

and

$$\dot{m}_2 = \dot{m}_1 = c_2 A_2 \rho_2 = \frac{c_2 A_2}{v_2} \quad \text{with} \quad v_2 = \frac{RT_2}{p_2} = 1.0566\frac{\mathrm{m}^3}{\mathrm{kg}}. \quad (4.109)$$

Finally, it applies

$$\frac{A_2}{A_1} = \frac{c_1 v_2}{c_2 v_1} = 0.1418. \quad (4.110)$$

4.31 The isentropic exponent follows

$$\kappa = \frac{c_p}{c_v} = \frac{c_p}{c_p - R} = 1.4003. \quad (4.111)$$

To determine the outlet temperature of the adiabatic but frictional turbine, the outlet temperature from a fictitious isentropic, i.e. adiabatic and frictionless, turbine is first determined:

$$T_{2s} = T_1\left(\frac{p_2}{p_1}\right)^{\frac{\kappa-1}{\kappa}} = 181.6427\,\mathrm{K}. \quad (4.112)$$

With the definition of the isentropic turbine efficiency

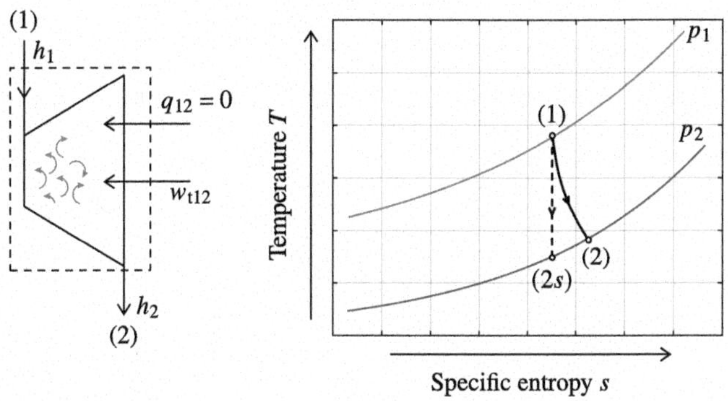

Fig. 4.16 Adiabatic turbine

$$\eta_{sT} = \frac{h_2 - h_1}{h_{2s} - h_1} = \frac{T_2 - T_1}{T_{2s} - T_1}, \tag{4.113}$$

one can then calculate the temperature at the outlet of the turbine, i.e.

$$T_2 = \eta_{sT} (T_{2s} - T_1) + T_1 = 190.1482\,\mathrm{K}. \tag{4.114}$$

The first law of thermodynamics obeys, cf. Fig. 4.16,

$$P_{t12} + \underbrace{\dot{Q}}_{=0} + \dot{m}h_1 = \dot{m}h_2. \tag{4.115}$$

respectively

$$P_{t12} = \dot{m} (h_2 - h_1) = \dot{m}c_p (T_2 - T_1). \tag{4.116}$$

The mass flow rate follows from the thermal equation of state[11], i.e.

$$\dot{m}RT_1 = p_1 \dot{V}_1. \tag{4.117}$$

Finally, one obtains

$$P_{t12} = \frac{p_1 \dot{V}_L}{RT_1} c_p (T_2 - T_1) = -8.6934\,\mathrm{kW}. \tag{4.118}$$

4.32

(a) Since it is a simple flow channel, no work is supplied from the outside, nor is any work released to the environment. The first law of thermodynamics for an open, steady state system is therefore in this case:

[11] Applied for state (1), as the volume flow rate is specified for this state.

$$\underbrace{\dot{Q}_{12} + P_{t,12}}_{=0} + \dot{m}\left(h_1 + \frac{1}{2}c_1^2\right) = \dot{m}\left(h_2 + \frac{1}{2}c_2^2\right). \tag{4.119}$$

Applying the caloric equation of state, we obtain

$$0 = c_p\,(T_2 - T_1) + \frac{1}{2}\left(c_2^2 - c_1^2\right). \tag{4.120}$$

Hence, the temperature at the outlet yields

$$T_2 = T_1 - \frac{c_2^2 - c_1^2}{2c_p} = 404.7809\,\text{K}. \tag{4.121}$$

To determine the specific entropy generation, the second law of thermodynamics is applied, i.e.

$$\dot{m}s_1 + \underbrace{\dot{S}_a}_{=0} + \dot{S}_i = \dot{m}s_2 \tag{4.122}$$

Due to the adiabatic mode of operation, it is $\dot{S}_a = 0$. The specific entropy generation then follows

$$s_i = s_2 - s_1. \tag{4.123}$$

The caloric equation of state for an isobaric change of state finally results in

$$s_i = s_2 - s_1 = c_p \ln\frac{T_2}{T_1} = 11.9289\frac{\text{J}}{\text{kgK}}. \tag{4.124}$$

(b) The thermodynamic mean temperature then is

$$T_m = \frac{T_2 - T_1}{\ln\frac{T_2}{T_1}} = 402.3857\,\text{K} \tag{4.125}$$

so that the specific dissipation follows

$$\psi_{12} = s_i T_m = 4800\frac{\text{J}}{\text{kg}}. \tag{4.126}$$

Alternatively, the specific dissipation can be calculated with the partial energy equation, i.e.

$$w_{t,12} = 0 = \underbrace{\int_1^2 v\,\mathrm{d}p}_{=0} + \psi_{12} + \frac{1}{2}\left(c_2^2 - c_1^2\right) \tag{4.127}$$

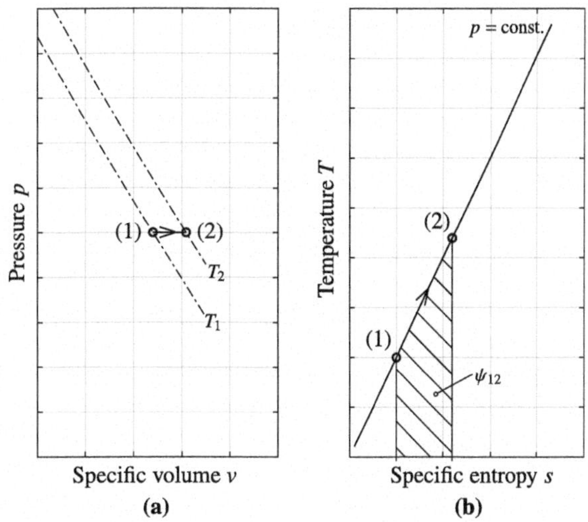

Fig. 4.17 Adiabatic channel: **a** p, v-diagram and **b** T, s-diagram

respectively

$$\psi_{12} = -\frac{1}{2}\left(c_2^2 - c_1^2\right) = 4800 \frac{\text{J}}{\text{kg}}. \tag{4.128}$$

(c) The ratio of outlet cross section A_2 to inlet cross section A_1 can be derived with the equation of continuity, i.e.

$$\dot{m}_1 = c_1 A_1 \rho_1 = \frac{c_1 A_1}{v_1} \tag{4.129}$$

and

$$\dot{m}_2 = \dot{m}_1 = c_2 A_2 \rho_2 = \frac{c_2 A_2}{v_2} \tag{4.130}$$

Finally, it applies for an isobaric change of state

$$\frac{A_2}{A_1} = \frac{c_1 v_2}{c_2 v_1} = \frac{c_1 T_2}{c_2 T_1} = 5.0598. \tag{4.131}$$

(d) The change of state is illustrated in Fig. 4.17. In these diagrams, a pressure of $p =$ 1 bar was arbitrarily chosen for representation. A gas constant of $R = 287 \frac{\text{J}}{\text{kgK}}$ was also used for the calculations. The p, v-diagram only shows a section so that the curvature of the isotherms is not visible. The same applies for the shape of the curved isobar in the T, s-diagram.

4.33

(a) It applies

$$\text{Ma} = \frac{c_2}{a_2} = 0.8 \tag{4.132}$$

so that the velocity yields

$$c_2 = 0.8a_2 = 0.8\sqrt{\kappa R T_2} \quad \text{with} \quad \kappa = \frac{c_p}{c_v} = \frac{c_p}{c_p - R} = 1.4003. \tag{4.133}$$

The energy balance (1) → (2) obeys

$$\underbrace{P_{t,12} + \dot{Q}_{12}}_{=0} + \dot{m}\left[h_1 + \frac{1}{2}c_1^2\right] = \dot{m}\left[h_2 + \frac{1}{2}c_2^2\right] \tag{4.134}$$

respectively with the caloric equation of state and[12] $c_1 = 0$

$$c_p\,(T_2 - T_1) + \frac{1}{2}c_2^2 = 0. \tag{4.135}$$

A combination of Eqs. 4.133 and 4.135 results in

$$T_2 = \frac{2c_p T_1}{0.8^2\kappa R + 2c_p} = 265.9364\,\text{K}. \tag{4.136}$$

The velocity then follows

$$c_2 = 0.8a_2 = 0.8\sqrt{\kappa R T_2} = 261.5334\,\text{m}\,\text{s}^{-1}. \tag{4.137}$$

The change of state (1) → (2) is adiabatic and free of friction, i.e. isentropic:

$$p_2 = p_1\left(\frac{T_2}{T_1}\right)^{\frac{\kappa}{\kappa-1}} = 6.5598 \times 10^4\,\text{Pa}. \tag{4.138}$$

Since state (2) is kept constant in the time span Δt by regulating the valve,[13] this means for the inflowing mass

$$\Delta m_{\text{in}} = \int_0^{\Delta t} \dot{m}_2\,\text{d}t = \int_0^{\Delta t} \rho_2 \dot{V}_2\,\text{d}t = \int_0^{\Delta t} \rho_2 c_2 A_2\,\text{d}t = \rho_2 c_2 A_2\,\Delta t. \tag{4.139}$$

With the thermal equation of state one gets

$$\Delta m_{\text{in}} = \frac{p_2 c_2 A_2\,\Delta t}{R T_2} = 899.1176\,\text{kg}. \tag{4.140}$$

---

[12] The environment is supposed to be in rest.

[13] The pressure p_2 remains constant, so that according to the isentropic equation the temperature T_2 also remains constant. So the specific volume v_2 and the velocity c_2 also stay constant.

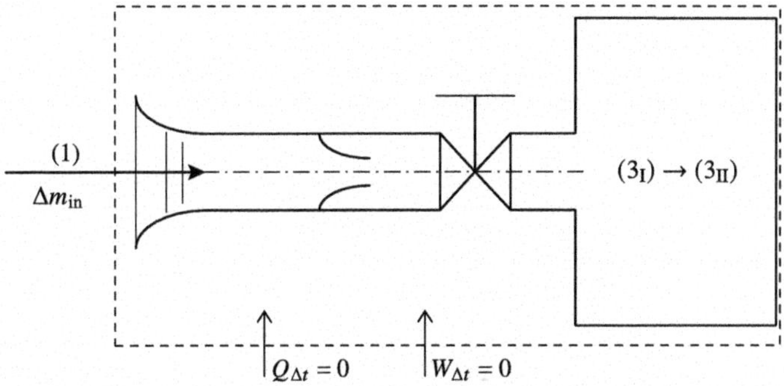

Fig. 4.18 Wind tunnel

(b) The system is balanced according to Fig. 4.18. Initially, the mass within the system obeys

$$m_{3_I} = \frac{V_3 p_{3_I}}{RT_{3_I}} = 1.3937 \times 10^3 \text{kg}. \tag{4.141}$$

After Δt the mass within the system is

$$m_{3_{II}} = m_{3_I} + \Delta m_{in} = 2.2928 \times 10^3 \text{kg}. \tag{4.142}$$

The transient energy balance[14] is

$$\underbrace{Q_{\Delta t} + W_{\Delta t}}_{=0} + \int_0^{\Delta t} (u_1 + p_1 v_1)\, \mathrm{d}m = U_{3_{II}} - U_{3_I}. \tag{4.143}$$

Since state (1) is constant it follows

$$(u_1 + p_1 v_1) \int_0^{\Delta t} \mathrm{d}m = (u_1 + p_1 v_1)\, \Delta m_{in} = U_{3_{II}} - U_{3_I}. \tag{4.144}$$

This results in

$$\begin{aligned} \Delta m_{in} u_1 + \Delta m_{in} p_1 v_1 &= \overbrace{\left(m_{3_I} + \Delta m_{in}\right)}^{=m_{3_{II}}} u_{3_{II}} - m_{3_I} u_{3_I} \\ &= m_{3_I} u_{3_{II}} + \Delta m_{in} u_{3_{II}} - m_{3_I} u_{3_I}. \end{aligned} \tag{4.145}$$

Rearranging leads to

[14] Outer energies are neglected for both the incoming energy and the system.

$$\Delta m_{\text{in}} p_1 v_1 = m_{3_{\text{I}}} \left(u_{3_{\text{II}}} - u_{3_{\text{I}}}\right) + \Delta m_{\text{in}} \left(u_{3_{\text{II}}} - u_1\right) \tag{4.146}$$

respectively with the caloric equation of state

$$\Delta m_{\text{in}} p_1 v_1 = m_{3_{\text{I}}} c_v \left(T_{3_{\text{II}}} - T_{3_{\text{I}}}\right) + \Delta m_{\text{in}} c_v \left(T_{3_{\text{II}}} - T_1\right) \tag{4.147}$$

with

$$c_v = c_p - R = 717 \frac{\text{J}}{\text{kgK}}. \tag{4.148}$$

Solving for $T_{3_{\text{II}}}$ reads as

$$T_{3_{\text{II}}} = \frac{\Delta m_{\text{in}} p_1 v_1 + c_v \left(m_{3_{\text{I}}} T_{3_{\text{I}}} + \Delta m_{\text{in}} T_1\right)}{c_v \left(m_{3_{\text{I}}} + \Delta m_{\text{in}}\right)} \tag{4.149}$$

respectively

$$T_{3_{\text{II}}} = \frac{\Delta m_{\text{in}} R T_1 + c_v \left(m_{3_{\text{I}}} T_{3_{\text{I}}} + \Delta m_{\text{in}} T_1\right)}{c_v \left(m_{3_{\text{I}}} + \Delta m_{\text{in}}\right)} = 347.0897\text{K}. \tag{4.150}$$

The pressure follows according to

$$p_{3_{\text{II}}} = \frac{m_{3_{\text{II}}} R T_{3_{\text{II}}}}{V_3} = 3.8067 \times 10^4 \text{Pa}. \tag{4.151}$$

(c) The entropy is balanced according to Fig. 4.19:

$$S_{\text{i},\Delta t} + \underbrace{S_{\text{a},\Delta t}}_{=0} + \int_0^{\Delta t} s_1 \, \text{d}m = S_{3_{\text{II}}} - S_{3_{\text{I}}}. \tag{4.152}$$

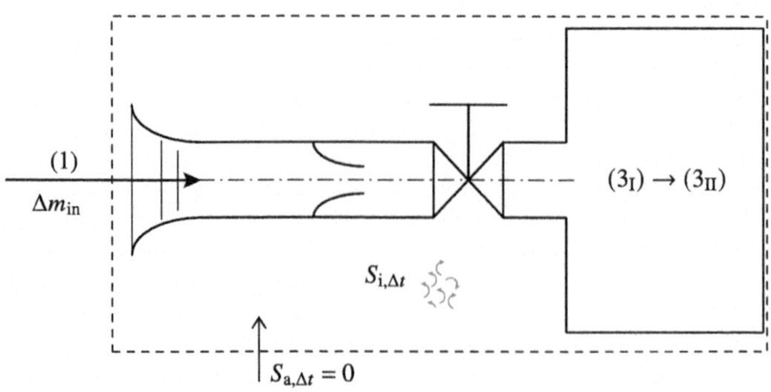

Fig. 4.19 Wind tunnel

Since state (1) is constant it follows

$$S_{\mathrm{i},\Delta t} + s_1 \int_0^{\Delta t} \mathrm{d}m = S_{\mathrm{i},\Delta t} + s_1 \Delta m_{\mathrm{in}} = S_{3_{\mathrm{II}}} - S_{3_{\mathrm{I}}}. \tag{4.153}$$

This results in

$$\begin{aligned} S_{\mathrm{i},\Delta t} + s_1 \Delta m_{\mathrm{in}} &= \overbrace{\left(m_{3_{\mathrm{I}}} + \Delta m_{\mathrm{in}}\right)}^{=m_{3_{\mathrm{II}}}} s_{3_{\mathrm{II}}} - m_{3_{\mathrm{I}}} s_{3_{\mathrm{I}}} \\ &= m_{3_{\mathrm{I}}} s_{3_{\mathrm{II}}} + \Delta m_{\mathrm{in}} s_{3_{\mathrm{II}}} - m_{3_{\mathrm{I}}} s_{3_{\mathrm{I}}}. \end{aligned} \tag{4.154}$$

Rearranging leads to

$$S_{\mathrm{i},\Delta t} = m_{3_{\mathrm{I}}} \left(s_{3_{\mathrm{II}}} - s_{3_{\mathrm{I}}}\right) + \Delta m_{\mathrm{in}} \left(s_{3_{\mathrm{II}}} - s_1\right) \tag{4.155}$$

respectively with the caloric equation of state

$$\begin{aligned} S_{\mathrm{i},\Delta t} &= m_{3_{\mathrm{I}}} \left(c_p \ln \frac{T_{3_{\mathrm{II}}}}{T_{3_{\mathrm{I}}}} - R \ln \frac{p_{3_{\mathrm{II}}}}{p_{3_{\mathrm{I}}}}\right) + \Delta m_{\mathrm{in}} \left(c_p \ln \frac{T_{3_{\mathrm{II}}}}{T_1} - R \ln \frac{p_{3_{\mathrm{II}}}}{p_1}\right) \\ &= 3.2742 \times 10^5 \, \mathrm{J\,K^{-1}}. \end{aligned} \tag{4.156}$$

The loss of exergy finally is

$$\Delta E_{\mathrm{x,V},\Delta t} = S_{\mathrm{i},\Delta t} T_{\mathrm{env}} = 9.8226 \times 10^4 \, \mathrm{kJ}. \tag{4.157}$$

4.34 The solution is depicted in Fig. 4.20.

- (1) → (2):
  The pressure must decrease, the temperature decreases, because the turbine releases energy, i.e.

Fig. 4.20 Changes of state

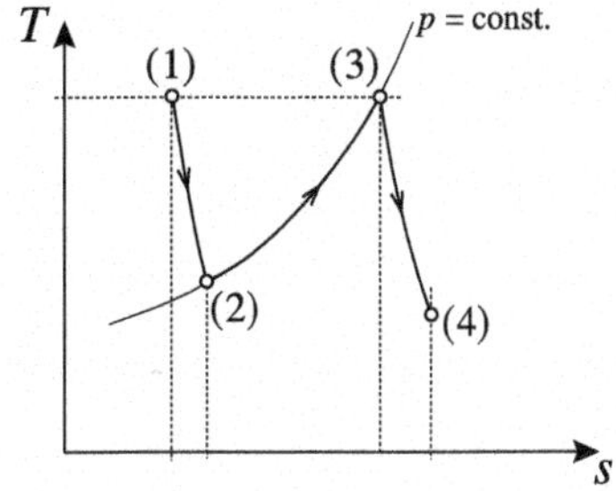

$$h_2 - h_1 = w_{t,12} + \underbrace{q_{12}}_{=0} < 0 \tag{4.158}$$

respectively

$$T_2 - T_1 < 0. \tag{4.159}$$

Since the turbine is adiabatic, the specific entropy cannot decrease. In contrast, with an isentropic efficiency of less than 1, the specific entropy must increase.

- (2) → (3):
  The change of state is isobaric and by supply of heat the initial temperature T_1 is reached again. The supply of heat increases the specific entropy. Since the outer energies are neglected and the change of state is isobaric, no dissipation occurs and no entropy is generated:

$$w_{t,23} = 0 = \underbrace{\int_2^3 v\,\mathrm{d}p}_{=0} + \psi_{23} \tag{4.160}$$

so that

$$\psi_{23} = 0 \quad \rightarrow \quad s_{i,23} = 0. \tag{4.161}$$

- (3) → (4):
  The pressure must decrease, the temperature decreases, because the turbine releases energy, i.e.

$$h_4 - h_3 = w_{t,34} + \underbrace{q_{34}}_{=0} < 0 \tag{4.162}$$

respectively

$$T_4 - T_3 < 0. \tag{4.163}$$

Since the turbine is adiabatic, the specific entropy cannot decrease. In contrast, with an isentropic efficiency of less than 1, the specific entropy must increase.

4.35 **(b)**
Answer (b) is correct, a detailed explanation has already been given in Problem 4.13. Isotherms in a T, s-diagram are horizontal lines. Answer (a) therefore is wrong. Specific work becomes visible in the p, v-diagram: The area under the change of state corresponds to the specific volume work of an open system. The projected area on the p-axis corresponds to the pressure work of an open system.
In the T, s-diagram, on the other hand, the sum of specific heat and specific dissipation can be identified as the area under the change of state. Therefore, answer (c) is incorrect.

Chapter 5
Exergy and Anergy

This chapter is dedicated to the thermodynamic evaluation of energy. It is examined to what extent different forms of energy can be converted into work. In particular, it is shown that, for example, thermal energy, i.e. heat, cannot be completely converted into mechanical work in cyclic processes. For example, every combustion engine releases part of the heat produced in the combustion chamber to the environment in the radiator, i.e. part of the thermal energy is not converted into mechanical energy. For this purpose, the energy, which is a conservation variable, is theoretically divided into a part that can be converted into work and is called exergy, and another part that cannot be further converted in a given environment and is called anergy. In the tasks dealt with in this chapter, the concept of exergy is further sharpened.

5.1 Problems

5.1 What is true for an adiabatic turbine?

(a) The specific entropy can never decrease.
(b) The specific exergy rises.
(c) Technical work must be supplied.
(d) None of the above statements is true.

5.2 Heat passes a wall from the inside of a house (1) to the environment (env) in a steady state. Which statement is true?

(a) Dissipation can be observed.
(b) The exergy gets fully lost on its way to the environment.
(c) There is no generation of entropy within the wall.
(d) None of the above statements is true.

5.3 Which of the areas marked in grey, cf. Fig. 5.1, shows the specific loss of exergy of an adiabatic turbine? Changes of potential/kinetic energies shall be ignored.

A. Schmidt, *Technical Thermodynamics Workbook for Engineers*,
https://doi.org/10.1007/978-3-031-50172-2_5

Fig. 5.1 Adiabatic turbine

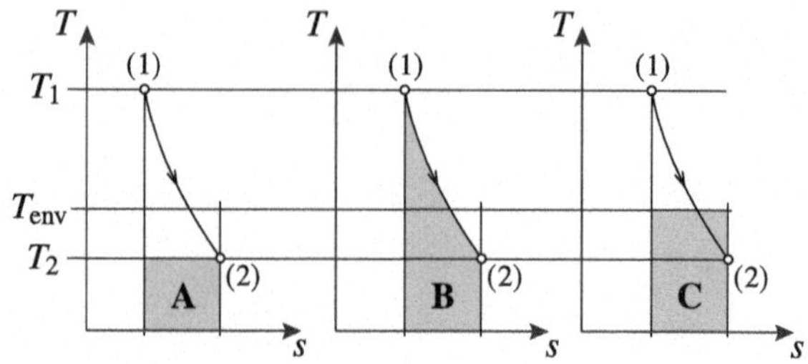

(a) A
(b) B
(c) C
(d) None of the areas shown.

5.4 Is it possible to increase the exergy of a closed, rigid system with an initial temperature of T_{env}? The fluid inside the system shall be an ideal gas.

(a) No, exergy can only decrease.
(b) Yes, only by cooling the system.
(c) Yes, by supply of stirrer work.
(d) None of the above statements is true.

5.5 What is true for the adiabatic throttling of an ideal gas in a steady state flow process when changes in potential/kinetic energies are neglected?

(a) Exergy rises.
(b) Temperature rises.
(c) Technical work must be supplied.
(d) None of the above statements is true.

5.6 An incompressible liquid passes a horizontal, adiabatic tube of constant cross section. On its way the pressure decreases. Which statement is true?

(a) Temperature drops.
(b) Anergy rises.
(c) Heat is supplied.
(d) None of the above statements is true.

5.7 An ideal gas passes stationary a heat exchanger without any pressure drop. Changes of outer energies can be ignored. The supplied flux of exergy equals a quarter of the supplied heat flux. Ambient temperature shall be 25 °C. What is the thermodynamic mean temperature of the gas?

(a) 112.38 °C
(b) 124.38 °C
(c) 385.53 °C
(d) 397.53 °C

5.8 What is true for steady state heat transfer?

(a) In real processes a loss of exergy occurs when heat is transferred since $\Delta T \neq 0$.
(b) The smaller the temperature difference at which heat is transferred, the greater the entropy generation.
(c) Energy is generated due to the heat transfer.
(d) None of the above statements applies.

5.9 Heat passes a wall from the inside of a house (1) to the environment (env). The wall shall be in steady state. Which statement is true?

(a) The absolute value of the exergy loss is equal to the absolute value of the exergy of the heat in state (1).
(b) This process describes a perpetuum mobile of the second kind.
(c) Dissipation occurs.
(d) None of the above statements applies.

5.10 Which statement is true for an adiabatic, horizontal diffuser?

(a) The exergy increases as it flows through the diffuser.
(b) The flow velocity increases.
(c) Entropy decreases.
(d) None of the above statements applies.

5.11 What applies to an adiabatic, horizontal nozzle?

(a) The specific entropy at the outlet is greater than or equal to the specific entropy at the inlet.
(b) The flow is slowed down.
(c) Heat is exchanged with the environment.
(d) None of the above statements applies.

5.12 A heat exchanger is operated in steady state with a flow of water and a flow of air, that are not mixing. The heat exchanger shall be adiabatic to the environment ($T_{\text{env}} = 280\,\text{K}$). The following information is known:

- Water flow: $\dot{m}_{\text{W}} = 2\frac{\text{kg}}{\text{s}}$, $T_{\text{W,in}} = 330\,\text{K}$, $T_{\text{W,out}} = 320\,\text{K}$, $c_{\text{W}} = 4.19\frac{\text{kJ}}{\text{kgK}}$
- Air flow: $\dot{m}_{\text{A}} = 2.5\frac{\text{kg}}{\text{s}}$, $T_{\text{A,in}} = 280\,\text{K}$, $c_p = 1.004\frac{\text{kJ}}{\text{kgK}}$

There shall be no pressure drop in the heat exchanger. Changes of kinetic/potential energies are to be ignored. Calculate the flux of loss of exergy!

5.13 In a closed vessel, 200 kg of steel ($c_{\text{St}} = 0.45\frac{\text{kJ}}{\text{kgK}}$) with an initial temperature of $T_{\text{St}} = 1200\,\text{K}$ are quenched in 1000 kg of water ($c_{\text{W}} = 4.19\frac{\text{kJ}}{\text{kgK}}$) with an initial temperature of $T_{\text{W}} = 288\,\text{K}$. Calculate the loss of exergy until the system reaches thermodynamic equilibrium! It is assumed that there is no heat exchange between the vessel and the environment. The mass of the vessel shall be neglected. Water as well as steel can be treated as incompressible. The environmental temperature is $T_{\text{env}} = 288\,\text{K}$.

5.14 An air/water heat exchanger adiabatic to the environment is operated stationary. The following conditions are known:

- Water (incompressible liquid): $T_{\text{W,in}} = 300\,\text{K}$, $p = \text{const.}$, $c_{\text{W}} = 4.19\frac{\text{kJ}}{\text{kgK}}$, $\dot{m}_{\text{W}} = 0.2\frac{\text{kg}}{\text{s}}$
- Air (ideal gas): $T_{\text{A,in}} = 500\,\text{K}$, $T_{\text{A,out}} = 400\,\text{K}$, $p = \text{const.}$, $c_p = 1.004\frac{\text{kJ}}{\text{kgK}}$, $\dot{m}_{\text{A}} = 1.1\frac{\text{kg}}{\text{s}}$
- Environment: $T_{\text{env}} = 300\,\text{K}$, $p_{\text{env}} = 1\,\text{bar}$

What is the ratio of exergy supplied to heat supplied on the water side in case changes of potential/kinetic energies are ignored?

5.15 An adiabatic tank ($V = 1\text{m}^3$ =const.) is filled with 1.4 kg of an ideal gas ($c_p = 1.004\frac{\text{kJ}}{\text{kgK}}$, $\kappa = 1.4$). The initial temperature at equilibrium is $\vartheta_1 = 20\,°\text{C}$. Now, 0.5 kW power are supplied over 45 s by an immersed electric heating element. In state (2) a new equilibrium is reached. What is the exergy loss during the change of state if ambient temperature is 20 °C?

5.16 A theoretical reversible thermal engine has an exergetic efficiency of 1. Which statement is true?

(a) The heat release corresponds exactly to the dissipation inside the machine.
(b) No heat is released.
(c) The heat release takes place at $T_0 \to T_{\text{env}}$.
(d) The Carnot efficiency in this case is also 1.

5.17 An incompressible, non-reactive body with a mass of 1 kg rests in the Earth's gravitational field two metres above the ground and is in thermal equilibrium with its environment. Which statement is true?

(a) Because of the equilibrium, the exergy of the mass is zero.
(b) Only because of its potential energy does the mass possess exergy.
(c) The exergy of the mass is smaller than its potential energy.
(d) None of the above statements applies.

5.2 Solutions

5.1 **(a)**
The change of state in an adiabatic turbine is shown in Fig. 5.2. To determine what occurs with the entropy in steady state, a balance is applied, i.e. the entropy supplied equals the entropy released:

$$\dot{m}s_1 + \dot{S}_{\text{i}} + \dot{S}_{\text{a}} = \dot{m}s_2 \tag{5.1}$$

respectively

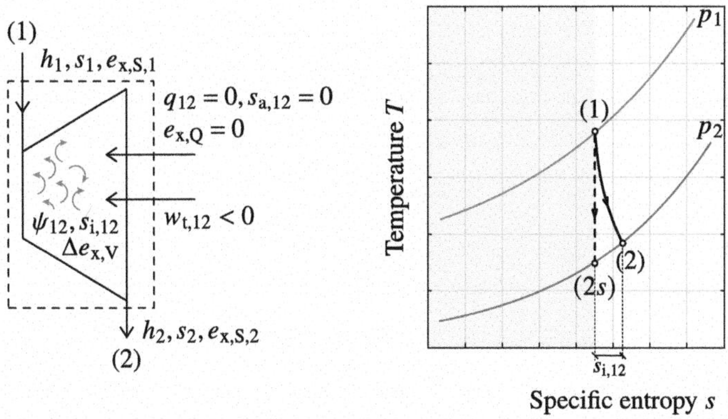

Fig. 5.2 Adiabatic turbine

$$s_2 - s_1 = \Delta s = s_{i,12} + s_{a,12}. \tag{5.2}$$

This equation simplifies further because the turbine is adiabatic, i.e.

$$s_2 - s_1 = \Delta s = s_{i,12} \geq 0. \tag{5.3}$$

Since the entropy generated can only become positive or zero, respectively, it is never possible for the specific entropy in an adiabatic turbine to decrease—answer (a) is therefore correct.

The balance of exergy obeys

$$\dot{m}e_{x,S,1} + \underbrace{\dot{E}_{x,Q}}_{=0} + P_t = \Delta\dot{E}_{x,V} + \dot{m}e_{x,S,2} \tag{5.4}$$

respectively

$$e_{x,S,2} - e_{x,S,1} = w_{t,12} - \Delta e_{x,V} = \underbrace{w_{t,12}}_{<0} - \underbrace{s_{i,12}T_{env}}_{\geq 0} < 0. \tag{5.5}$$

Due to the expansion, i.e. $\mathrm{d}p < 0$, of the fluid, the turbine emits work,[1] i.e.

$$w_{t,12} = \underbrace{\int_1^2 v\,\mathrm{d}p}_{<0} + \psi_{12} + \Delta e_{a,12} < 0. \tag{5.6}$$

[1] Possible dissipation within the turbine is driven by the pressure work and therefore reduces the power release of the turbine. Compared to the pressure work, the external work plays a minor role.

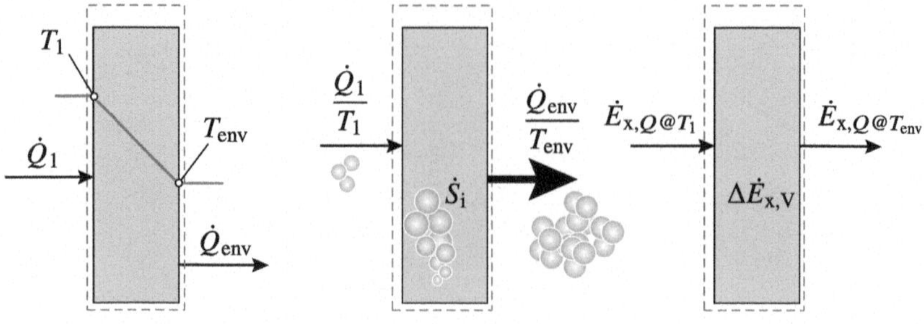

Fig. 5.3 Heat flux passing a wall: **a** Energy balance, **b** Entropy balance, **c** Exergy balance

Since the turbine is applied to release power, answer (c) is incorrect. Furthermore, working capability, i.e. exergy, is lost due to possible irreversibilities. Exergy therefore decreases, cf. Eq. 5.5, and can never increase. Answer (b) is therefore wrong.

5.2 **(b)**

The thermodynamics of a heat flux passing a plane wall have already been discussed in Problem 2.2. The first law of thermodynamics obeys, cf. Fig. 5.3a,

$$\dot{Q}_1 = \dot{Q}_{env} = \dot{Q}. \tag{5.7}$$

The balance of entropy follows

$$\frac{\dot{Q}}{T_1} + \dot{S}_i = \frac{\dot{Q}}{T_{env}} \tag{5.8}$$

so that the entropy generation is

$$\dot{S}_i = \dot{Q}\left(\frac{1}{T_{env}} - \frac{1}{T_1}\right). \tag{5.9}$$

Now we want to supplement an exergetic consideration, see also Fig. 5.3c. Just like energy and entropy, exergy is an extensive state value that can be balanced, i.e. counted. However, only energy is a conservation quantity,[2] so its value remains constant; energy can neither be created nor destroyed. When balancing entropy, it must be noted that there can be a source term for entropy that is linked to dissipation or inhomogeneities that initiate irreversible balancing processes. If entropy is generated, this is associated with a loss of working capability, i.e. exergy. So while there can be source terms with entropy, there are sinks with regard to exergy. This is the reason why neither entropy nor exergy are conservation quantities.

Let us now turn to the exergy balance according to Fig. 5.3c: On the left side of the wall, exergy enters the system because the heat flux is at a temperature above

[2] The other conservation quantities are mass and momentum.

ambient temperature. By means of a thermal engine between reservoirs T_1 and T_{env}, this heat could be converted into work. The maximum work is therefore the exergy

$$\dot{E}_{\text{x},Q@T_1} = \dot{Q}\left(1 - \frac{T_{\text{env}}}{T_1}\right). \tag{5.10}$$

Since entropy is generated inside the wall, there is a loss of exergy

$$\Delta\dot{E}_{\text{x,V}} = \dot{S}_{\text{i}} T_{\text{env}} = \dot{Q}\left(1 - \frac{T_{\text{env}}}{T_1}\right). \tag{5.11}$$

The heat flux that leaves the wall cannot be further converted in a thermal engine because there is no longer a temperature potential: The heat occurs at ambient temperature. Therefore the exergy is

$$\dot{E}_{\text{x},Q@T_{\text{env}}} = \dot{Q}\left(1 - \frac{T_{\text{env}}}{T_{\text{env}}}\right) = 0. \tag{5.12}$$

For the steady state case, the exergy balance is that entering exergy equals exiting exergy,[3] i.e.

$$\dot{E}_{\text{x},Q@T_1} = \dot{E}_{\text{x},Q@T_{\text{env}}} + \Delta\dot{E}_{\text{x,V}} = \Delta\dot{E}_{\text{x,V}} \tag{5.13}$$

Consequently, in this case, the exergy gets fully lost on its way to the environment. Answer (b) is correct. Although entropy is generated in the system, there is no dissipation. The cause of entropy generation is a disturbance in the temperature distribution: due to the temperature inhomogeneity, an irreversible balancing process occurs, which is linked to entropy generation. Answer (c) is therefore wrong. For dissipation to occur, a mechanical process is required in which particles of matter move at different velocities and friction occurs. In this case, the wall is at rest, so there are no mechanical processes taking place that could cause dissipation. Therefore, answer (a) is wrong.

5.3 **(c)**
All three graphs shown illustrate the change of state of an adiabatic, non-isentropic turbine. The specific dissipation is responsible for the increase of the specific entropy, since

$$\text{d}s = \delta s_{\text{a}} + \delta s_{\text{i}} = \underbrace{\frac{\delta q}{T}}_{=0} + \frac{\delta\psi}{T} > 0 \tag{5.14}$$

respectively

$$s_2 - s_1 = s_{\text{i},12} = \frac{\psi_{12}}{T_{\text{m}}} > 0. \tag{5.15}$$

For the specific exergy loss,

[3] Note that the exergy loss is listed on the exiting exergy side.

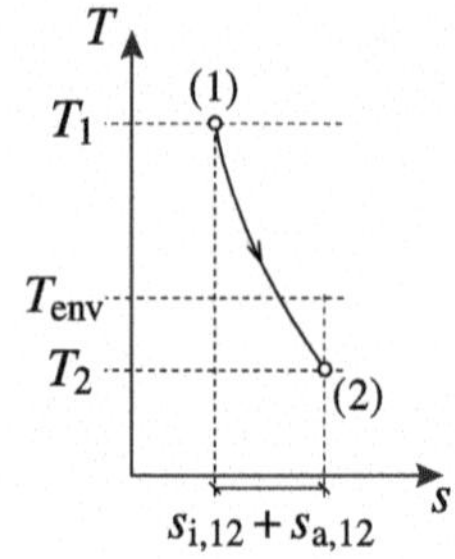

Fig. 5.4 Diabatic turbine

$$\Delta e_{x,V} = s_{i,12} T_{env} \tag{5.16}$$

always applies. For the case of an adiabatic change of state, one can thus visualise the specific exergy loss very simply graphically, since

$$\Delta e_{x,V} = (s_2 - s_1)\, T_{env}. \tag{5.17}$$

The specific exergy loss therefore corresponds to a rectangle with edge lengths $(s_2 - s_1)$ and T_{env}. This is correctly shown in Graph C, so answer (c) applies.
In the case of a diabatic, frictional change of state, it is not possible to visualise the specific exergy loss so easily, see Fig. 5.4. This is because the change in specific entropy is then based on two mechanisms, i.e.

$$s_2 - s_1 = s_{i,12} + s_{a,12} > 0. \tag{5.18}$$

Thus, $s_{i,12}$ cannot be unambiguously deduced from a $T,\ s$-diagram, since this diagram only depicts the total change in specific entropy. A rectangular structure $s_{i,12} \cdot T_{env}$ for the representation of the specific exergy loss does not succeed here.In order to draw the specific exergy loss as a rectangle in the $T,\ s$-diagram, $s_{i,12}$ must be determined.

5.4 **(c)**
Figure 5.5 shows an overview for a closed system. By exchanging the process values of work W_{12} and heat Q_{12} with the environment, the energetic state of the system changes. The first law of thermodynamics, see Fig. 5.5a, obeys

$$W_{12} + Q_{12} = U_2 - U_1 + \Delta E_{a,12}. \tag{5.19}$$

For the case shown, according to the partial energy equation, the work W_{12} is composed as follows

$$W_{12} = W_{V,12} + W_{mech,12} + \Psi_{12}. \tag{5.20}$$

The work exchanged across the system boundary is composed of volume work and mechanical work to change the position and velocity of the centre of gravity. In addition, part of the work dissipates through friction. The second law of thermodynamics, i.e. the entropy balance, cf. Fig. 5.5b, reads as follows

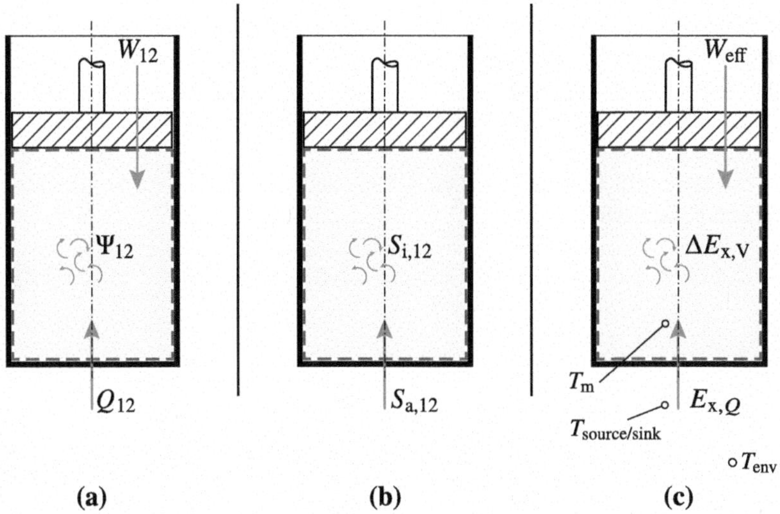

Fig. 5.5 Closed system: **a** Energy **b** Entropy **c** Exergy balance

$$S_{a,12} + S_{i,12} = S_2 - S_1 . \tag{5.21}$$

The state value entropy inside the system can change through entropy generation due to friction processes in the system $S_{i,12}$ and through entropy exchanged with the heat $S_{a,12}$. Entropy generation can only be positive, respectively zero in the best case. Negative entropy generation is not possible. If there is only $S_{i,12}$ and no $S_{a,12}$, the entropy in the system can therefore never decrease. $S_{a,12}$ depends on the direction of the heat exchange. If the system is heated, entropy $S_{a,12}$ is supplied to the system. In the opposite case, i.e. in cooling mode, entropy $S_{a,12}$ is removed from the system.

Now we want to apply the exergy balance for the closed system, see Fig. 5.5c. In state (1), let the exergy in the system be $E_{x,1}$. By supplying heat, which, if the temperature within the system at which the heat is transferred is greater than ambient temperature, i.e. $T_m > T_{env}$,[4] contains working capability, i.e. exergy $E_{x,Q} > 0$. Consequently, the exergy of the system increases. Anyhow, the sign of exergy of heat is summarised in Fig. 7.12 and depends from the direction of heat transfer and the temperature range compared to ambient temperature.
Exergy can get lost inside the system if entropy is generated, e.g. by friction processes. Conclusively, this reduces the exergy within the system. Furthermore, the exergy inside the system changes due to the work exchanged at the system boundary: The effective work W_{eff} is relevant, since the work W_{12} that crosses the system boundary is not completely usable. For example, when the system expands, part of the work W_{12} is utilised to compress the external environment. In the reverse case

[4] Heat is transferred into the system when the environment of the system has a higher temperature than the system under consideration, i.e. $T_{source/sink} > T_m$. However, the ambient temperature T_{env} is the reference level for determining the working capability of a system, i.e. the exergy.

of compressing the system, the expanding environment supports this compression. This is summarised by the effective work[5]

$$W_{\text{eff}} = W_{12} + p_{\text{env}}\,(V_2 - V_1) \tag{5.22}$$

which is relevant for the exergy balance. Such an exergy balance therefore leads to the exergy of the system in state (2)

$$E_{\text{x},2} = E_{\text{x},1} + E_{\text{x},Q} - \Delta E_{\text{x,V}} + W_{\text{eff}} \tag{5.23}$$

respectively

$$E_{\text{x},2} - E_{\text{x},1} = \underbrace{E_{\text{x},Q} + W_{\text{eff}}}_{\text{IN}} - \underbrace{\Delta E_{\text{x,V}}}_{\text{OUT}}\,. \tag{5.24}$$

As this equation shows, the exergy of a closed system can increase, decrease or remain constant. This depends on the entering ($E_{\text{x},Q}$, W_{eff}) and exiting ($\Delta E_{\text{x,V}}$) exergies. Answer (a) is therefore incorrect. If heat for instance is released to an environment with $T_{\text{sink}} = T_{\text{env}}$,[6] i.e. the system is cooled, $E_{\text{x},Q} < 0$, i.e. the cooling is responsible for a decrease in exergy. Answer (b) does not apply.

Now we want to investigate what happens when stirring work is supplied to the rigid system. In this case, the system should be adiabatic in order to focus exclusively on the influence of the work. The effective work then is

$$W_{\text{eff}} = W_{12} \tag{5.25}$$

since the volume is constant. According to the partial energy equation, it further is

$$W_{\text{eff}} = W_{12} = \underbrace{W_{\text{V},12}}_{=0} + \underbrace{W_{\text{mech},12}}_{=0} + \Psi_{12} = \Psi_{12}. \tag{5.26}$$

The stirrer work is fully dissipated. The change of exergy therefore is

$$\begin{aligned} E_{\text{x},2} - E_{\text{x},1} &= \Psi_{12} - \Delta E_{\text{x,V}} = \Psi_{12} - S_{\text{i},12} T_{\text{env}} = \Psi_{12} - \frac{\Psi_{12}}{T_{\text{m}}} T_{\text{env}} \\ &= \Psi_{12}\left(1 - \frac{T_{\text{env}}}{T_{\text{m}}}\right). \end{aligned} \tag{5.27}$$

[5] If the system expands and releases work, i.e. $W_{12} < 0$, the volume work on the environment $p_{\text{env}}\,(V_2 - V_1) > 0$. Overall, therefore, less work will effectively be usable at the piston, i.e. $W_{\text{eff}} > W_{12}$. If the system is compressed and consumes work, i.e. $W_{12} > 0$ then the volume work on the environment $p_{\text{env}}\,(V_2 - V_1) < 0$. In total, therefore, less work is effectively required on the piston, i.e. $W_{\text{eff}} < W_{12}$.

[6] This means that the thermodynamic mean temperature of the system $T_{\text{m}} > T_{\text{env}}$, see also Problem 7.16.

According to the first law of thermodynamics

$$W_{12} = \Psi_{12} = U_2 - U_1 = mc_v\,(T_2 - T_1) > 0 \tag{5.28}$$

The temperature T_2 is greater than the initial temperature $T_1 = T_{\text{env}}$, so that the thermodynamic mean temperature from (1) to (2) is $T_{\text{m}} > T_{\text{env}}$. Consequently, it yields

$$E_{\text{x},2} - E_{\text{x},1} = \Psi_{12}\left(1 - \frac{T_{\text{env}}}{T_{\text{m}}}\right) > 0 \tag{5.29}$$

respectively in specific notation

$$e_{\text{x},2} - e_{\text{x},1} = \psi_{12}\left(1 - \frac{T_{\text{env}}}{T_{\text{m}}}\right) > 0. \tag{5.30}$$

By supplying stirrer work the exergy of a closed system can be increased. Answer (c) is correct.

Equation 5.30 can also be determined via the exergy of a closed, isochoric system, cf. Appendix E, i.e.

$$\begin{aligned} e_{\text{x},1} &= u_1 - u_{\text{env}} + \underbrace{\frac{1}{2}c_1^2 + gz_1}_{=0} + T_{\text{env}}\,(s_{\text{env}} - s_1) - p_{\text{env}}\,\underbrace{(v_{\text{env}} - v_1)}_{=0} \\ &= c_v\,(T_1 - T_{\text{env}}) + T_{\text{env}}c_v \ln\frac{T_{\text{env}}}{T_1} = e_{\text{x},1}\,(T) \end{aligned} \tag{5.31}$$

and

$$e_{\text{x},2} = c_v\,(T_2 - T_{\text{env}}) + T_{\text{env}}c_v \ln\frac{T_{\text{env}}}{T_2} = e_{\text{x},2}\,(T) \tag{5.32}$$

The difference of Eqs. 5.32 and 5.31 gives Eq. 5.30, since

$$T_{\text{m}} = \frac{T_2 - T_1}{\ln\frac{T_2}{T_1}}. \tag{5.33}$$

The temperature-dependent specific exergy of the closed system is shown in Fig. 5.6. We will now examine three cases:

- Case a: $T_{\text{m}} < T_{\text{env}}$
  By supply of stirrer work, the temperature increases from $T_{1\text{a}}$ to $T_{2\text{a}}$ according to Eq. 5.28. The system from (1) to (2) has a thermodynamic mean temperature $T_{\text{m,a}}$. As Eq. 5.31 has shown, exergy solely exists due to its temperature. Work cannot be gained because the volume of the system shall not change and outer energies are not relevant. In this case, the environment is always the warm reservoir and the system is the cold reservoir. In state (1), there is a greater thermal potential to

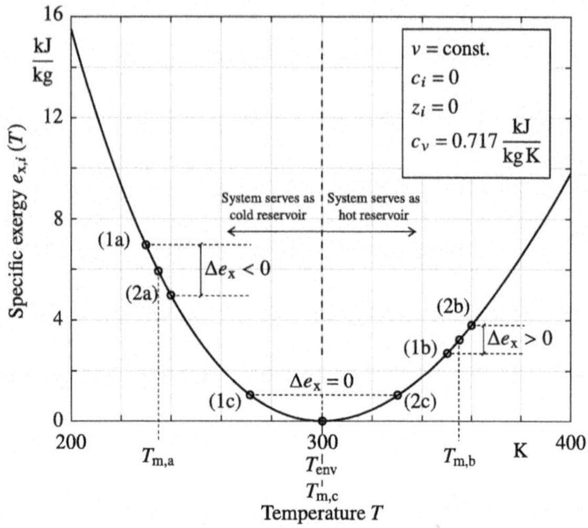

Fig. 5.6 Specific exergy of a closed, isochoric system

the environment than in state (2), so the specific exergy decreases from (1) to (2), see Fig. 5.7.

- Case b: $T_\text{m} > T_\text{env}$
  Same as in the previous case the temperature increases from $T_{1\text{b}}$ to $T_{2\text{b}}$ according to Eq. 5.28. Its thermodynamic mean temperature is $T_{\text{m,b}}$. However, in this case, the environment is the cold reservoir and the system is the warm reservoir. In state (1), there is a smaller thermal potential to the environment than in state (2), so the specific exergy increases from (1) to (2), see Fig. 5.8.
- Case c: $T_\text{m} = T_\text{env}$
  Once again, the temperature increases from $T_{1\text{c}}$ to $T_{2\text{c}}$ according to Eq. 5.28. Exergy only arises through a thermal potential. In state (1), the environment is the warm reservoir and the system is the cold reservoir. This changes in state (2), because then the environment is the cold reservoir and the system is the warm reservoir. The thermodynamic mean temperature for this case is exactly ambient temperature, i.e. $T_{\text{m,c}} = T_\text{env}$: According to Eq. 5.31 the specific exergy must remain constant.

5.5 **(d)**
The thermodynamics of an adiabatic throttle has already been explained, cf. Problem 4.12, and is summarised in Fig. 5.9. The explicit application of the energy and entropy balance is omitted here, as this has already been done in detail in Problem 4.12. There it has been shown that for an adiabatic throttling of an ideal gas with neglect of the outer energies, the change of state is isothermal. Answer (b) is therefore not correct. It has also been shown that the throttle is work-isolated, i.e. that no work is exchanged across the system boundary. Answer (c) does not apply.

In addition, an exergetic balance should now be conducted, see Fig. 5.9. We now want to formulate how large the energy flux $\dot{m}e_{\text{s},S,2}$ is that leaves the system boundary with the fluid, i.e.

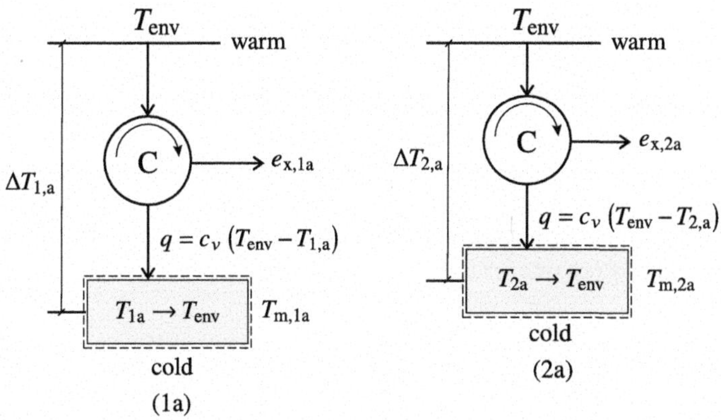

Fig. 5.7 Exergy of the closed system in states (1a) and (2a)

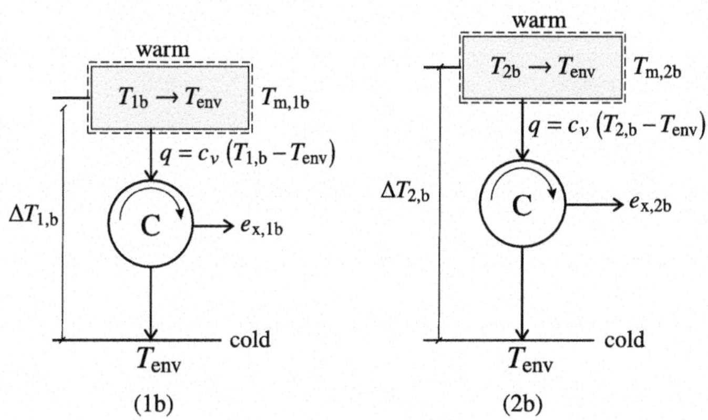

Fig. 5.8 Exergy of the closed system in states (1b) and (2b)

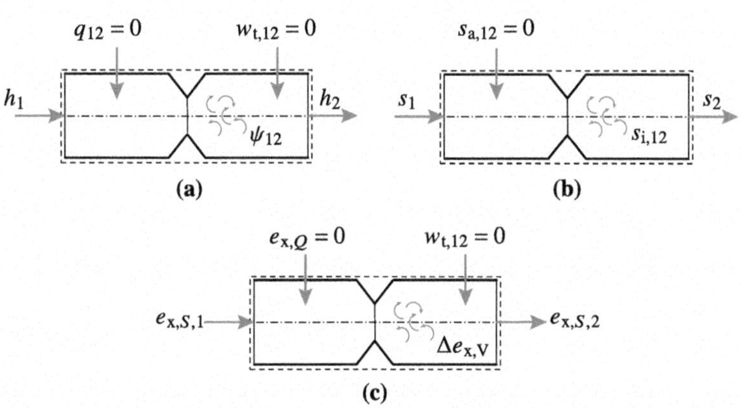

Fig. 5.9 Adiabatic throttle: **a** Energy **b** Entropy **c** Exergy balance

$$\dot{m}e_{x,S,2} = \dot{m}e_{x,S,1} + \underbrace{\dot{E}_{x,Q}}_{=\dot{m}e_{x,Q}} + \underbrace{P_t}_{=\dot{m}w_t} - \Delta\dot{E}_{x,V} \tag{5.34}$$

The basis is the exergetic state of the entering fluid $\dot{m}e_{x,S,1}$. As it flows through, the fluid continues to absorb exergy (+) through $\dot{E}_{x,Q}$ and P_t. Furthermore, exergy is decreased (−) by irreversibilities, i.e. $\Delta\dot{E}_{x,V}$. This balance is expressed by Eq. 5.34. Now we want to decide whether the exergy of the fluid increases or decreases. To do this, the equation is rewritten, i.e.

$$\dot{m}\left(e_{x,S,2} - e_{x,S,1}\right) = \dot{E}_{x,Q} + P_t - \Delta\dot{E}_{x,V} \tag{5.35}$$

respectively

$$e_{x,S,2} - e_{x,S,1} = e_{x,Q} + w_t - \Delta e_{x,V}. \tag{5.36}$$

Since the throttle is adiabatic and work-insulated, the equation simplifies to

$$e_{x,S,2} - e_{x,S,1} = -\Delta e_{x,V} = -s_{i,12}T_{env} < 0. \tag{5.37}$$

In other words, the exergy of the fluid must decrease. Answer (a) is therefore wrong. Alternatively, the change in exergy of the fluid flow can also be calculated as follows, see Appendix E:

$$\begin{aligned} e_{x,S,2} - e_{x,S,1} &= \underbrace{h_2 - h_1}_{=0} + T_{env}\left(s_1 - s_2\right) \\ &= T_{env}\left(c_p \ln\frac{T_1}{T_2} - R\ln\frac{p_1}{p_2}\right) = -T_{env}R\ln\frac{p_1}{p_2} < 0. \end{aligned} \tag{5.38}$$

5.6 **(b)**
In order to clarify what happens to the temperature, the first law of thermodynamics is first applied, i.e.

$$P_t + \dot{Q} + \dot{m}h_1 = \dot{m}h_2 \tag{5.39}$$

respectively

$$w_t + q = h_2 - h_1. \tag{5.40}$$

The potential energy is neglected because the flow channel is horizontal. Since the fluid is incompressible, $v_1 = v_2 = v$. It follows from the continuity equation

$$\dot{m}_1 = \frac{c_1 A}{v} = \frac{c_2 A}{v} = \dot{m}_2 \tag{5.41}$$

that the velocity must remain constant at a constant cross-section. The kinetic energy is therefore also neglected in the energy balance. A flow channel is work-isolated, in addition the process is adiabatic, so Eq. 5.40 simplifies:

$$0 = h_2 - h_1. \tag{5.42}$$

Applying the caloric equation of state for incompressible fluids, yields

$$\underbrace{c\,(T_2 - T_1)}_{>0} + \underbrace{v\,(p_2 - p_1)}_{<0} = 0. \tag{5.43}$$

Since, according to the task, the pressure decreases, the temperature must increase according to Eq. 5.43. Answer (a) must be wrong.

It follows from the partial energy equation

$$w_\mathrm{t} = 0 = \underbrace{\int_1^2 v\,\mathrm{d}p}_{<0} + \underbrace{\psi_{12}}_{>0} \tag{5.44}$$

that dissipation $\psi_{12} > 0$ must occur. As a result, entropy is generated, i.e.

$$s_{\mathrm{i},12} = \frac{\psi_{12}}{T_\mathrm{m}} > 0. \tag{5.45}$$

The entropy generated is responsible for an exergy loss according

$$\Delta e_{\mathrm{x,V}} = s_{\mathrm{i},12} T_\mathrm{env} > 0. \tag{5.46}$$

Since energy is the sum of exergy and anergy and remains constant, the anergy must increase when the exergy decreases due to exergy lost. Answer (b) is therefore correct. Answer (c) does not make any sense because the channel is supposed to be adiabatic.

5.7 **(b)**
First, an exergy balance is applied according to Fig. 5.10. The exergy supplied in the steady state case corresponds to the exergy released, i.e.

$$\underbrace{\dot{m}e_{\mathrm{x},S,1} + \dot{E}_{\mathrm{x},Q@T_\mathrm{m}}}_{\mathrm{IN}} = \underbrace{\Delta\dot{E}_{\mathrm{x,V}} + \dot{m}e_{\mathrm{x},S,2}}_{\mathrm{OUT}}. \tag{5.47}$$

Since the change of state is isobaric and outer energies are negligible, the following occurs according to the partial energy equation

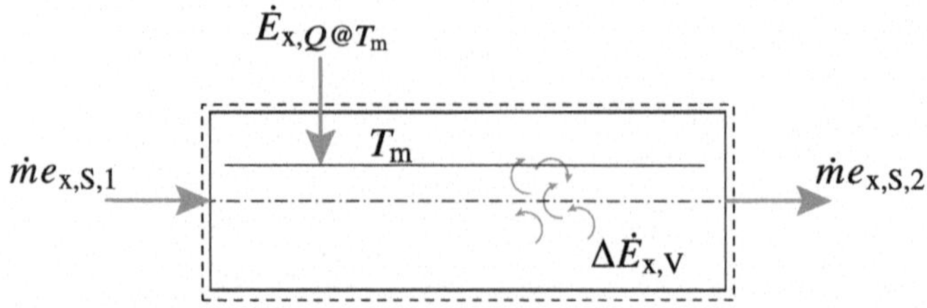

Fig. 5.10 Exergy balance heat exchanger

$$w_t = 0 = \underbrace{\int_1^2 v\,dp}_{=0} + \psi_{12} + \underbrace{\Delta e_{a,12}}_{=0} \tag{5.48}$$

no dissipation occurs, i.e. $\psi_{12} = 0$. If no dissipation occurs, no entropy is generated,[7] i.e. $s_{i,12} = 0$ and thus no exergy gets lost, i.e. $e_{x,V} = 0$. The exergy balance simplifies very much and the supplied exergy is only caused by the supplied heat, i.e.

$$\dot{m}e_{x,S,2} - \dot{m}e_{x,S,1} = \dot{E}_{x,Q@T_m}. \tag{5.49}$$

The following applies to the exergy of heat

$$\dot{E}_{x,Q@T_m} = \dot{Q}\left(1 - \frac{T_{env}}{T_m}\right) \tag{5.50}$$

respectively

$$\frac{1}{4} = 1 - \frac{T_{env}}{T_m} \quad \text{with} \quad T_{env} = 298.15\,\text{K}. \tag{5.51}$$

The thermodynamic mean temperature thus yields

$$T_m = \frac{4}{3}T_{env} = 397.5333\,\text{K}. \tag{5.52}$$

Answer (b) is correct.

5.8 **(a)**
The exergy balance for heat transfer through a wall has already been presented in Problem 5.2. In the case that the temperature on one side is T_1 and on the other $T_2 < T_1$, then according to Fig. 5.11 for the entropy generation yields

[7] Entropy generation through heat transfer into the system does not play a role due to the system boundary, which only considers the fluid and not, for example, the wall of the heat exchanger.

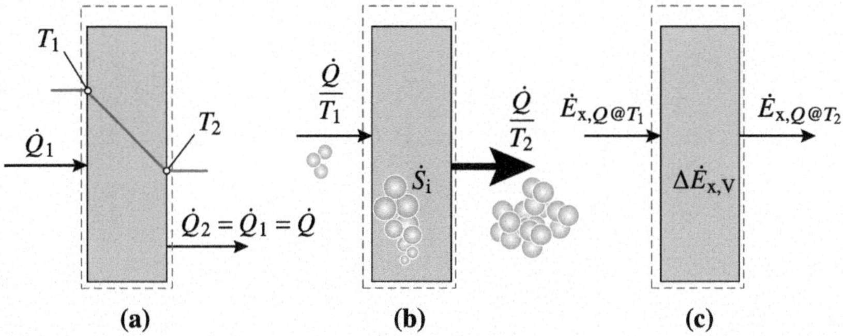

Fig. 5.11 Heat flux passing a wall: **a** Energy balance, **b** Entropy balance, **c** Exergy balance

$$\dot{S}_i = \dot{Q}\left(\frac{1}{T_2} - \frac{1}{T_1}\right) = \dot{Q}\frac{\Delta T}{T_1 T_2} > 0 \quad \text{with} \quad \Delta T = T_1 - T_2 > 0. \tag{5.53}$$

The exergy loss is therefore equal to

$$\Delta\dot{E}_{x,V} = T_{env}\dot{S}_i = T_{env}\dot{Q}\frac{\Delta T}{T_1 T_2} > 0. \tag{5.54}$$

According to heat and mass transfer, the heat flux $\dot{Q}$ can be expressed by

$$\dot{Q} = kA\,(T_1 - T_2) = kA\,\Delta T > 0. \tag{5.55}$$

The exergy loss is therefore

$$\Delta\dot{E}_{x,V} = T_{env}kA\frac{\Delta T^2}{T_1 T_2} > 0. \tag{5.56}$$

Answer (a) is correct. Furthermore, the greater ΔT, the more entropy is generated, i.e.

$$\dot{S}_i = kA\frac{\Delta T^2}{T_1 T_2}. \tag{5.57}$$

Answer (b) is wrong. Energy, however, is a conservation quantity, thus it can not be generated. Answer (c) does not apply.

5.9 **(a)**
As already shown in Problem 5.2, see Fig. 5.3, the exergy of the incoming heat flux is

$$\dot{E}_{x,Q@T_1} = \dot{Q}\left(1 - \frac{T_{env}}{T_1}\right) > 0 \quad \text{with} \quad T_1 > T_{env}. \tag{5.58}$$

The exergy of the outgoing heat flux is zero because the heat is now at ambient temperature, i.e.

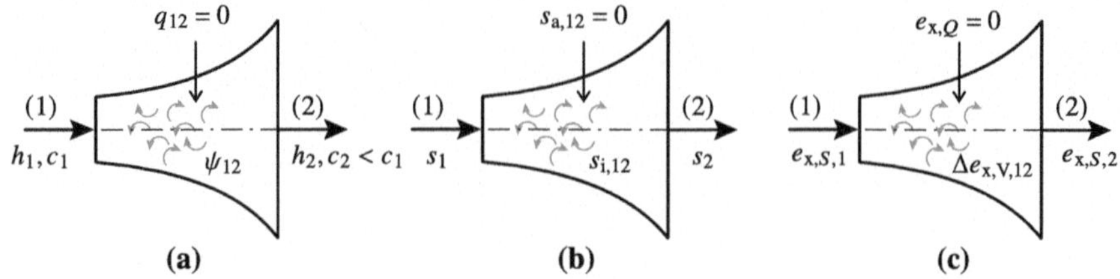

Fig. 5.12 Adiabatic diffuser: **a** Energy balance, **b** Entropy balance, **c** Exergy balance

$$\dot{E}_{x,\,Q@T_{env}} = \dot{Q}\left(1 - \frac{T_{env}}{T_{env}}\right) = 0. \tag{5.59}$$

In other words, all the entering exergy is lost—answer (a) is therefore correct. A perpetual mobile of the 2nd kind is a fictitious machine that violates the second law of thermodynamics. Consequently, such a machine cannot exist. Answer (b) is wrong because the problem described does not violate the principles of thermodynamics. Dissipation requires friction, e.g. by flowing fluid particles. In the problem described, no friction occurs because the wall is perfectly at rest. Answer (c) is therefore also wrong. The entropy generated is due to heat transfer at $\Delta T > 0$.

5.10 **(d)**
A diffuser is utilised to reduce the flow velocity by increasing the cross-section, see Fig. 5.12. Hence, answer (b) does not apply. In technical diffusers, the kinetic energy is used to increase the pressure. However, with very large dissipation there are also decelerated flows with pressure decrease. This is, however, technically undesirable in most cases, see [2]. The balance of exergy for the adiabatic diffuser obeys

$$\underbrace{\dot{m}e_{x,S,1} + \overbrace{\dot{E}_{x,Q}}^{=0}}_{\text{IN}} = \underbrace{\dot{m}e_{x,S,2} + \Delta\dot{E}_{x,V}}_{\text{OUT}} \tag{5.60}$$

so that

$$\dot{m}e_{x,S,2} - \dot{m}e_{x,S,1} = -\Delta\dot{E}_{x,V} = -T_{env}\dot{S}_i \leq 0. \tag{5.61}$$

Consequently, the exergy can only decrease or, in the best case, remain constant. Answer (a) is therefore wrong. Finally, we apply the entropy balance, i.e.

$$\underbrace{\dot{m}s_1 + \overbrace{\dot{S}_a}^{=0} + \dot{S}_a}_{\text{IN}} = \underbrace{\dot{m}s_2}_{\text{OUT}} \tag{5.62}$$

so that

$$\dot{m}s_2 - \dot{m}s_1 = \dot{S}_i = \frac{\dot{\Psi}}{T_m} \geq 0. \tag{5.63}$$

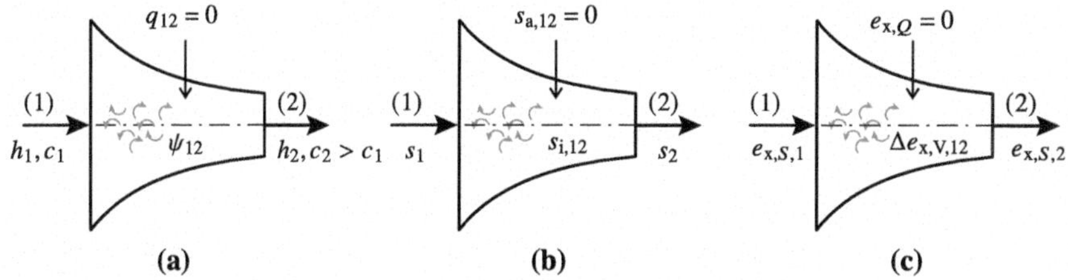

Fig. 5.13 Adiabatic nozzle: **a** Energy balance, **b** Entropy balance, **c** Exergy balance

Consequently, entropy can only increase or, in the best case, remain constant. Answer (c) is therefore also wrong. Answer (d) therefore applies.

5.11 **(a)**
A nozzle is the inverse of the diffuser, see Problem 5.10. In a nozzle, the kinetic energy is increased while the pressure decreases. The velocity of the fluid therefore increases, see Fig. 5.13. Answer (b) is incorrect. Since the nozzle is assumed to be adiabatic, no heat is exchanged with the environment. Answer (c) is therefore also incorrect.
Finally, we apply the entropy balance, i.e.

$$\underbrace{\dot{m}s_1 + \overbrace{\dot{S}_a}^{=0} + \dot{S}_a}_{\text{IN}} = \underbrace{\dot{m}s_2}_{\text{OUT}} \tag{5.64}$$

so that

$$\dot{m}s_2 - \dot{m}s_1 = \dot{S}_i = \frac{\dot{\Psi}}{T_m} \geq 0. \tag{5.65}$$

Consequently, entropy can only increase or, in the best case, remain constant. Answer (a) applies.

5.12 For the steady state case, the first law of thermodynamics is applied,[8] cf. Fig. 5.14, i.e.

$$\dot{m}_A h_{A,in} + \dot{m}_W h_{W,in} = \dot{m}_A h_{A,out} + \dot{m}_W h_{W,out} \tag{5.66}$$

respectively

$$\dot{m}_A \left(h_{A,in} - h_{A,out}\right) = \dot{m}_W \left(h_{W,out} - h_{W,in}\right). \tag{5.67}$$

In the next step, the caloric equations of state for an ideal gas and an incompressible liquid are applied, so that

[8] Although a heat flux $\dot{Q}$ is transferred, it does not cross the selected system boundary and therefore does not appear in the energy balance. When applying the energy balance, it is irrelevant whether the unit is operated in parallel flow or counterflow.

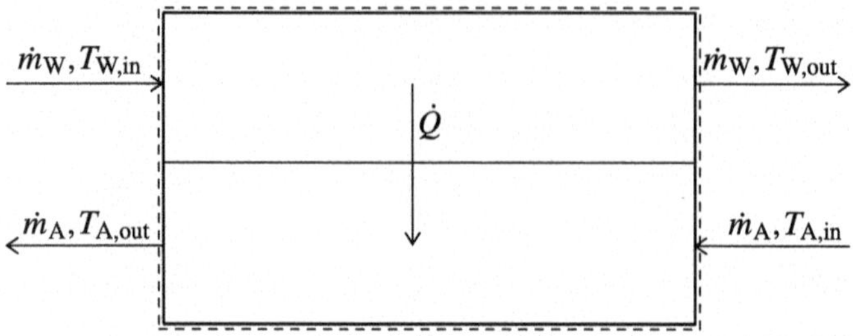

Fig. 5.14 Heat exchanger

$$\dot{m}_A c_p \left(T_{A,in} - T_{A,out}\right) = \dot{m}_W \left[c_W \left(T_{W,out} - T_{W,in}\right) + v_W \underbrace{\left(p_{W,out} - p_{W,in}\right)}_{=0} \right]. \quad (5.68)$$

This equation can be solved for the outlet temperature of the air[9]

$$T_{A,out} = T_{A,in} + \frac{\dot{m}_W c_W \left(T_{W,in} - T_{W,out}\right)}{\dot{m}_A c_p} = 313.3865\,\text{K}. \quad (5.69)$$

The balance of entropy obeys

$$\dot{m}_A s_{A,in} + \dot{m}_W s_{W,in} + \dot{S}_i + \underbrace{\dot{S}_a}_{=0} = \dot{m}_A s_{A,out} + \dot{m}_W s_{W,out} \quad (5.70)$$

so that the generated entropy inside the system boundary obeys

$$\dot{S}_i = \dot{m}_A \left(s_{A,out} - s_{A,in}\right) + \dot{m}_W \left(s_{W,out} - s_{W,in}\right). \quad (5.71)$$

Applying the caloric equations of state for ideal gas and incompressible liquid under isobaric conditions, yields

$$\dot{S}_i = \dot{m}_W c_W \ln \frac{T_{W,out}}{T_{W,in}} + \dot{m}_A c_p \ln \frac{T_{A,out}}{T_{A,in}}. \quad (5.72)$$

Finally, the loss of exergy can be calculated according to

$$\Delta \dot{E}_{x,V} = \dot{S}_i T_{env} = 6.990\,\text{kW}. \quad (5.73)$$

5.13 Due to the quenching process, a thermodynamic balancing process takes place that is irreversible in one direction, i.e. a state of equilibrium. Therefore, the change

[9] Obviously, the heat exchanger can be operated both in parallel flow and in counterflow, cf. Problem 4.21.

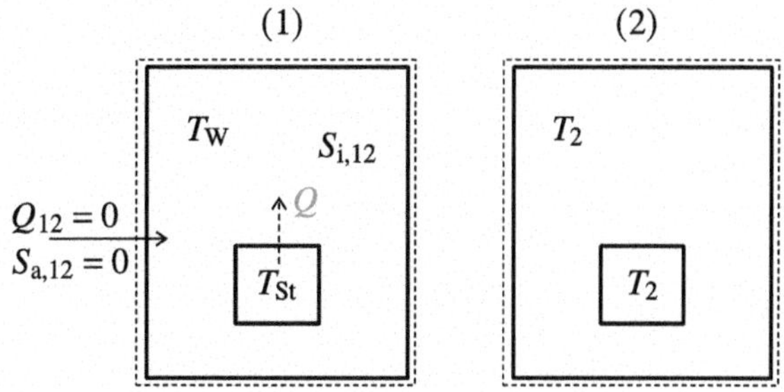

Fig. 5.15 Quenching process

of state is associated with generation of entropy and thus a loss of exergy.
The first law of thermodynamics, see Fig. 5.15, obeys

$$Q_{12} + W_{12} = 0 = \Delta U = \Delta U_{St} + \Delta U_W. \tag{5.74}$$

Applying the caloric equation of state, results in

$$0 = m_{St}c_{St}\left(T_2 - T_{St}\right) + m_W c_{W<}\left(T_2 - T_W\right). \tag{5.75}$$

Solving for the equilibrium temperature leads to

$$T_2 = \frac{m_W c_W T_W + m_{St}c_{St}T_{St}}{m_W c_W + m_{St}c_{St}} = 307.1776\,\text{K}. \tag{5.76}$$

Once the equilibrium temperature is known, the entropy generated can be determined from the entropy balance,[10] i.e.

$$S_{i,12} + \underbrace{S_{a,12}}_{=0} = \Delta S = \Delta S_{St} + \Delta S_W \tag{5.77}$$

respectively

$$S_{i,12} = m_{St}\left(s_2 - s_{St}\right) + m_W\left(s_2 - s_W\right). \tag{5.78}$$

With the caloric equation of state for incompressible solids/fluids it is

$$S_{i,12} = m_W c_W \ln\frac{T_2}{T_W} + m_{St}c_{St}\ln\frac{T_2}{T_{St}} = 1.4747 \times 10^5\,\text{J}\,\text{K}^{-1}. \tag{5.79}$$

Finally, the loss of exergy for the balancing process yields

$$\Delta E_{x,V} = S_{i,12}T_{env} = 4.2472 \times 10^7\,\text{J} = 4.2472 \times 10^4\,\text{kJ}. \tag{5.80}$$

[10] The cause for the generation of entropy lies in the internal heat exchange between steel/water and not in mechanical dissipation.

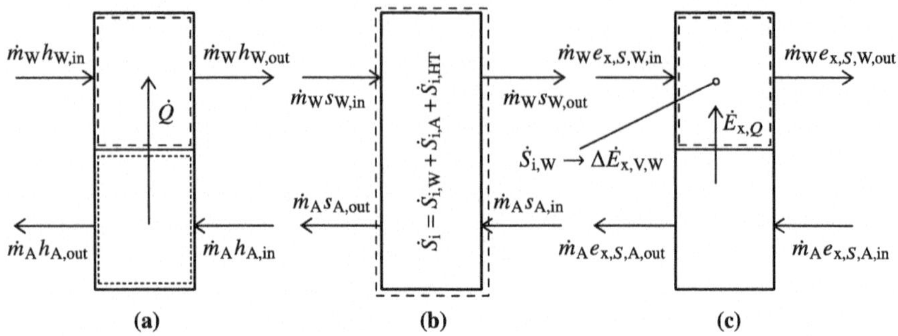

Fig. 5.16 Heat exchanger: **a** Energy balance, **b** Entropy balance, **c** Exergy balance

5.14 This task deals with a heat exchanger analogous to Problem 5.12. First one determines the outlet temperature of the water. For this purpose, the first law of thermodynamics is applied to the system *air* according to Fig. 5.16, i.e.

$$\dot{m}_A h_{A,in} = \dot{m}_A h_{A,out} + \dot{Q}. \tag{5.81}$$

With the caloric equation of state one gets the transferred flux of heat

$$\dot{Q} = \dot{m}_A c_p \left(T_{A,in} - T_{A,out}\right) = 110.44\,\text{kW}. \tag{5.82}$$

The outlet temperature can now be determined with the first law of thermodynamics for the *water*

$$\dot{m}_W h_{W,in} + \dot{Q} = \dot{m}_W h_{W,out}. \tag{5.83}$$

Since there is no pressure loss, we obtain

$$T_{W,out} = T_{W,in} + \frac{\dot{Q}}{c_W \dot{m}_W} = 431.79\,\text{K}. \tag{5.84}$$

Alternatively, this temperature can also be determined via an overall energy balance of the heat exchanger.

Finally, we apply an exergetic consideration to the water system. For this purpose, we determine how much exergy the water has absorbed, i.e.

$$\begin{aligned}\Delta\dot{E}_{x,W} &= \dot{m}_W \left(e_{x,S,W,out} - e_{x,S,W,in}\right)\\ &= \dot{m}_W \left[c_W \left(T_{W,out} - T_{W,in}\right) - T_{env} c_W \ln\frac{T_{W,out}}{T_{W,in}}\right] = 18.891\,\text{kW}.\end{aligned} \tag{5.85}$$

Therefore, the ratio of absorbed exergy to supplied heat is as follows

$$\xi = \frac{\Delta \dot{E}_{x,W}}{\dot{Q}} = 1 - \frac{T_{env}}{T_{m,W}} = 0.1711. \tag{5.86}$$

The exergy balance

$$\dot{m}_W e_{x,S,W,in} + \dot{E}_{x,Q} = \dot{m}_W e_{x,S,W,out} + \Delta \dot{E}_{x,V,W} \tag{5.87}$$

shows that the increase in exergy of the water is composed as follows:

$$\Delta \dot{E}_{x,W} = \dot{m}_W \left(e_{x,S,W,out} - e_{x,S,W,in}\right) = \dot{E}_{x,Q} - \underbrace{\Delta \dot{E}_{x,V,W}}_{=0} = \dot{E}_{x,Q}. \tag{5.88}$$

Because $\mathrm{d}p = 0$ and $\mathrm{d}e_a = 0$, it follows from the partial energy equation that $\psi_W = 0$. Thus, no entropy is generated in the water, i.e. $s_{i,W} = 0$. If no entropy is generated, there is also no exergetic loss, i.e. $\Delta e_{x,V,W} = 0$. According to Eq. 5.88, the change in exergy of the water is justified by the supplied heat solely.

5.15 First we start with the obvious and determine the specific heat capacity c_v, i.e.

$$c_v = \frac{c_p}{\kappa} = 717.1429 \frac{\mathrm{J}}{\mathrm{kgK}}. \tag{5.89}$$

The first law of thermodynamics for a closed system is as follows

$$Q_{12} + W_{12} = U_2 - U_1. \tag{5.90}$$

The supplied (electrical) work can be determined by the applied and constant heating power and the duration of operation

$$W_{12} = P_{el} \cdot \Delta t. \tag{5.91}$$

With the help of the caloric equation of state for an ideal gas, the energy balance results in

$$P_{el} \cdot \Delta t = m c_v \left(T_2 - T_1\right). \tag{5.92}$$

Hence, the temperature in state (2) is

$$T_2 = T_1 + \frac{P_{el} \cdot \Delta t}{m c_v} = 315.5604\,\mathrm{K}. \tag{5.93}$$

From the thermal equation of state, the pressure in state (1) and in state (2) can be easily determined, since the volume of the tank is supposed to be constant, i.e.

$$p_1 = \frac{m R T_1}{V} = 1.1773\,\mathrm{bar} \tag{5.94}$$

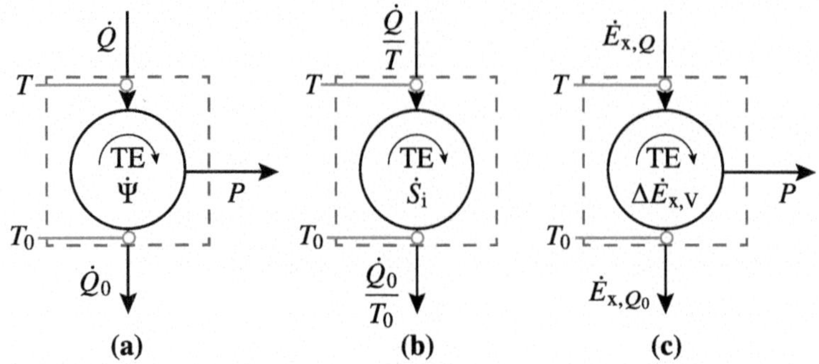

Fig. 5.17 Thermal engine: **a** Energy balance, **b** Entropy balance, **c** Exergy balance

and

$$p_2 = \frac{mRT_2}{V} = 1.2673\,\text{bar}. \tag{5.95}$$

From the entropy balance follows

$$S_i + \underbrace{S_a}_{=0} = S_2 - S_1 = m\,(s_2 - s_1) \tag{5.96}$$

so that

$$S_i = m\left[c_p \ln\frac{T_2}{T_1} - R\ln\frac{p_2}{p_1}\right] = 73.9602\,\text{J}\,\text{K}^{-1}. \tag{5.97}$$

Finally, the loss of exergy yields

$$\Delta E_{x,V} = T_{env} S_i = 2.1681 \times 10^4\,\text{J}. \tag{5.98}$$

Alternatively, an exergy balance can be applied, i.e.

$$E_{x,2} = E_{x,1} + \underbrace{W_{eff,12}}_{=W_{12}} - \Delta E_{x,V} + \underbrace{E_{x,Q}}_{=0} \tag{5.99}$$

so that

$$\begin{aligned} \Delta E_{x,V} &= E_{x,1} - E_{x,2} + W_{12} \\ &= m\left[c_v\,(T_1 - T_2) + T_{env}\left(c_p \ln\frac{T_2}{T_1} - R\ln\frac{p_2}{p_1}\right)\right] + W_{12} = 2.1681 \times 10^4\,\text{J}. \end{aligned} \tag{5.100}$$

5.16 (**c**)
An overview of the thermal engine is shown in Fig. 5.17. The exergetic efficiency of such a machine is defined as

$$\eta_{\mathrm{ex}} = \frac{P}{\dot{E}_{\mathrm{x},Q}}. \tag{5.101}$$

According to Fig. 5.17c the exergy balance is

$$\underbrace{\dot{E}_{\mathrm{x},Q}}_{\mathrm{IN}} = \underbrace{\dot{E}_{\mathrm{x},Q_0} + P + \Delta\dot{E}_{\mathrm{x,V}}}_{\mathrm{OUT}}. \tag{5.102}$$

Since the machine works reversibly, i.e. $\Delta\dot{E}_{\mathrm{x,V}} = 0$, the following is obtained

$$P = \dot{E}_{\mathrm{x},Q} - \dot{E}_{\mathrm{x},Q_0}. \tag{5.103}$$

Therefore, one gets for the exergetic efficiency

$$\eta_{\mathrm{ex}} = \frac{P}{\dot{E}_{\mathrm{x},Q}} = 1 - \frac{\dot{E}_{\mathrm{x},Q_0}}{\dot{E}_{\mathrm{x},Q}}. \tag{5.104}$$

For an exergetic efficiency of 1, the exergy of the released heat $\dot{Q}_0$ must therefore be zero, i.e.

$$\dot{E}_{\mathrm{x},Q_0} = \dot{Q}_0\left(1 - \frac{T_{\mathrm{env}}}{T_0}\right) = 0. \tag{5.105}$$

This is achieved[11] when $T_0 \to T_{\mathrm{env}}$. Answer (c) is correct. Answer (a) is incorrect because dissipation does not occur in a reversible machine. According to the second law of thermodynamics, a reversible machine must also give off heat, as otherwise the entropy in the machine would increase over time. Answer (b) is wrong. The balance of entropy obeys

$$\underbrace{\frac{\dot{Q}}{T} + \overbrace{\dot{S}_{\mathrm{i}}}^{=0}}_{\mathrm{IN}} = \underbrace{\frac{\dot{Q}_0}{T_0}}_{\mathrm{OUT}}. \tag{5.106}$$

Hence, the released heat muss be

$$\dot{Q}_0 = \dot{Q}\frac{T_0}{T}. \tag{5.107}$$

According to the energy balance it is

$$P = \dot{Q} - \dot{Q}_0 = \dot{Q} - \dot{Q}\frac{T_0}{T} \tag{5.108}$$

[11] For the machine to release heat to the environment, T_0 must always be greater than T_{env}. If T_0 is allowed to strive towards T_{env}, the ever decreasing temperature potential must be compensated for by an increasing heat-transferring surface.

so that the thermal efficiency follows[12]

$$\eta_{\mathrm{th}} = \eta_{\mathrm{C}} = \frac{P}{\dot{Q}} = 1 - \frac{T_0}{T} < 1. \tag{5.109}$$

Answer (d) is wrong.

5.17 **(b)**
According to the formulary in Appendix E, the exergy of a closed system obeys

$$E_{\mathrm{x},1} = m\left[u_1 - u_{\mathrm{env}} + \frac{1}{2}c_1^2 + gz_1 + T_{\mathrm{env}}\left(s_{\mathrm{env}} - s_1\right) - p_{\mathrm{env}}\left(v_{\mathrm{env}} - v_1\right)\right]. \tag{5.110}$$

Since it is assumed that the body is incompressible, i.e. $v_{\mathrm{env}} = v_1$, and in rest the following applies

$$E_{\mathrm{x},1} = m\left[c\left(T_1 - T_{\mathrm{env}}\right) + gz_1 + T_{\mathrm{env}}c\ln\frac{T_{\mathrm{env}}}{T_1}\right]. \tag{5.111}$$

Furthermore, the body is in thermal equilibrium with the environment, i.e. $T_1 = T_{\mathrm{env}}$, so that the exergy of the body is

$$E_{\mathrm{x},1} = mgz_1. \tag{5.112}$$

The exergy of the body therefore corresponds exactly to its potential energy. Answer (b) is therefore correct and the other answer options are eliminated.

[12] Since it is a reversible machine, the thermal efficiency equals Carnot efficiency.

Chapter 6
Thermodynamic Cycles

The concept of energetic transformation, which has already been dealt with in the previous chapters, is closely linked to thermodynamic cycles. These are discussed in further detail in this chapter. Basically, cyclic processes can be divided into clockwise and counterclockwise cycles. Both have in common that they operate between two temperature reservoirs. The tasks presented here address, on the one hand, thermal engines that partially convert a heat flux taken from the warm reservoir into mechanical work and release the unconverted part of the supplied thermal energy as heat to the cold reservoir. These processes are thermodynamically evaluated and usually described with a thermal efficiency. On the other hand, the tasks deal with cooling machines and heat pumps, respectively, that extract heat from the cold reservoir and transfer it to the warm reservoir while consuming mechanical work.

6.1 Problems

6.1 How can the exergetic efficiency of a thermal engine be improved?

(a) Increase the temperature at what heat is released.
(b) Heat release should be realised close to ambient temperature and irreversibilities should be avoided.
(c) No heat release to the environment.
(d) None of the above statements is true.

6.2 The owner of a fridge (coefficient of performance 1.1) decides to operate the fridge as a heat pump within the same temperature range. What is true for this heat pump?

(a) It is impossible to operate the fridge as a heat pump.
(b) The released heat flux is 10 % of the supplied power.

A. Schmidt, *Technical Thermodynamics Workbook for Engineers*,
https://doi.org/10.1007/978-3-031-50172-2_6

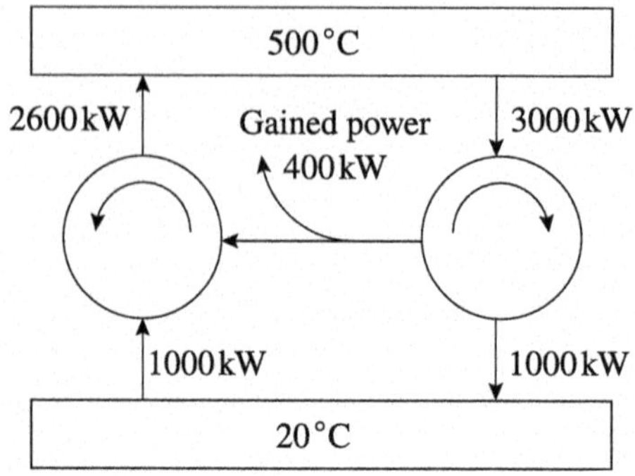

Fig. 6.1 Combined cycles

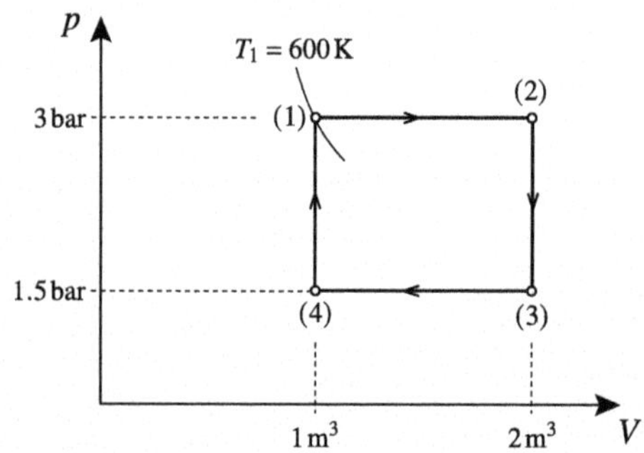

Fig. 6.2 Thermodynamic cycle

(c) To calculate the coefficient of performance of the heat pump, ambient temperature is missing.
(d) None of the above statements is true.

6.3 Two reservoirs with temperatures of 20 °C and 500 °C are given. A heat pump and a thermal engine are to be operated as shown in Fig. 6.1. Which statement is true?

(a) The process can be operated in this way.
(b) The process only violates the 1st law of thermodynamics.
(c) The process violates the 1st and 2nd law of thermodynamics.
(d) The process only violates the 2nd law of thermodynamics.
(e) The process violates mass conservation.

6.4 An ideal gas in a closed system goes through a cycle according to the qualitative Fig. 6.2. What is the temperature in state (3)?

(a) 1200 K
(b) 1000 K
(c) 600 K
(d) 450 K

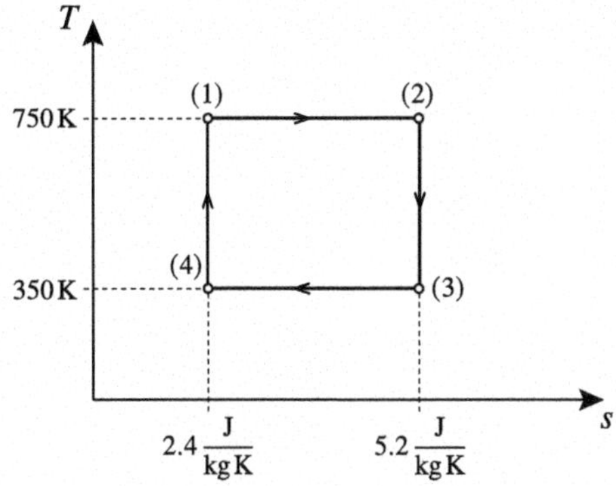

Fig. 6.3 Thermodynamic cycle

6.5 A reversible thermal engine, operated with an ideal gas, follows Fig. 6.3. What is the amount of specific net work released by the thermal engine?

(a) $2.10\frac{\text{kJ}}{\text{kg}}$
(b) $1.12\frac{\text{kJ}}{\text{kg}}$
(c) $0.98\frac{\text{kJ}}{\text{kg}}$
(d) $0.96\frac{\text{kJ}}{\text{kg}}$

6.6 In a thermal engine, 10 kW of thermal power is converted into 4 kW of mechanical power. The heat release takes place to an environment with $T_{\text{env}} = 300\,\text{K}$. The exergetic efficiency is 80 %. At what temperature is the heat supplied?

(a) 450 K
(b) 500 K
(c) 550 K
(d) 600 K

6.7 A reversible Stirling engine, operated with an ideal gas, follows Fig. 6.4. What is the thermal efficiency?

(a) 0.3904
(b) 0.5333
(c) 0.6123
(d) 1

6.8 The coefficient of performance of a cooling machine is 4. The heat release of the cooling machine is $\dot{Q} = 2\,\text{kW}$. How much power P needs to be supplied?

6.9 A reversible heat pump takes in ambient air with $\vartheta_{\text{env}} = 12\,°\text{C}$. A partial flow $\dot{m}_{\text{H}} = 1.2\frac{\text{kg}}{\text{s}}$ of this air is heated isobarically to $\vartheta_{\text{H}} = 65\,°\text{C}$. The other partial flow serves as a heat source and is cooled isobarically to $\vartheta_{\text{C}} = 3\,°\text{C}$, cf. Fig. 6.5. The specific heat capacity of air shall be constant and is $c_p = 1004\frac{\text{J}}{\text{kgK}}$. What is the required power P in case changes of potential/kinetic energies are ignored?

Fig. 6.4 Ideal stirling cycle

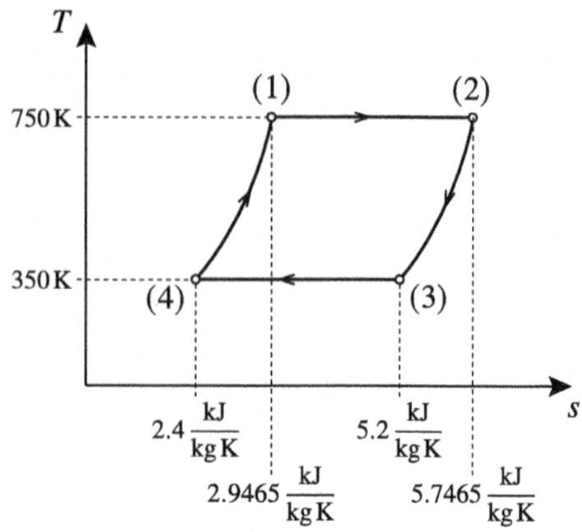

Fig. 6.5 Heat pump

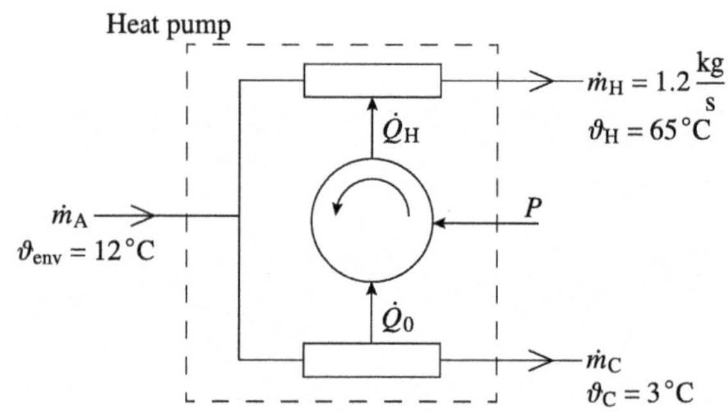

Fig. 6.6 Cycle

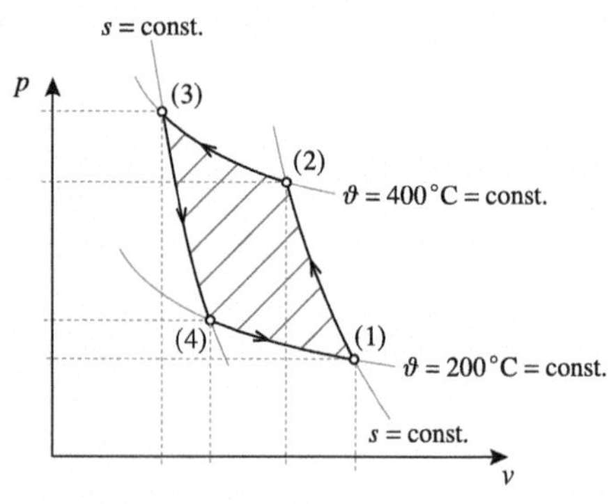

6.10 Let a reversible cyclic process be given, see Fig. 6.6, which is operated with an ideal gas. What is the coefficient of performance if the process is to be operated as a heat pump?

6.11 The heat source of a thermal engine is an air flow of $\dot{m} = 0.85\,\frac{\text{kg}}{\text{s}}$ at $\vartheta_1 = 1000\,°\text{C}$ which can be cooled isobarically to $\vartheta_2 = 25\,°\text{C}$, see Fig. 6.7. Calculate the maximum power P that the thermal engine can release in best case. The heat sink

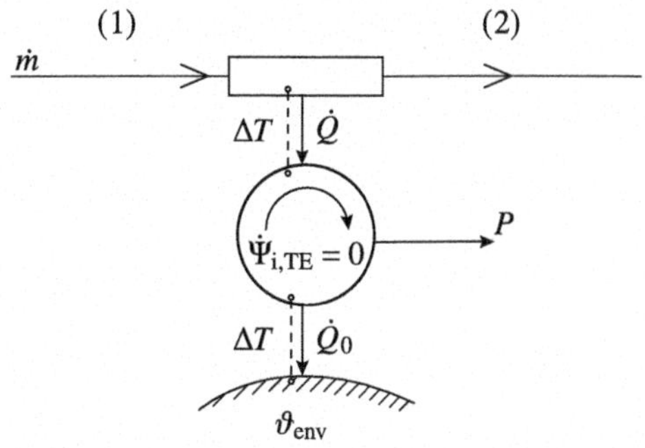

Fig. 6.7 Thermal engine

shall have a constant temperature of $\vartheta_{env} = 10\,°C$. For heat transfer, a local potential of $\Delta T = 5\,K$ is always assumed. It shall be $c_p = 1.05\frac{kJ}{kgK}$.

6.12 A thermal engine is supplied with a heat flux $\dot{Q}_1$ at a constant temperature $\vartheta_1 = 250\,°C$ and a second heat flux $\dot{Q}_2 = 3 \cdot \dot{Q}_1$ at a constant temperature $\vartheta_2 = 500\,°C$. At a constant temperature $\vartheta_0 = 10\,°C$ heat is released. What is the thermal efficiency η_{th} in best case?

6.13 A Stirling process is used to convert a solar thermal heat flux of 10 kW into electricity. The waste heat from the cyclic process is used to heat a swimming pool. The solar thermal heat is available at 300 °C. The swimming pool is to be kept constantly warm at 30 °C. The Stirling cyclic process consists of the following changes of state:

- (1) → (2) Isothermal expansion (heat supply)
- (2) → (3) Isochoric, regenerative heat release (adiabatic to the outside)
- (3) → (4) Isothermal compression (heat release)
- (4) → (1) Isochoric, regenerative heat supply (adiabatic to the outside)

The working fluid is carbon dioxide. Carbon dioxide can be regarded as an ideal gas. Ambient state: 300 K, 1 bar

(a) Draw a black box diagram of the process, and label all heat fluxes and powers.
(b) What maximum electrical power and how much waste heat would an idealised Carnot cyclic process deliver for the given hot/cold reservoirs?

For heat transfer to and from the cyclic process, a driving temperature difference of 10 K is required. The real process generates only 80 % of the maximum electrical power (see task part b).

(c) What electrical current and how much waste heat does the real process deliver?
(d) Determine for the real process:

(1) How much exergy is lost in solar thermal heat transfer due to the required driving temperature difference?

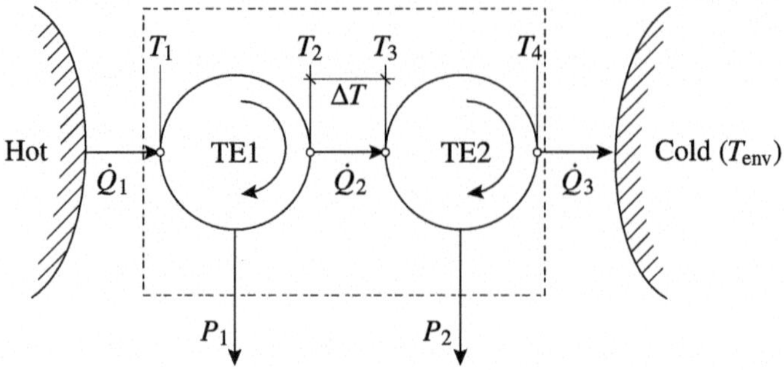

Fig. 6.8 Thermal engines

(2) How much exergy is lost during heat transfer to the swimming pool water due to the required driving temperature difference?
(3) The total exergy loss of the process.
(4) How to explain the difference between the total exergy loss and the exergy losses due to heat transfer?

6.14 Two reversible thermal engines, i.e. TE1 and TE2, are operated in serial connection, see Fig. 6.8. The power release of TE1 is $P_1 = 30\,\text{kW}$. The heat supply is $\dot{Q}_1 = 100\,\text{kW}$ at $T_1 = 500\,\text{K}$. For the heat transfer between the two engines a temperature difference of $\Delta T = 5\,\text{K}$ is required, see Fig. 6.8.

(a) What is the overall thermal efficiency, in case the temperature of the heat release is $T_4 = 300\,\text{K}$?
(b) What is the exergy loss $\Delta \dot{E}_{x,V}$ of the entire plant,[1] see Fig. 6.8, when $T_{env} = 295\,\text{K}$?

6.15 A closed cyclic process with an ideal gas as the working fluid ($c_p = 5.196\frac{\text{kJ}}{\text{kgK}}$) undergoes the following changes of state, cf. Fig 6.9.

- (1) $\rightarrow$ (2) Irreversible, adiabatic compression of the working fluid from state (1) ($\vartheta_1 = 10\,°\text{C}$, $p_1 = 1.0\,\text{bar}$) to $p_2 = 5.7\,\text{bar}$
- (2) $\rightarrow$ (3) Isobaric heating to $\vartheta_3 = 700\,°\text{C}$ in a heat exchanger. The hot air flow used for heating ($\dot{m}_a = 204.5\frac{\text{kg}}{\text{h}}$) thus cools isobarically from $\vartheta_{a,in} = 1200\,°\text{C}$ to $\vartheta_{a,out} = 850\,°\text{C}$. The warm air may also be considered as an ideal gas ($c_{p,a} = 1.005\frac{\text{kJ}}{\text{kgK}}$).
- (3) $\rightarrow$ (4) Irreversible, adiabatic expansion to state (4) ($\vartheta_4 = 290\,°\text{C}$, $p_4 = 1.0\,\text{bar}$). The entropy change caused by irreversibilities is $s_{i34} = 0.773\frac{\text{kJ}}{\text{kgK}}$.
- (4) $\rightarrow$ (1) Reversible, isobaric cooling to initial state

[1] Shown by the dashed line.

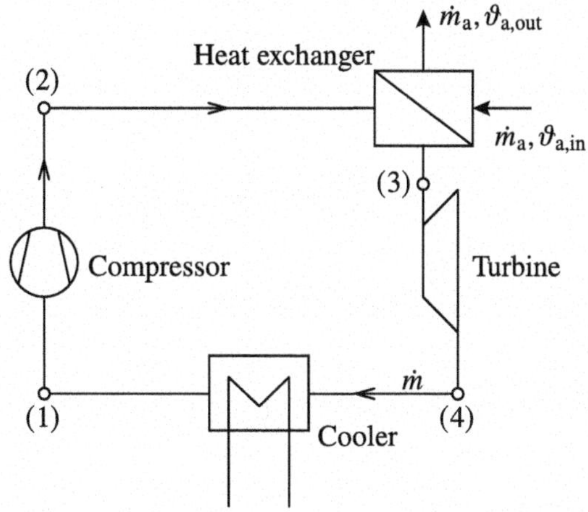

Fig. 6.9 Thermodynamic cycle

Assumptions:

- The ambient temperature is $\vartheta_{env} = 19.85\,°C$
- The isentropic efficiency of the compressor is $\eta_{s,V} = 0.9$
- Changes in kinetic and potential energies are not to be taken into account

(a) Determine the gas constant R of the working fluid.
(b) Calculate the mass flux $\dot{m}$ of the working fluid.
(c) What is the isentropic efficiency of the turbine?
(d) Calculate the flux of exergy loss $\Delta \dot{E}_{x,V}$ occurring in the heat exchanger.

6.2 Solutions

6.1 **(b)**
The thermal engine has already been dealt with in Problem 5.16. Nevertheless, we will now summarise the most important findings, see also Fig. 6.10. A thermal engine operates between two temperature reservoirs and uses the resulting temperature potential, i.e. a heat flux $\dot{Q}$ follows the temperature gradient $T_h \rightarrow T_c$, see Fig. 6.10a. However, the supplied heat flux $\dot{Q}$ is converted into mechanical power P. The first law of thermodynamics obeys

$$\dot{Q} = \dot{Q}_0 + P. \tag{6.1}$$

The balance of entropy reads as

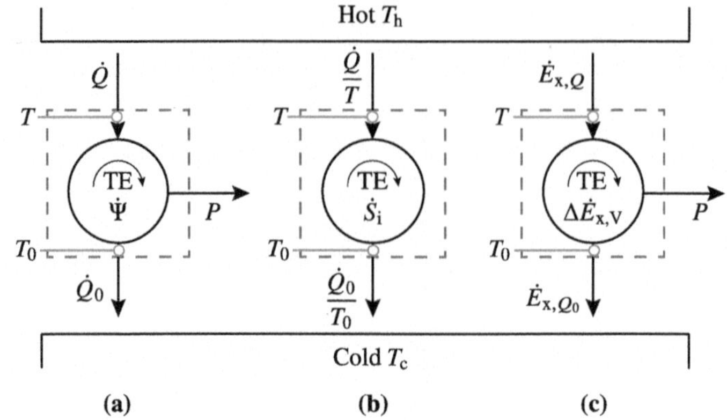

Fig. 6.10 Thermal engine: **a** Energy balance, **b** Entropy balance, **c** Exergy balance

$$\overbrace{\frac{\dot{Q}}{T} + \underbrace{\dot{S}_i}_{\geq 0}}^{\text{IN}} = \overbrace{\frac{\dot{Q}_0}{T_0}}^{\text{OUT}} \quad \text{with} \quad T > T_0. \tag{6.2}$$

It can be seen that even in the reversible case, a heat flux $\dot{Q}_0$ is required to keep the entropy in the machine constant. If no heat flux were released, the entropy fluxes into the machine would cause the state value entropy to increase with time. If a state value changes in time, the state is not steady state. Mind that $T > T_0$. Consequently, even the best thermal engine requires a cooling. Hence, answer (c) is wrong.

The exergetic balance is

$$\underbrace{\dot{E}_{x,Q}}_{\text{IN}} = \underbrace{\dot{E}_{x,Q_0} + P + \Delta\dot{E}_{x,V}}_{\text{OUT}} \tag{6.3}$$

respectively

$$P = \dot{E}_{x,Q} - \dot{E}_{x,Q_0} - \Delta\dot{E}_{x,V}. \tag{6.4}$$

Therefore, one gets for the exergetic efficiency

$$\eta_{ex} = \frac{P}{\dot{E}_{x,Q}} = 1 - \frac{\dot{E}_{x,Q_0}}{\dot{E}_{x,Q}} - \frac{\Delta\dot{E}_{x,V}}{\dot{E}_{x,Q}}. \tag{6.5}$$

Let us now first focus on the heat release $\dot{Q}_0$. The following applies to the related exergy $\dot{E}_{x,Q_0}$

$$\dot{E}_{x,Q_0} = \dot{Q}_0 \left(1 - \frac{T_{env}}{T_0}\right). \tag{6.6}$$

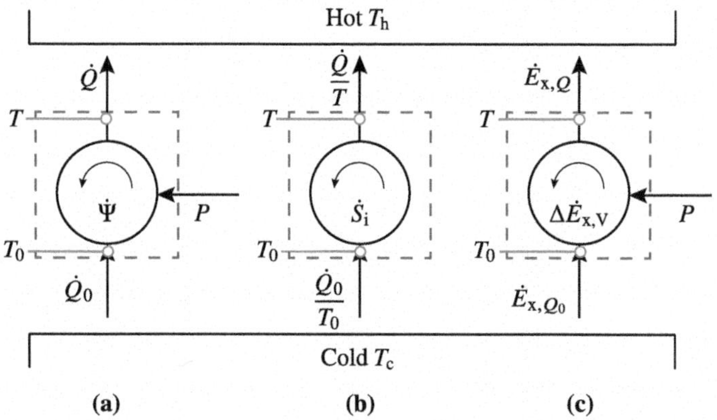

Fig. 6.11 Counterclockwise cycle: **a** Energy balance, **b** Entropy balance, **c** Exergy balance

According to this equation, the exergy of the released heat increases when T_0 is increased, i.e. according to Eq. 6.5, the exergetic efficiency decreases. Thus, a lot of exergy is released unused to the environment with the heat flux $\dot{Q}_0$. This reduces the exergetic efficiency. Answer (a) is therefore incorrect.

Let us now examine answer (b). It applies[2]

$$\lim_{T_0 \to T_{env}} \dot{E}_{x,Q_0} = \lim_{T_0 \to T_{env}} \dot{Q}_0 \left(1 - \frac{T_{env}}{T_0}\right) = 0. \tag{6.7}$$

According to Eq. 6.5, this means that a heat release at $T_0 \to T_{env}$ increases the exergetic efficiency. In the case that irreversibilities are avoided, $\Delta\dot{E}_{x,V} = 0$. This also leads to an increase in exergetic efficiency, cf. Eq. 6.5. Answer (b) therefore is correct.

6.2 **(d)**
A fridge represents a counterclockwise cycle as shown in Fig. 6.11. Analogous to the thermal engine, counterclockwise cycles also operate between two temperature levels. Unlike the thermal engine, however, a heat flux $\dot{Q}_0$ is taken from the cold container and raised to a higher temperature level against the temperature potential, i.e. against the natural flow direction of the heat. To realise a heat flux against the temperature gradient, power needs to be supplied to the machine.[3] According to the first law of thermodynamics this results in

$$\dot{Q} = \dot{Q}_0 + P. \tag{6.8}$$

[2] Since heat must be released from the machine, T_0 approaches T_{env} from above.

[3] The entropy balance will provide the physical explanation.

The balance of entropy reads as

$$\frac{\dot{Q}}{T} = \frac{\dot{Q}_0}{T_0} + \underbrace{\dot{S}_\mathrm{i}}_{\geq 0} \quad \text{with} \quad T > T_0. \tag{6.9}$$

Thus, to satisfy the entropy balance, $\dot{Q}$ must be greater than $\dot{Q}_0$. According to the first law of thermodynamics, this can be achieved if power P is supplied to the process from outside. This is the thermodynamic explanation why a refrigerator consumes electrical power.

Finally, the balance of exergy yields

$$\underbrace{\dot{E}_{\mathrm{x},Q_0} + P}_{\text{IN}} = \underbrace{\dot{E}_{\mathrm{x},Q} + \Delta \dot{E}_{\mathrm{x,V}}}_{\text{OUT}}. \tag{6.10}$$

There are two possible technical applications for the counterclockwise cycle:

- In case the cooling power $\dot{Q}_0$ is of interest, the machine is denoted fridge or cooling machine, respectively. The coefficient of performance is

$$\varepsilon_\mathrm{CM} = \frac{\text{Benefit}}{\text{Effort}} = \frac{\dot{Q}_0}{P}. \tag{6.11}$$

  In best case it is $T = T_\mathrm{env}$.
- If one wants to utilise the released heat $\dot{Q}$ for heating, one speaks of a heat pump. The technical design is exactly the same as for the cooling machine.[4] Anyhow, the coefficient of performance for this process obeys

$$\varepsilon_\mathrm{HP} = \frac{\text{Benefit}}{\text{Effort}} = \frac{\dot{Q}}{P} = \frac{\dot{Q}_0 + P}{P} = \varepsilon_\mathrm{CM} + 1. \tag{6.12}$$

  In best case it is $T_0 = T_\mathrm{env}$.

In this Problem, the coefficient of performance of the fridge is

$$\varepsilon_\mathrm{CM} = 1.1 \tag{6.13}$$

so that the coefficient of performance (COP) for the heat pump with the same boundary conditions equals

$$\varepsilon_\mathrm{HP} = \varepsilon_\mathrm{CM} + 1 = 2.1 = \frac{\dot{Q}}{P}. \tag{6.14}$$

[4] Obviously, answer (a) does not apply.

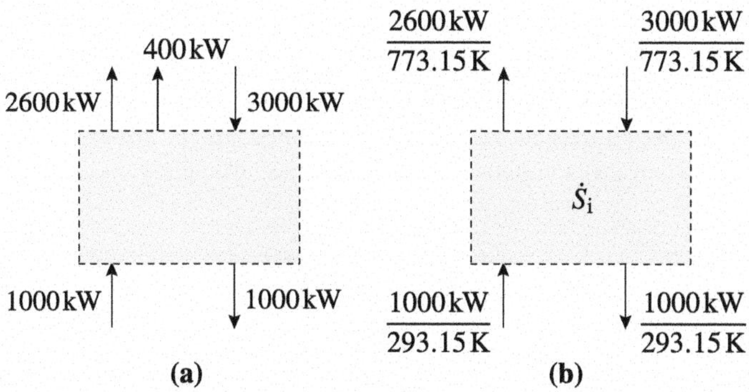

Fig. 6.12 System boundary for Problem 6.3

Consequently, answer (b) is wrong, $P = 2.1 \cdot \dot{Q}$. Furthermore, ambient temperature is not required to calculate the COP of the heat pump, cf. Eq. 6.12. Answer (c) is also not correct.

6.3 **(d)**
Let us check whether the first law of thermodynamics is violated according to the system boundary shown in Fig. 6.12a. In steady state, the energy inflows must be balanced by the energy outflows:

$$\underbrace{1000\,\text{kW} + 3000\,\text{kW}}_{\text{IN}} = \underbrace{1000\,\text{kW} + 2600 + 400\,\text{kW}}_{\text{OUT}} \tag{6.15}$$

This equation is true, the first law of thermodynamics is not violated.

The next step is to check the entropy balance, see Fig. 6.12b:

$$\underbrace{\frac{1000\,\text{kW}}{293.15\,\text{K}} + \frac{3000\,\text{kW}}{773.15\,\text{K}}}_{\text{IN}} + \dot{S}_i = \underbrace{\frac{1000\,\text{kW}}{293.15\,\text{K}} + \frac{2600\,\text{kW}}{773.15\,\text{K}}}_{\text{OUT}} \tag{6.16}$$

so that

$$\dot{S}_i = -\frac{400\,\text{kW}}{773.15\,\text{K}} < 0. \tag{6.17}$$

Consequently, the second law of thermodynamics is violated, the process can not be operated. Answer (d) is true. Since it is a closed system, the mass in the system remains constant, so it does not violate conservation of mass.

Alternatively, the thermal efficiency can be determined. The thermal fluxes are:

$$\dot{Q}_{\text{in},@T_{\max}} = 3000\,\text{kW} - 2600\,\text{kW} = 400\,\text{kW} = P \tag{6.18}$$

and

$$\dot{Q}_{\text{out},@T_{\min}} = 1000\,\text{kW} - 1000\,\text{kW} = 0. \tag{6.19}$$

In other words, the entire heat is converted into mechanical power, i.e. the second law of thermodynamics is violated,[5] and the thermal efficiency yields

$$\begin{aligned} \eta_{\text{th}} &= \frac{P}{\dot{Q}_{\text{in}}} = \frac{400\,\text{kW}}{400\,\text{kW}} = 1 \\ &> \underbrace{1 - \frac{T_{\min}}{T_{\max}}}_{=\eta_{\text{C}}} = 0.6208. \end{aligned} \tag{6.20}$$

and exceed Carnot efficiency. The machine is a perpetuum mobile of the 2nd kind.

6.4 **(c)**
For state (1) it applies

$$p_1 V_1 = mRT_1 \rightarrow mR = \frac{p_1 V_1}{T_1}. \tag{6.21}$$

State (2) obeys

$$p_1 2V_1 = mRT_2 \rightarrow T_2 = \frac{2p_1 V_1}{mR} = 2T_1. \tag{6.22}$$

Consequently, state (3) follows

$$\frac{1}{2} p_1 2V_1 = mRT_3 \rightarrow T_3 = \frac{p_1 V_1}{mR} = T_1 = 600\,\text{K}. \tag{6.23}$$

Answer (c) is correct.

6.5 **(b)**
Since it is a reversible process, entropy changes are only possible by heat release or supply, i.e.

$$\text{d}s = \frac{\delta q}{T} + \underbrace{\frac{\delta \psi}{T}}_{=0}. \tag{6.24}$$

Thus, for the specific entropy decrease from (3) to (4), specific heat must be released, cf. Fig. 6.13. Integration of Eq. 6.24 yields[6]

$$s_4 - s_3 = \int_3^4 \frac{\delta q}{T} = \frac{q_{34}}{T} < 0. \tag{6.25}$$

[5] Any thermal engine in steady state operation requires a cooling.

[6] Mind that the change of state is isothermal, i.e. $T_3 = T_4$.

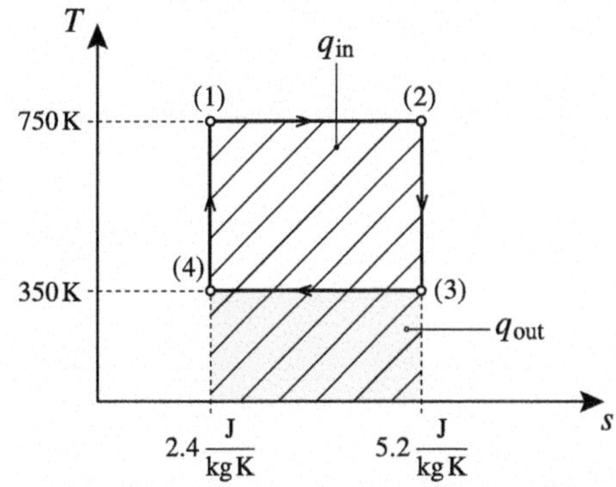

Fig. 6.13 Reversible thermodynamic cycle

The released heat therefore obeys

$$q_{out} = |q_{34}| = |T_3 (s_4 - s_3)| = 1.12 \frac{\text{kJ}}{\text{kg}}. \tag{6.26}$$

6.6 **(d)**
To answer this question, we refer to Fig. 6.10 from Problem 6.1. However, for this Problem it is

$$\dot{Q} = 10\,\text{kW} \tag{6.27}$$

and

$$P = 4\,\text{kW}. \tag{6.28}$$

Hence, according to the first law of thermodynamics it yields

$$\dot{Q}_0 = \dot{Q} - P = 6\,\text{kW}. \tag{6.29}$$

The thermal efficiency therefore is

$$\eta_{th} = \frac{\text{Benefit}}{\text{Effort}} = \frac{P}{\dot{Q}} = 0.4. \tag{6.30}$$

The exergetic efficiency follows

$$\eta_{ex} = \frac{\text{ex. Benefit}}{\text{ex. Effort}} = \frac{P}{\dot{E}_{x,Q}} = \frac{P}{\dot{Q}} \frac{\dot{Q}}{\dot{E}_{x,Q}} = \eta_{th} \frac{\dot{Q}}{\dot{E}_{x,Q}} = \eta_{th} \frac{1}{1 - \frac{T_{env}}{T}}. \tag{6.31}$$

Hence, it applies

$$\eta_{ex} = 0.8 = \frac{\eta_{th}}{1 - \frac{T_{env}}{T}} \tag{6.32}$$

respectively

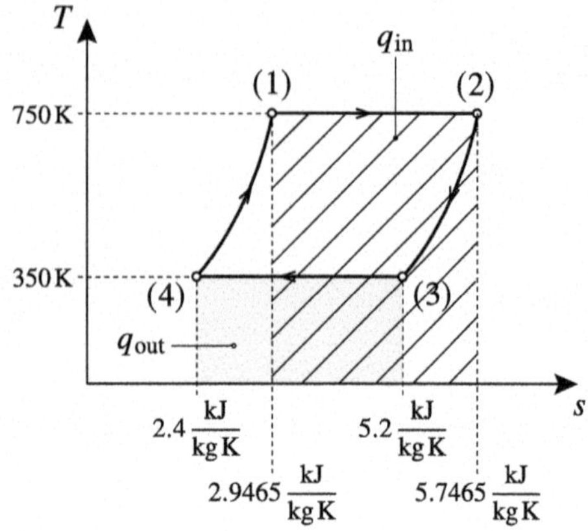

Fig. 6.14 Ideal stirling cycle

$$T = \frac{T_{env}}{1 - \frac{\eta_{th}}{\eta_{ex}}} = 600\,\text{K}. \tag{6.33}$$

6.7 **(b)**
Entropy changes within a reversible thermal engine, i.e. in this case with the Stirling engine, are only due to heat supply or release, see Fig. 6.14. The special feature of the Stirling process is the regenerative heat transfer from (4) to (1) and from (2) to (3): Heat release from (2) to (3) takes place within the machine by means of a thermal accumulator. This amount of heat is therefore not released to the environment via the system boundary. Moreover, in the reversible case, this (charged) quantity of heat corresponds exactly to the required supply of heat from (4) to (1)—the thermal accumulator is therefore discharged again. For the calculation of the thermal, reversible efficiency, however, only the heat exchanged with the environment is relevant, i.e.

$$\eta_{th} = 1 - \frac{q_{out}}{q_{in}} \tag{6.34}$$

with

$$q_{out} = |T_3 \cdot (s_4 - s_3)| = 980 \frac{\text{kJ}}{\text{kg}} \tag{6.35}$$

and

$$q_{in} = T_1 \cdot (s_2 - s_1) = 2100 \frac{\text{kJ}}{\text{kg}}. \tag{6.36}$$

Finally, the thermal efficiency is

$$\eta_{th} = 1 - \frac{q_{out}}{q_{in}} = 0.5333. \tag{6.37}$$

6.8 To solve this problem, we refer to the schematic diagram (black box) of a counterclockwise cycle shown in Fig. 6.11. The following applies to the COP of a cooling machine

$$\varepsilon_{\mathrm{CM}} = \frac{\dot{Q}_0}{P} = 4. \tag{6.38}$$

According to the first law of thermodynamics it is

$$\dot{Q}_0 + P = \dot{Q} \quad \text{with} \quad \dot{Q} = 2\,\mathrm{kW}. \tag{6.39}$$

Hence, it yields

$$\varepsilon_{\mathrm{CM}} = \frac{\dot{Q} - P}{P} = \frac{\dot{Q}}{P} - 1. \tag{6.40}$$

Solving for P results in

$$P = \frac{\dot{Q}}{\varepsilon_{\mathrm{CM}} + 1} = 0.4\,\mathrm{kW}. \tag{6.41}$$

6.9 First, the first law of thermodynamics is applied for the given system boundary, cf. Fig. 6.5, i.e.[7]

$$\dot{m}_{\mathrm{A}} h_{\mathrm{env}} + P = \dot{m}_{\mathrm{C}} h_{\mathrm{C}} + \dot{m}_{\mathrm{H}} h_{\mathrm{H}} \tag{6.42}$$

with

$$\dot{m}_{\mathrm{A}} = \dot{m}_{\mathrm{C}} + \dot{m}_{\mathrm{H}}. \tag{6.43}$$

Solving for P results in

$$P = \dot{m}_{\mathrm{H}} \left(h_{\mathrm{H}} - h_{\mathrm{env}}\right) - \dot{m}_{\mathrm{C}} \left(h_{\mathrm{env}} - h_{\mathrm{C}}\right) \tag{6.44}$$

respectively with the caloric equation of state

$$P = \dot{m}_{\mathrm{H}} c_p \left(T_{\mathrm{H}} - T_{\mathrm{env}}\right) - \dot{m}_{\mathrm{C}} c_p \left(T_{\mathrm{env}} - T_{\mathrm{C}}\right). \tag{6.45}$$

Unfortunately, the mass flux $\dot{m}_{\mathrm{C}}$ is missing, so that a second balance needs to be done. In the next step, for the given system boundary, cf. Fig. 6.5, the entropy is balanced, i.e.

$$\overbrace{\underbrace{\dot{S}_{\mathrm{i}} + \dot{S}_{\mathrm{a}}}_{=0} + \dot{m}_{\mathrm{A}} s_{\mathrm{env}}}^{\mathrm{IN}} = \overbrace{\dot{m}_{\mathrm{C}} s_{\mathrm{C}} + \dot{m}_{\mathrm{H}} s_{\mathrm{H}}}^{\mathrm{OUT}} \tag{6.46}$$

respectively with the mass balance according to Eq. 6.43

$$0 = \dot{m}_{\mathrm{H}} \left(s_{\mathrm{A}} - s_{\mathrm{env}}\right) + \dot{m}_{\mathrm{C}} \left(s_{\mathrm{C}} - s_{\mathrm{env}}\right). \tag{6.47}$$

Applying the caloric equation of state results in

[7] $\dot{Q}_{\mathrm{H}}$ and $\dot{Q}_0$ do not appear in the balance because they do not cross the system boundary.

$$0 = \dot{m}_{\mathrm{H}} c_p \ln \frac{T_{\mathrm{H}}}{T_{\mathrm{env}}} + \dot{m}_{\mathrm{C}} c_p \ln \frac{T_{\mathrm{C}}}{T_{\mathrm{env}}}. \tag{6.48}$$

Solving for the mass flux $\dot{m}_{\mathrm{C}}$ yields

$$\dot{m}_{\mathrm{C}} = -\dot{m}_{\mathrm{H}} \frac{\ln \frac{T_{\mathrm{H}}}{T_{\mathrm{env}}}}{\ln \frac{T_{\mathrm{C}}}{T_{\mathrm{env}}}} = 6.3786 \frac{\mathrm{kg}}{\mathrm{s}}. \tag{6.49}$$

Finally, the required power P is

$$P = \dot{m}_{\mathrm{H}} c_p \left(T_{\mathrm{H}} - T_{\mathrm{env}}\right) - \dot{m}_{\mathrm{C}} c_p \left(T_{\mathrm{env}} - T_{\mathrm{C}}\right) = 6.2174 \times 10^3\,\mathrm{W}. \tag{6.50}$$

6.10 The following changes of state can be taken from the diagram:

- (1)→(2):
  Isentropic, reversible compression, i.e. $s_1 = s_2$, no heat exchange, since

$$\mathrm{d}s = 0 = \frac{\delta q}{T} + \underbrace{\frac{\delta \psi}{T}}_{=0} \rightarrow \delta q = 0 \tag{6.51}$$

- (2)→(3):
  Isothermal, reversible compression, i.e. $T_2 = T_3$, heat release, since

$$\underbrace{\delta w}_{>0} + \delta q = \mathrm{d}u = 0 \rightarrow \delta q < 0 \tag{6.52}$$

  with

$$q_{\mathrm{out}} = \left|T_2 \left(s_3 - s_2\right)\right| = T_2 \left(s_2 - s_3\right) \tag{6.53}$$

- (3)→(4):
  Isentropic, reversible expansion, i.e. $s_3 = s_4$, no heat exchange, since

$$\mathrm{d}s = 0 = \frac{\delta q}{T} + \underbrace{\frac{\delta \psi}{T}}_{=0} \rightarrow \delta q = 0 \tag{6.54}$$

- (4)→(1):
  Isothermal, reversible expansion, i.e. $T_4 = T_1$, heat supply, since

$$\underbrace{\delta w}_{<0} + \delta q = \mathrm{d}u = 0 \rightarrow \delta q > 0 \tag{6.55}$$

  with

$$q_{\mathrm{in}} = T_1 \left(s_1 - s_4\right) \tag{6.56}$$

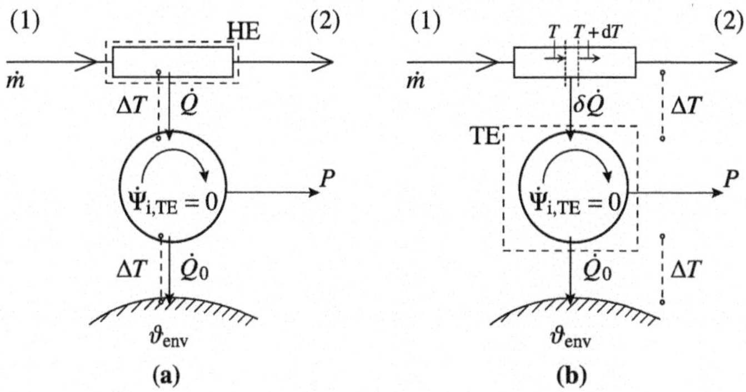

Fig. 6.15 Thermal engine: **a** Balance HE, **b** Balance TE

The first law of thermodynamics for the heat pump obeys

$$q_{in} + w = q_{out} \tag{6.57}$$

and the COP follows

$$\begin{aligned}\varepsilon_{HP} &= \frac{q_{out}}{w}\\ &= \frac{q_{out}}{q_{out}-q_{in}}\\ &= \frac{T_2\,(s_2-s_3)}{T_2\,(s_2-s_3)-T_1\,(s_1-s_4)}.\end{aligned} \tag{6.58}$$

Finally, since $s_1 = s_2$ and $s_3 = s_4$, it is

$$\varepsilon_{HP} = \frac{T_2}{T_2 - T_1} = 3.3657. \tag{6.59}$$

6.11 Let us start with an energy balance with respect to system boundary HE according to Fig. 6.15a, i.e.

$$\underbrace{\dot m h_1}_{\text{IN}} = \underbrace{\dot m h_2 + \dot Q}_{\text{OUT}}. \tag{6.60}$$

Hence, with the caloric equation of state the entire heat flux $\dot Q$ follows

$$\dot Q = -\dot m c_p\,(T_2 - T_1) = 870.1875\,\text{kW}. \tag{6.61}$$

Since the temperature in the heat exchanger is not constant, a differential approach is followed for the heat flux into the thermal engine, see Fig. 6.15b:

$$\dot m h\,(T) = \delta\dot Q + \dot m h\,(T + \mathrm{d}T) \tag{6.62}$$

respectively

$$\delta \dot{Q} = -\dot{m} c_p \, \mathrm{d}T. \tag{6.63}$$

Now the TE system is being further investigated. The differential local heat flux $\delta \dot{Q}$ brings an entropy flux into the machine. To calculate this, the temperature at which the heat flux enters the system must be known. To ensure that heat flows into the machine, this temperature must always be $\Delta T = 5\,\mathrm{K}$ lower than the local temperature in the heat exchanger (HE), i.e. the total entropy entering the system obeys

$$\dot{S}_{\mathrm{in}} = \int_1^2 \frac{\delta \dot{Q}}{T - \Delta T} \tag{6.64}$$

respectively

$$\dot{S}_{\mathrm{in}} = -\int_1^2 \frac{\dot{m} c_p \, \mathrm{d}T}{T - \Delta T} = -\dot{m} c_p \ln \frac{T_2 - \Delta T}{T_1 - \Delta T}. \tag{6.65}$$

The entropy flux from the machine can be calculated more easily, since the temperature of the sink does not vary locally, but has a constant ambient temperature T_{env}. In order to supply heat to this sink, the temperature in the machine, at the point where heat is supplied, must always be $\Delta T = 5\,\mathrm{K}$ greater than the sink temperature. The entropy flux released by the machine therefore is

$$\dot{S}_{\mathrm{out}} = \frac{\dot{Q}_0}{T_{\mathrm{env}} + \Delta T}. \tag{6.66}$$

Finally, an overall entropy balance for the thermal engine is conducted. Since the released power is in demand in the best case, the machine must be reversible, i.e. $\dot{S}_{\mathrm{i}} = 0$. The entropy balance obeys

$$\underbrace{\dot{S}_{\mathrm{out}}}_{\mathrm{OUT}} = \overbrace{\dot{S}_{\mathrm{in}} + \underbrace{\dot{S}_{\mathrm{i}}}_{=0}}^{\mathrm{IN}} \tag{6.67}$$

respectively

$$\frac{\dot{Q}_0}{T_{\mathrm{env}} + \Delta T} = -\dot{m} c_p \ln \frac{T_2 - \Delta T}{T_1 - \Delta T}. \tag{6.68}$$

Solving for the release heat flux $\dot{Q}_0$ brings

$$\dot{Q}_0 = -\dot{m} c_p \left(T_{\mathrm{env}} + \Delta T \right) \ln \frac{T_2 - \Delta T}{T_1 - \Delta T} = 376.6646\,\mathrm{kW}. \tag{6.69}$$

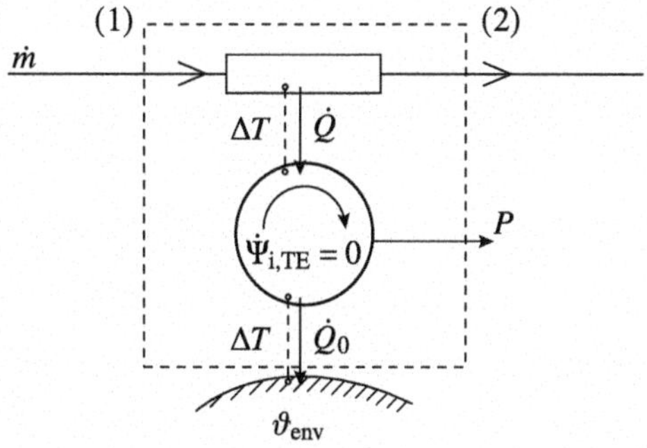

Fig. 6.16 Thermal engine: Overall balance

The first law of thermodynamics for the thermal engine follows

$$\underbrace{\dot{Q}_0 + P}_{\text{OUT}} = \underbrace{\dot{Q}}_{\text{IN}} . \tag{6.70}$$

Finally, the released power P in best case is

$$P = \dot{Q} - \dot{Q}_0 = 493.5229\,\text{kW}. \tag{6.71}$$

<u>Additional information:</u>
For the heat *supply*, an local $\Delta T = 5\,\text{K}$ is always fixed, i.e. at any position within the heat exchanger. This specification describes the assumed heat transfer approach. If the thermodynamic mean temperature inside the heat exchanger were used instead, the general premise $\Delta T = 5\,\text{K}$ would not be considered, leading to a different result:

$$\begin{aligned} \dot{S}_{\text{in,alt}} &= \frac{\dot{Q}}{T_\text{m} - \Delta T} = \frac{-\dot{m}c_p\,(T_2 - T_1)}{\frac{T_2 - T_1}{\ln \frac{T_2}{T_1}} - \Delta T} \\ &\neq \underbrace{-\dot{m}c_p \ln \frac{T_2 - \Delta T}{T_1 - \Delta T}}_{\text{see Eq. 6.65}} \end{aligned} \tag{6.72}$$

<u>Verification:</u> The results obtained are now examined[8] in an overall balance according to Fig. 6.16. The correctness of the balances can be easily checked by inserting the calculated numerical values.

- Energy balance

$$\dot{m}h_1 = \dot{m}h_2 + P + \dot{Q}_0 \tag{6.73}$$

$$\rightarrow P + \dot{Q}_0 = \dot{m}c_p\,(T_1 - T_2) \tag{6.74}$$

[8] This is not asked in the Problem definition and is only intended to check the results.

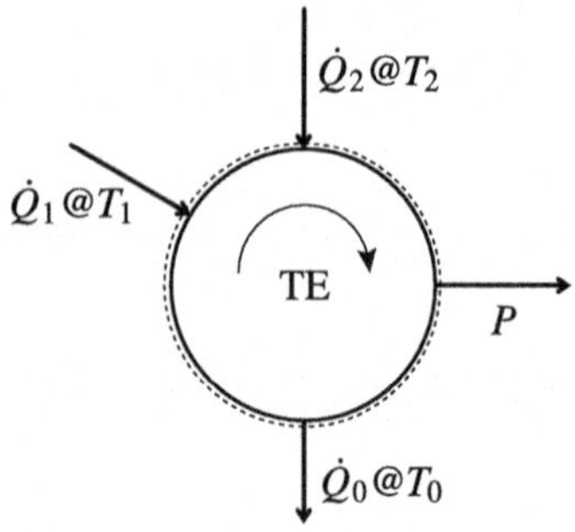

Fig. 6.17 Thermal engine

- Entropy balance[9]

$$\dot{m}s_1 + \dot{S}_\mathrm{i} = \dot{m}s_2 + \frac{\dot{Q}_0}{T_\mathrm{env}} \tag{6.75}$$

$$\rightarrow \dot{S}_\mathrm{i} = \dot{m}c_p \ln\frac{T_2}{T_1} + \frac{\dot{Q}_0}{T_\mathrm{env}} \tag{6.76}$$

- Exergy balance

$$\dot{m}e_{\mathrm{x},S,1} = \dot{m}e_{\mathrm{x},S,2} + \dot{S}_\mathrm{i}T_\mathrm{env} + P + \underbrace{\dot{E}_{\mathrm{x},Q_0@T_\mathrm{env}}}_{=0} \tag{6.77}$$

$$\rightarrow \dot{m}\left[c_p\left(T_1 - T_2\right) + T_\mathrm{env}c_p \ln\frac{T_2}{T_1}\right] = \dot{S}_\mathrm{i}T_\mathrm{env} + P \tag{6.78}$$

6.12 Figure 6.17 shows the thermal engine described. First law of thermodynamics obeys

$$\underbrace{\dot{Q}_1 + \dot{Q}_2}_{\mathrm{IN}} = \underbrace{\dot{Q}_0 + P}_{\mathrm{OUT}}. \tag{6.79}$$

The balance of entropy follows

$$\underbrace{\frac{\dot{Q}_1}{T_1} + \frac{\dot{Q}_2}{T_2} + \dot{S}_\mathrm{i}}_{\mathrm{IN}} = \underbrace{\frac{\dot{Q}_0}{T_0}}_{\mathrm{OUT}} \quad \text{with} \quad \dot{Q}_2 = 3\dot{Q}_1. \tag{6.80}$$

Since the machine is to be studied for the best case, we assume reversible operation, i.e. $\dot{S}_\mathrm{i} = 0$. Hence, it is

$$\dot{Q}_0 = \dot{Q}_1\frac{T_0}{T_1} + 3\dot{Q}_1\frac{T_0}{T_2}. \tag{6.81}$$

[9] Although the actual thermal engine is reversible, due to heat transfer with $\Delta T > 0$ entropy is generated within the chosen system.

The thermal efficiency yields

$$\eta_{\text{th}} = \frac{\text{Benefit}}{\text{Effort}} = \frac{P}{\dot{Q}_1 + \dot{Q}_2} \tag{6.82}$$

respectively with Eq. 6.79

$$\eta_{\text{th}} = \frac{\dot{Q}_1 + \dot{Q}_2 - \dot{Q}_0}{\dot{Q}_1 + \dot{Q}_2} = 1 - \frac{\dot{Q}_0}{\dot{Q}_1 + \dot{Q}_2}. \tag{6.83}$$

Finally, substitution of $\dot{Q}_0$ according to Eq. 6.81 results in

$$\eta_{\text{th}} = 1 - \frac{\dot{Q}_1 \frac{T_0}{T_1} + 3\dot{Q}_1 \frac{T_0}{T_2}}{4\dot{Q}_1} = 1 - \frac{\frac{T_0}{T_1} + 3\frac{T_0}{T_2}}{4} = 0.5900. \tag{6.84}$$

6.13

(a) This is a clockwise cycle as shown in Fig. 6.18.
(b) In best case, the machine is a Carnot machine, i.e. there are no irreversibilities inside the machine as well as in the heat transfer. In this case $\Delta T = 0$ and the efficiency is only determined by T_{min} and T_{max}, i.e.

$$\eta_{\text{C}} = \frac{P_{\text{el,max}}}{\dot{Q}} = 1 - \frac{T_{\text{min}}}{T_{\text{max}}}. \tag{6.85}$$

Hence, the maximum electrical power yields

$$P_{\text{el,max}} = \dot{Q}\left(1 - \frac{T_{\text{min}}}{T_{\text{max}}}\right) = 4.7108\,\text{kW}. \tag{6.86}$$

The first law of thermodynamics for the Stirling engine obeys

$$\dot{Q} = P_{\text{el,max}} + \dot{Q}_0 \tag{6.87}$$

so that

$$\dot{Q}_0 = \dot{Q} - P_{\text{el,max}} = 5.2892\,\text{kW}. \tag{6.88}$$

(c) Due to irreversibilites the power reduces to

$$P_{\text{el}} = 0.8 P_{\text{el,max}} = 3.7686\,\text{kW}. \tag{6.89}$$

The heat flux for the swimming pool yields

$$\dot{Q}_0 = \dot{Q} - P_{\text{el}} = 6.2314\,\text{kW}. \tag{6.90}$$

(d) Now the real process is to be examined in detail.

(1) An entropy balance for the heat transfer into the Stirling engine, cf. Fig. 6.19a yields

$$\frac{\dot{Q}}{T_{\max}} + \dot{S}_{\mathrm{i,HT_1}} = \frac{\dot{Q}}{T_{\max} - \Delta T}. \tag{6.91}$$

Thus, the generated entropy due to the heat transfer is

$$\dot{S}_{\mathrm{i,HT_1}} = \frac{\dot{Q}}{T_{\max} - \Delta T} - \frac{\dot{Q}}{T_{\max}} = 3.0982 \times 10^{-4}\,\mathrm{kW\,K^{-1}} \tag{6.92}$$

and the exergy loss follows

$$\dot{E}_{\mathrm{x,V,HT_1}} = \dot{S}_{\mathrm{i,HT_1}} T_{\mathrm{env}} = 0.0929\,\mathrm{kW}. \tag{6.93}$$

(2) An entropy balance for the heat transfer out of the Stirling engine, cf. Fig. 6.19b yields

$$\frac{\dot{Q}_0}{T_{\min} + \Delta T} + \dot{S}_{\mathrm{i,HT_2}} = \frac{\dot{Q}_0}{T_{\min}}. \tag{6.94}$$

Thus, the generated entropy due to the heat transfer is

$$\dot{S}_{\mathrm{i,HT_2}} = \frac{\dot{Q}_0}{T_{\min}} - \frac{\dot{Q}_0}{T_{\min} + \Delta T} = 6.5641 \times 10^{-4}\,\mathrm{kW\,K^{-1}} \tag{6.95}$$

and the exergy loss follows

$$\dot{E}_{\mathrm{x,V,HT_2}} = \dot{S}_{\mathrm{i,HT_2}} T_{\mathrm{env}} = 0.1969\,\mathrm{kW}. \tag{6.96}$$

(3) An entropy balance of the overall Stirling engine, cf. Fig. 6.19c yields

$$\frac{\dot{Q}}{T_{\max}} + \dot{S}_{\mathrm{i,total}} = \frac{\dot{Q}_0}{T_{\min}}. \tag{6.97}$$

Thus, the generated entropy due to the heat transfer is

$$\dot{S}_{\mathrm{i,total}} = \frac{\dot{Q}_0}{T_{\min}} - \frac{\dot{Q}}{T_{\max}} = 0.0031\,\mathrm{kW\,K^{-1}} \tag{6.98}$$

and the exergy loss follows

$$\dot{E}_{\mathrm{x,V,total}} = \dot{S}_{\mathrm{i,total}} T_{\mathrm{env}} = 0.9324\,\mathrm{kW}. \tag{6.99}$$

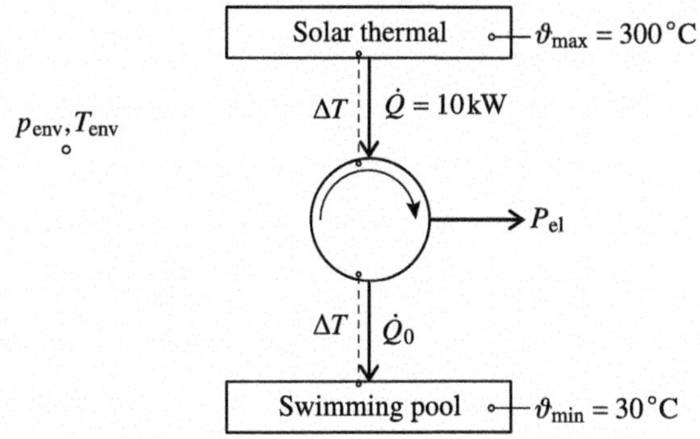

Fig. 6.18 Stirling engine

(4) The entropy generated within the entire system, see Fig. 6.19c, is composed of the entropy generated by heat transfer and by irreversibilities within the Stirling machine, i.e.

$$\dot{S}_{i,total} = \dot{S}_{i,HT_1} + \dot{S}_{i,HT_2} + \dot{S}_{i,SM}. \tag{6.100}$$

Accordingly,[10] the following applies to the exergy losses

$$\dot{E}_{x,V,total} = \dot{E}_{x,V,HT_1} + \dot{E}_{x,V,HT_2} + \dot{E}_{x,V,SM}. \tag{6.101}$$

I.e. due to irreversibilities in the machine there is a loss of exergy according to

$$\dot{E}_{x,V,SM} = \dot{E}_{x,V,total} - \dot{E}_{x,V,HT_1} - \dot{E}_{x,V,HT_2} = 0.6425\,kW. \tag{6.102}$$

6.14

(a) The heat flux $\dot{Q}_2$ can be determined by an energetic balance for the first thermal engine TE1, i.e.

$$\dot{Q}_2 = \dot{Q}_1 - P_1 = 70\,kW. \tag{6.103}$$

Finally, temperature T_2 at which TE1 releases the heat follows from an entropy balance for the reversible thermal engine TE1, i.e.

$$\frac{\dot{Q}_1}{T_1} + \underbrace{\dot{S}_{i,TE1}}_{=0} = \frac{\dot{Q}_2}{T_2} \tag{6.104}$$

so that

[10] By mulitplying with ambient temperature T_{env}.

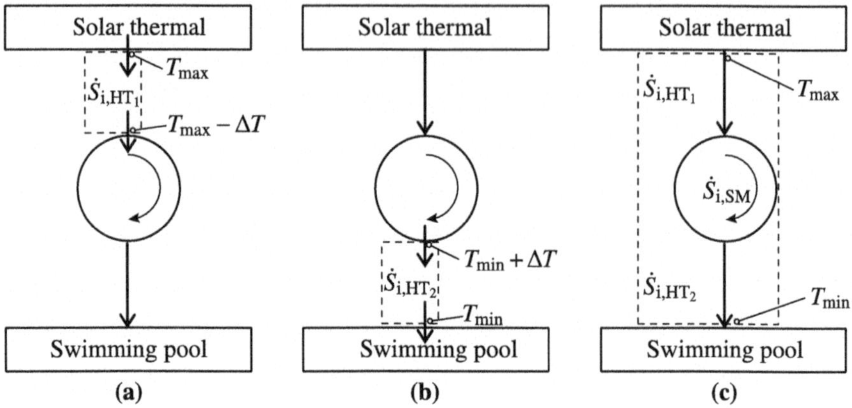

Fig. 6.19 Stirling engine—system boundaries

$$T_2 = T_1 \frac{\dot{Q}_2}{\dot{Q}_1} = 350\,\text{K}. \tag{6.105}$$

To transfer the heat flux $\dot{Q}_2$ *into* thermal engine TE2, T_3 must be smaller than T_2 by ΔT, i.e.

$$T_3 = T_2 - \Delta T = 345\,\text{K} \tag{6.106}$$

The entropy balance for the reversible thermal engine TE2 obeys

$$\frac{\dot{Q}_2}{T_3} + \underbrace{\dot{S}_{i,TE2}}_{=0} = \frac{\dot{Q}_3}{T_4} \tag{6.107}$$

so that

$$\dot{Q}_3 = T_4 \frac{\dot{Q}_2}{T_3} = 60.8696\,\text{kW}. \tag{6.108}$$

Finally, the released power P_2 of TE2 yields from an energy balance for TE2, i.e.

$$P_2 = \dot{Q}_2 - \dot{Q}_3 = 9.1304\,\text{kW}. \tag{6.109}$$

The overall thermal efficiency thus amounts to

$$\eta_{th} = \frac{\text{Benefit}}{\text{Effort}} = \frac{P_1 + P_2}{\dot{Q}_1} = 0.3913. \tag{6.110}$$

(b) Now the overall system, see Fig. 6.8, is examined. Although TE1 and TE2 are reversible by themselves, $\dot{S}_i$ is generated in the overall system due to the internal heat transfer with $\Delta T > 0$. Therefore, the total entropy balance is:

$$\frac{\dot{Q}_1}{T_1} + \dot{S}_i = \frac{\dot{Q}_3}{T_4}. \tag{6.111}$$

The generated flux of entropy therefore results in

$$\dot{S}_i = \frac{\dot{Q}_3}{T_4} - \frac{\dot{Q}_1}{T_1} = 0.0029\,\mathrm{kW\,K^{-1}}. \tag{6.112}$$

Finally, the generation of entropy causes an exergetic loss according to

$$\Delta\dot{E}_{x,V} = T_{env}\dot{S}_i = 0.8551\,\mathrm{kW}. \tag{6.113}$$

6.15

(a) The gas constant follows from an entropy balance for the adiabatic turbine, since the specific entropy generation is given, i.e.

$$\dot{m}s_3 + \dot{S}_i + \underbrace{\dot{S}_a}_{=0} = \dot{m}s_4 \tag{6.114}$$

so that

$$s_{i34} = s_4 - s_4. \tag{6.115}$$

Applying the caloric equation of state yields

$$s_{i34} = s_4 - s_4 = c_p \ln\frac{T_4}{T_3} - R\ln\frac{p_4}{p_3}. \tag{6.116}$$

Solving for the gas constant R finally leads to

$$R = \frac{c_p \ln\frac{T_4}{T_3} - s_{i34}}{\ln\frac{p_4}{p_3}} = 2.0771\frac{\mathrm{kJ}}{\mathrm{kgK}}. \tag{6.117}$$

(b) The mass flux $\dot{m}$ follows from the energy balance of the heat exchanger, i.e.

$$\underbrace{\dot{m}_a h_{a,in} + \dot{m}h_2}_{\text{IN}} = \underbrace{\dot{m}_a h_{a,out} + \dot{m}h_3}_{\text{OUT}} \tag{6.118}$$

so that

$$\dot{m} = \frac{\dot{m}_a\left(h_{a,in} - h_{a,out}\right)}{h_3 - h_2}. \tag{6.119}$$

With the caloric equation of state it yields

$$\dot{m} = \frac{\dot{m}_{\mathrm{a}} c_{p,\mathrm{a}} \left(\vartheta_{\mathrm{a,in}} - \vartheta_{\mathrm{a,out}}\right)}{c_p \left(\vartheta_3 - \vartheta_2\right)}. \tag{6.120}$$

In this equation, the temperature ϑ_2 is unknown, which is now calculated using the isentropic efficiency of the compressor, i.e.

$$\eta_{s\mathrm{V}} = 0.9 = \frac{T_{2s} - T_1}{T_2 - T_1}. \tag{6.121}$$

T_{2s} follows an isentropic change of state

$$T_{2s} = T_1 \left(\frac{p_2}{p_1}\right)^{\frac{\kappa-1}{\kappa}} = 567.7819\,\mathrm{K} \quad \text{with} \quad \kappa = \frac{c_p}{c_p - R} = 1.666. \tag{6.122}$$

Hence, temperature T_2 is

$$T_2 = T_1 + \frac{T_{2s} - T_1}{\eta_{s\mathrm{V}}} = 599.4076\,\mathrm{K}. \tag{6.123}$$

Finally, the mass flow rate obeys

$$\dot{m} = \frac{\dot{m}_{\mathrm{a}} c_{p,\mathrm{a}} \left(\vartheta_{\mathrm{a,in}} - \vartheta_{\mathrm{a,out}}\right)}{c_p \left(\vartheta_3 - \vartheta_2\right)} = 0.0103 \frac{\mathrm{kg}}{\mathrm{s}}. \tag{6.124}$$

(c) The isentropic efficiency of the turbine is

$$\eta_{s\mathrm{T}} = \frac{T_4 - T_3}{T_{4s} - T_3} = 0.8404 \quad \text{with} \quad T_{4s} = T_3 \left(\frac{p_4}{p_3}\right)^{\frac{\kappa-1}{\kappa}} = 485.3051\,\mathrm{K}. \tag{6.125}$$

(d) The exergy loss can be determined with the help of an entropy balance of the heat exchanger, see Fig. 6.20a. The balance obeys

$$\underbrace{\dot{m} s_2 + \dot{m}_{\mathrm{a}} s_{\mathrm{n}} + \dot{S}_{\mathrm{i}} + \overbrace{\dot{S}_{\mathrm{a}}}^{=0}}_{\mathrm{IN}} = \underbrace{\dot{m} s_3 + \dot{m}_{\mathrm{a}} s_{\mathrm{out}}}_{\mathrm{OUT}} \tag{6.126}$$

so that

$$\dot{S}_{\mathrm{i}} = \dot{m}_{\mathrm{a}} \left(s_{\mathrm{out}} - s_{\mathrm{in}}\right) + \dot{m} \left(s_3 - s_3\right). \tag{6.127}$$

Since the heat exchanger is isobaric on both the air and fluid side, the following applies

$$\dot{S}_{\mathrm{i}} = \dot{m}_{\mathrm{a}} c_{p,\mathrm{a}} \ln \frac{T_{\mathrm{a,out}}}{T_{\mathrm{a,in}}} + \dot{m} c_p \ln \frac{T_3}{T_2} = 0.0104\,\mathrm{kW\,K^{-1}}. \tag{6.128}$$

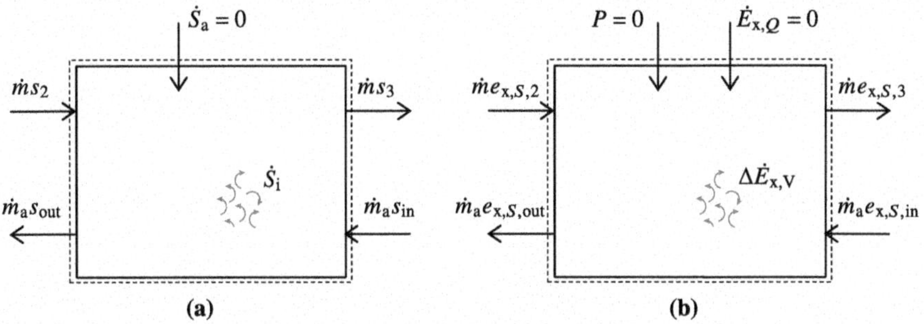

Fig. 6.20 Heat exchanger: **a** Entropy balance **b** Exergy balance

Finally, the loss of exergy is

$$\Delta\dot{E}_{x,V} = \dot{S}_i T_{env} = 3.0535\,\text{kW}. \tag{6.129}$$

Alternatively, the loss of exergy can be derived from an exergetic balance, see Fig. 6.20b, i.e.

$$\underbrace{\dot{m}_a e_{x,S,in} + \dot{m}e_{x,S,2} + \overbrace{\dot{E}_{x,Q} + P}^{=0}}_{\text{IN}} = \underbrace{\Delta\dot{E}_{x,V} + \dot{m}_a e_{x,S,out} + \dot{m}e_{x,S,3}}_{\text{OUT}} \tag{6.130}$$

so that

$$\Delta\dot{E}_{x,V} = \dot{m}_a\left(e_{x,S,in} - e_{x,S,out}\right) + \dot{m}\left(e_{x,S,2} - e_{x,S,3}\right). \tag{6.131}$$

It further applies

$$e_{x,S,in} - e_{x,S,out} = h_{a,in} - h_{a,out} - T_{env}\left(s_{in} - s_{out}\right) \tag{6.132}$$

and

$$e_{x,S,2} - e_{x,S,3} = h_2 - h_3 - T_{env}\left(s_2 - s_3\right). \tag{6.133}$$

Finally, it is

$$\Delta\dot{E}_{x,V} = \dot{m}_a T_{env} c_{p,a} \ln\frac{T_{a,out}}{T_{a,in}} + \dot{m} T_{env} c_p \ln\frac{T_3}{T_2} = 3.0535\,\text{kW}. \tag{6.134}$$

Chapter 7
Real Fluids

The first chapters of this book have addressed ideal fluids, i.e. ideal gases and incompressible liquids. Now the focus is turned to real fluids. These fluids behave differently in many respects from the ideal fluids known so far. Furthermore, the chapter covers the phase change of real fluids and their thermodynamic evaluation. The previously presented thermal and caloric equations of state can no longer be applied. Instead, other methods, e.g. the steam table, are utilised. The tasks presented here therefore address technical components that are already known from the previous tasks, but are now operated with real fluids. Thermodynamic cycle processes such as a steam power plant are considered as well as heat pumps and cooling machines, respectively, that use refrigerants as the working fluid.

7.1 Problems

7.1 What applies to adiabatic throttling of water in saturated steam state? Outer energies are to be neglected.

(a) Pressure rises.
(b) Entropy increases.
(c) Saturated steam can not be throttled.
(d) None of the above statements is true.

7.2 The changes of state in a compression heat pump are shown schematically in a log p, h-diagram, see Fig. 7.1. Which distance represents the specific heat exchanged in the condenser?

(a) A
(b) B
(c) C
(d) None of the graphs shown.

A. Schmidt, *Technical Thermodynamics Workbook for Engineers*,
https://doi.org/10.1007/978-3-031-50172-2_7

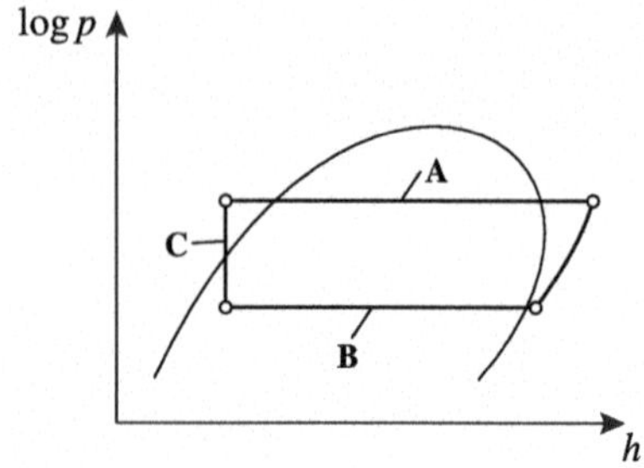

Fig. 7.1 Compression heat pump

7.3 Which measure can be applied to increase the thermal efficiency of a steam power process?

(a) Intermediate superheating.
(b) Reduction of the combustion temperature.
(c) Reduction of the isentropic turbine efficiency.
(d) None of the above statements is true.

7.4 What does a negative Joule–Thomson coefficient δ_h mean for adiabatic throttle neglecting the outer energies?

(a) An increase in pressure.
(b) An increase in fluid temperature during throttling.
(c) No dissipation occurs.
(d) None of the above statements is true.

7.5 In a h, s-diagram, which characteristic do the isotherms of strongly superheated vapour have at low pressures? The fluid is water.

(a) They coincide with the isobars.
(b) Vertical course.
(c) Horizontal course.
(d) None of the above statements is true.

7.6 Saturated vapour of 200 °C is adiabatically throttled to 1 bar. Outer energies are negligible. What happens? The fluid is water.

(a) Temperature remains constant.
(b) Temperature rises.
(c) Temperature decreases.
(d) None of the above statements is true.

7.7 Just boiling water is to be vaporised completely isobaric, reversible at 200 °C. Outer energies shall be ignored. What is the specific heat required for this change of state?

(a) $1321.6\,\frac{\text{kJ}}{\text{kg}}$
(b) $1571.8\,\frac{\text{kJ}}{\text{kg}}$

(c) $1939.7\,\frac{\text{kJ}}{\text{kg}}$
(d) $2341.9\,\frac{\text{kJ}}{\text{kg}}$

7.8 What applies to the adiabatic throttle of water in saturated steam state? Outer energies are to be neglected.

(a) Temperature rises.
(b) Pressure rises.
(c) Entropy decreases.
(d) None of the above statements is true.

7.9 The following state values are known from a refrigerant at 20 °C: Specific entropy of the boiling liquid $s' = 0.2924\,\frac{\text{kJ}}{\text{kg K}}$, specific entropy of the saturated vapour $s'' = 0.9102\,\frac{\text{kJ}}{\text{kg K}}$. What is the enthalpy of vaporisation at this temperature?

(a) $181.1\,\frac{\text{kJ}}{\text{kg}}$
(b) $210.2\,\frac{\text{kJ}}{\text{kg}}$
(c) $400.6\,\frac{\text{kJ}}{\text{kg}}$
(d) $2501.1\,\frac{\text{kJ}}{\text{kg}}$

7.10 A real gas is throttled adiabatically. Outer energies are to be neglected. Which statement applies?

(a) The change of state is reversible.
(b) The temperature can rise, fall or remain constant depending on the gas.
(c) According to the first law of thermodynamics, entropy remains constant.
(d) None of the above statements is true.

7.11 Saturated vapour of 1 bar is isentropically compressed to 4 bar. Outer energies are negligible. The fluid is water. Which statement applies?

(a) Temperature remains constant.
(b) After the compression, water is present in a superheated state.
(c) Specific entropy must decrease.
(d) None of the above statements is true.

7.12 How much heat is released when water condenses very slowly isothermally at 150 °C from ($''$) to ($'$)? Outer energies shall be disregarded.

(a) $1231.6\,\frac{\text{kJ}}{\text{kg}}$
(b) $1671.3\,\frac{\text{kJ}}{\text{kg}}$
(c) $2113.7\,\frac{\text{kJ}}{\text{kg}}$
(d) $2441.9\,\frac{\text{kJ}}{\text{kg}}$

7.13 What applies to the critical point of water?

(a) Above the critical temperature, no liquefaction is possible by increasing the pressure.

(b) Water exists in all three states of aggregation.
(c) The enthalpy of vaporisation becomes maximum at the critical point.
(d) None of the above statements is true.

7.14 A fluid flows through a horizontal, adiabatic nozzle. What is true?

(a) The entropy of the fluid decreases.
(b) The pressure decreases.
(c) The technical power is negative.
(d) None of the above statements is true.

7.15 The Joule–Thomson coefficient of a gas is $\delta_h = -0.5\,\frac{\text{K}}{\text{bar}}$. What can be expected from an adiabatic throttle if the external energy is neglected?

(a) The temperature of the gas increases.
(b) The specific entropy of the gas decreases.
(c) The specific exergy of the gas increases.
(d) None of the above statements is true.

7.16 Superheated vapour enters a heat exchanger and is cooled down until it leaves the apparatus as saturated vapour. A pressure loss is measured during this process. The thermodynamic mean temperature is above ambient temperature. Which statement is correct?

(a) If the change in outer energies is negligible, dissipation must occur in the apparatus.
(b) The exergy of the vapour increases as it flows through the apparatus.
(c) Since heat is released by the vapour, the pressure must decrease according to the thermal equation of state.
(d) None of the above statements is true.

7.17 Wet steam from state (1) is isochorically, reversibly expanded. In state (2), let $x_2 > 0$. The working fluid is water. Which sketch, see Fig. 7.2, shows qualitatively the change of state and the specific heat exchanged?

(a) A
(b) B
(c) C
(d) D

7.18 What is the basis for increasing the thermal efficiency of a steam power process through regenerative feedwater preheating?

(a) Joule–Thomson effect of the fluid used.
(b) Clausius-Clapeyron effect.
(c) Increased thermodynamic mean temperature at the external supply of heat.
(d) None of the above statements is true.

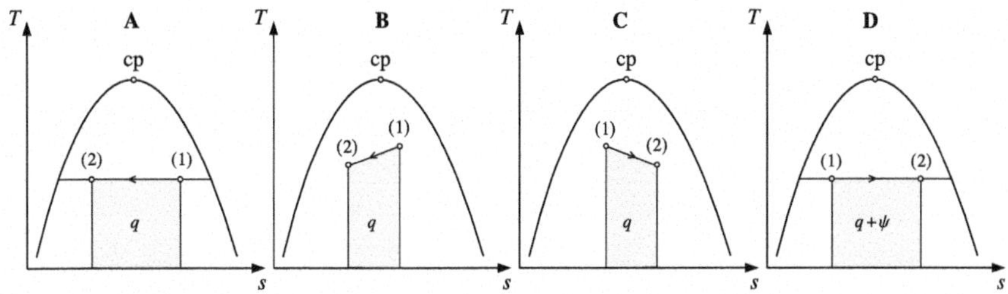

Fig. 7.2 Isochore, reversible expansion

7.19 What is true about the triple point of water?

(a) At the triple point, pressure and temperature are identical for all three states of aggregation.
(b) All three states of aggregation have the same specific enthalpy in this state.
(c) All three states of aggregation have the same density in that state.
(d) None of the above statements is true.

7.20 In a horizontal nozzle, vapour of state (1) ($p_1 = 20\,\text{bar}$, $\vartheta_1 = 500\,°\text{C}$) is to be expanded adiabatically polytropically with constant polytropic exponent $n = 1.2$ to a pressure of $p_2 = 15\,\text{bar}$. The inlet velocity of the vapour is $c_1 = 10\,\frac{\text{m}}{\text{s}}$. In state (2) the velocity is $c_2 = 389.21\,\frac{\text{m}}{\text{s}}$. Calculate the specific dissipation ψ_{12} in the nozzle!

7.21 To determine the vapour content of wet steam, it passes an adiabatic throttle. Changes in the kinetic/potential energies are negligible. The following data were measured:

- pressure when entering the throttle: $p_1 = 10.026\,\text{bar}$
- pressure at the outlet of the throttle: $p_2 = 1\,\text{bar}$
- temperature at the outlet: $\vartheta_2 = 125\,°\text{C}$.

Determine the vapour content x_1 before throttling.

7.22 Water is to be considered as a real fluid. What is the slope of the vapour pressure curve $\frac{dp}{dT}$ at $\vartheta = 50\,°\text{C}$?

7.23 What is the specific technical work done by wet steam of $h_1 = 2575.6\,\frac{\text{kJ}}{\text{kg}}$, $s_1 = 6.1404\,\frac{\text{kJ}}{\text{kg K}}$ when adiabatically, isentropically expanded to $p_2 = 0.84609\,\text{bar}$? Determine the pressure in state (1). Changes of kinetic/potential energy are negligible.

7.24 In a horizontal nozzle, water vapour of state (1) ($p_1 = 20\,\text{bar}$, $\vartheta_1 = 500\,°\text{C}$) is to be expanded adiabatically polytropically with constant polytropic exponent $n = 1.2$ to a pressure of $p_2 = 15\,\text{bar}$. The inlet velocity of the vapour is $c_1 = 10\,\frac{\text{m}}{\text{s}}$. In state (2) the velocity is $c_2 = 389.21\,\frac{\text{m}}{\text{s}}$. Calculate the specific dissipation ψ_{12} in the nozzle. What is the thermodynamic mean temperature?

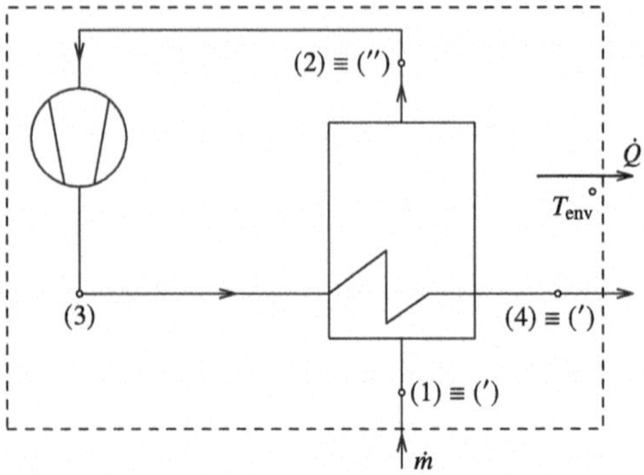

Fig. 7.3 Process

7.25 Water in saturated steam state with a pressure of 4 bar is to be compressed to 8 bar in an adiabatic, polytropic compressor ($n = 1.45$). What is the specific pressure work y_{12} of the compressor?

7.26 Water is subjected to a process according to Fig. 7.3. The pressure in state (1) is 4 bar, the pressure in state (4) is 8 bar. Let the mass flux be $\dot{m} = 3500 \frac{\text{kg}}{\text{h}}$. A total heat flux of 76.936 kW is released to the environment ($T_{env} = 293.15$ K) across the sketched system boundary. What is the flux of exergy that gets lost within the system boundary?

7.27 An adiabatic, polytropic turbine expands vapour from state (1) ($p_1 = 170$ bar, $\vartheta_1 = 540\,°$C) to state (2) ($p_2 = 35$ bar, $\vartheta_2 = 320\,°$C). What is the isentropic efficiency $\eta_{s,\text{T}}$ of the turbine?

7.28 In a vertical cylinder, which is closed by a freely movable piston, there is 5 kg wet steam under a constant pressure of 4 bar. To determine their specific vapour content x_1, heat is slowly supplied to the cylinder until the temperature of the vapour changes. The working medium is water and the amount of heat was determined to be 4000 kJ. What is the steam content x_1?

7.29 A simple steam power process with water as the process fluid is to be analysed. The water mass flow is $200 \frac{\text{kg}}{\text{s}}$. The isentropic efficiency of the feedwater pump is 70 %. The vaporisation takes place at a pressure of 70 bar. After expansion, wet steam is present at a pressure of 1 bar with a vapour content of 0.99. Condensation takes place by heat transfer to river water. The river water mass flow of $\dot{m}_{\text{W}} = 11865 \frac{\text{kg}}{\text{s}}$ with an average specific heat capacity of $c_{\text{W}} = 4.185 \frac{\text{kJ}}{\text{kg K}}$ is heated in the condenser from $\vartheta_{env} = \vartheta_{\text{w,in}} = 10\,°$C to $\vartheta_{\text{w,out}} = 20\,°$C. The river water shall be treated as an incompressible fluid, cf. Fig. 7.4. The following assumptions shall be made:

- Pressure losses are negligible
- Kinetic and potential energies may be neglected
- The system components have no heat losses to the environment

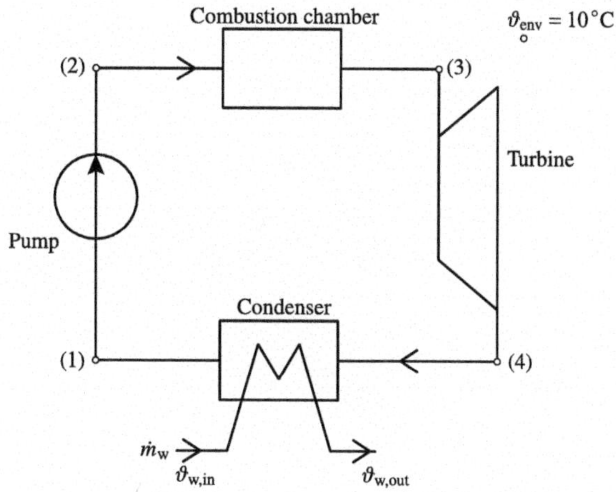

Fig. 7.4 Steam power process

(a) Determine the specific enthalpy of the state behind the turbine. To what temperature must the vapour be superheated to obtain a turbine power of 200 MW?
(b) Determine the temperature at the inlet to the feedwater pump and the specific enthalpy at its outlet.
(c) What is the thermal efficiency of the process? Name 3 measures by which it could be increased.
(d) What is the flux of exergy loss $\Delta \dot{E}_{x,V}$ in the condenser?

7.2 Solutions

7.1 **(b)**
Analogous to the ideal gas, the pressure of a vapour/steam is *reduced* by a throttle. Answer (a) is therefore ruled out. To understand what causes the pressure loss in the throttle,cf. Fig. 7.5, the part energy equation is stated, i.e.

$$w_{t,12} = 0 = \int_1^2 v\,dp + \psi_{12} + \underbrace{\Delta e_{a,12}}_{=0} \tag{7.1}$$

so that

$$\boxed{\int_1^2 v\,dp = -\psi_{12} \leq 0} \tag{7.2}$$

This means that energy must be dissipated within the throttle to reduce the pressure. If the throttle worked reversibly, the change of state would be isobaric under the given premises. The first law of thermodynamics, see Fig. 7.5a, obeys

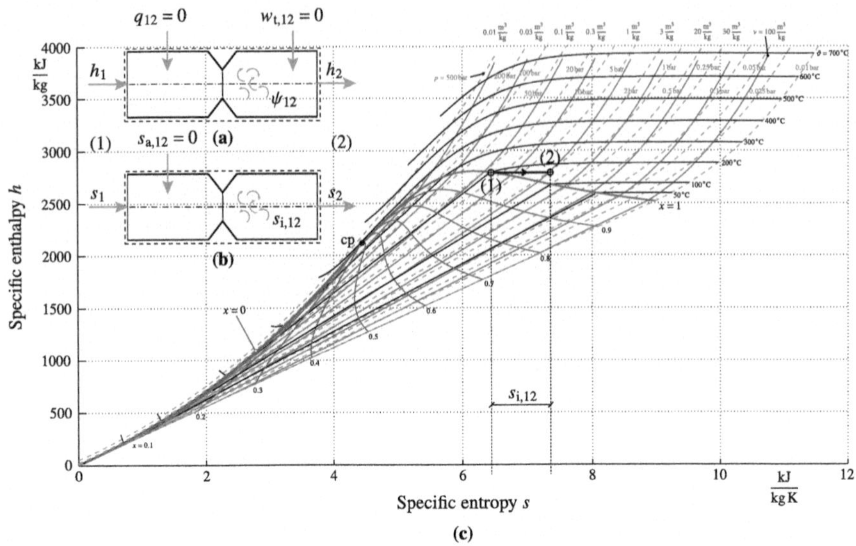

Fig. 7.5 Adiabatic throttling of water (saturated steam)

$$h_1 + \underbrace{w_{t,12} + q_{12}}_{=0} = h_2 \tag{7.3}$$

so that

$$\boxed{h_1 = h_2} \tag{7.4}$$

The change of state is therefore isenthalp. Next, the entropy balance, cf. Fig. 7.5b, follows

$$s_1 + s_{i,12} + \underbrace{s_{a,12}}_{=0} = s_2 \tag{7.5}$$

so that

$$\boxed{s_2 - s_1 = s_{i,12} = \int_1^2 \frac{\delta \psi}{T} = \frac{\psi_{12}}{T_m} \geq 0} \tag{7.6}$$

Due to the required dissipation, the specific entropy must therefore increase—answer (b) is correct.

So far, we have not placed any constraints on the fluid in any balance equation, i.e. no equation of state has been applied. Thus, the equations given are valid for ideal gases and real fluids, i.e. answer (c) does not apply.

Figure 7.5c shows an example of the throttling of saturated steam. State (1) must lie on the isoline $x = 1$. According to the entropy balance, cf. Eq. 7.6, the specific entropy rises, so that state (2) is to the right of state (1). The more specific dissipation occurs within the throttle, the further state (2) moves to the right. Equation 7.4 has shown that the specific enthalpy remains constant, so the change of state (1) to (2) is horizontal in the h, s-diagram. The h, s-diagram shows how the pressure decreases with the described change of state. From the isotherms in the state diagram, it can also be seen how the temperature for saturated steam *decreases* with isenthalp throttling.[1] Finally, state (2) is superheated vapour since it is to the right of the $x = 1$ isoline.

The major deviation from the adiabatic throttling of an ideal gas can be seen in the temperature behavior. For an ideal gas, the caloric equation of state can be applied for the energy balance, see Eq. 7.4, i.e.

$$h_2 - h_1 = c_p (T_2 - T_1) = 0 \tag{7.7}$$

so that the change of state runs isothermally with $T_1 = T_2$.

7.2 **(a)**
The basic design of the compression heat pump is shown in Fig. 7.6a. Figure 7.6b depicts the corresponding changes of state in a $\log p, h$-diagram, respectively. In this example, vapour from state (1) is adiabatically compressed to a high pressure p_{high}. Thus, the first law of thermodynamics[2] for the compressor follows

$$h_1 + w_t + \underbrace{q_{12}}_{=0} = h_2 \tag{7.8}$$

so that

$$h_2 - h_1 = w_t > 0. \tag{7.9}$$

Hence, the required specific compressor work can be represented as a horizontal distance in the $\log p, h$-diagram. Likewise, the higher pressure level compared to state (1) can be seen in the diagram.

As the specific enthalpy increases from (1) to (2), so does the temperature of the fluid, which can now release heat at the high pressure level. This takes place in the condenser from (2) to (3). In the chosen representation, cf. Fig. 7.6b, the heat release should be isobaric, so that the change of state is represented by a horizontal line. The energy balance is

$$h_2 + \underbrace{w_{t,23}}_{=0} + q_{23} = h_3 \tag{7.10}$$

so that

$$h_3 - h_2 = q_{23} < 0 \quad \text{and} \quad q_{\text{out}} = |q_{23}| \,. \tag{7.11}$$

Answer (a) is correct, the specific heat exchanged in the condenser is represented by the distance A, see Fig. 7.6b.

---

[1] In the case shown, saturated steam with $\vartheta_1 = 200\,°\text{C}$ is adiabatically throttled to about 2 bar. The temperature in state (2) is then approximately 161.2 °C.

[2] Outer energies shall be disregarded.

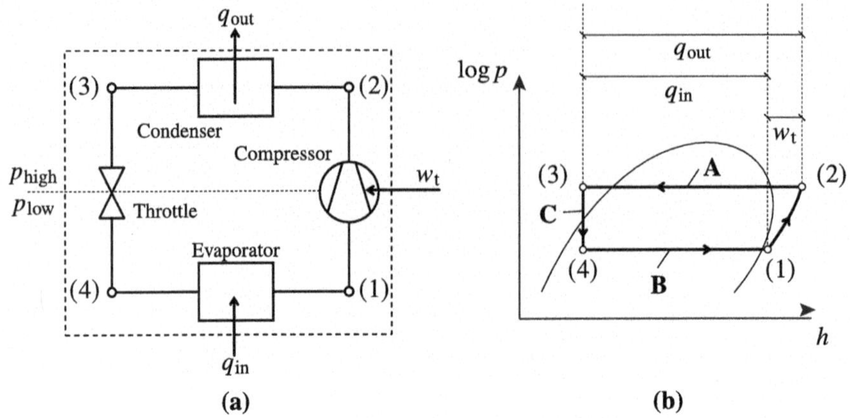

Fig. 7.6 Compression heat pump

In order to run through a cyclic process, the fluid, which is a supercooled liquid in state (3) after heat release in the selected example, must be expanded to the lower pressure level. This is done from (3) to (4) in an adiabatic throttle. Neglecting the outer energies, the throttling is isenthalp, see Problem 7.1, i.e. the change of state runs vertically down in the log p, h-diagram until p_{low} is reached.

To close the cyclic process, heat is now supplied[3] from (4) to (1) in the evaporator. The first law of thermodynamics obeys

$$h_4 + \underbrace{w_{t,41}}_{=0} + q_{41} = h_1 \tag{7.12}$$

so that

$$h_1 - h_4 = q_{41} > 0 \quad \text{and} \quad q_{in} = q_{41}. \tag{7.13}$$

In the example, the heat supply should be isobar, so that the change of state follows a horizontal straight line. The required specific heat can thus be represented as a horizontal distance in the log p, h-diagram.

By grabbing the process values in the log p, h-diagram, the first law of thermodynamics for the overall system, see Fig. 7.6a,

$$\boxed{\underbrace{q_{in} + w_t}_{\text{IN}} = \underbrace{q_{out}}_{\text{OUT}}} \tag{7.14}$$

can easily be visualised.

In order for the heat pump to release as much heat as possible, the supplied specific heat q_{in} at the change of state (4) after (1) should be maximised. Therefore,

[3] In the chosen example, the supply of heat should be isobaric.

it is advantageous to *vaporise* a fluid, since the phase change is associated with a high thermal energy demand, expressed by the specific enthalpy of vaporisation Δh_v, i.e. the energy density is high. For the heat pump to be able to absorb heat at a low temperature level, the fluid in the evaporator must have a lower temperature than the environment, i.e. the vaporisation must take place at a very low temperature level. As can be seen from the steam table, cf. Appendix A, for example, the pressure is set very low for this purpose, so that vaporisation can take place at low temperatures.

7.3 **(a)**

A steam power plant is a clockwise cycle, i.e. a thermal engine. Heat is supplied at a high temperature level, partially converted into mechanical/electrical power, and the unconverted part is released to the environment as heat at a low temperature level. The second law of thermodynamics has shown that even in the best case, i.e. in which all components operate reversibly, it is *impossible* to convert the supplied heat *completely* into power. Part of the supplied heat *must* be released, see Problem 5.16. In the best case, a thermal engine operates as a Carnot engine, i.e. the achievable thermal efficiency becomes maximum, i.e.

$$\eta_{th} = 1 - \frac{T_{min}}{T_{max}}. \tag{7.15}$$

According to Carnot, heat should be supplied to a thermal engine at the highest possible temperature T_{max}. The required heat output, on the other hand, should be at a low temperature level T_{min}. The lower temperature level cannot be influenced and is usually imposed by the environment. Technical efforts are being made to increase the maximum temperature of the supply of heat. Answer (b) does not apply.

A steam power plant with intermediate superheating is shown schematically in Fig. 7.7a. After the supply of heat from (2) to (5), the fluid in the high-pressure turbine is not completely expanded to the lowest pressure level p_{LP}, but only to an intermediate pressure p_{MP}. After leaving the high-pressure turbine, heat is supplied to the working fluid again, this time at a high temperature level, until the maximum permissible temperature is reached. Now the fluid is expanded to the lower process pressure in state (8) by a low-pressure turbine. Subsequently, state (1) is reached by heat release[4] at low pressure or temperature level, respectively. A pump increases the pressure to state (2) and the supply of heat takes place as described. The supply of heat $q_{in,1}$ takes place at a comparatively low average temperature because of the low temperature of the supercooled liquid in state (1). This contradicts Carnot's demand to supply heat at the highest possible temperature. The second supply of heat $q_{in,2}$ in the re-heater, however, takes place at a high temperature—quite in the spirit of Carnot: If $q_{in,1}$ and $q_{in,2}$ are combined, the average temperature of the supplied heat is raised. According to Carnot, the efficiency must therefore increase. This measure is called *carnotisation*. Answer (a) is correct.

The highest efficiency for a thermal engine is achieved when the changes of state are reversible, i.e. without dissipation. The smaller the isentropic turbine efficiency,

[4] Note that according to the second law of thermodynamics, heat must be released.

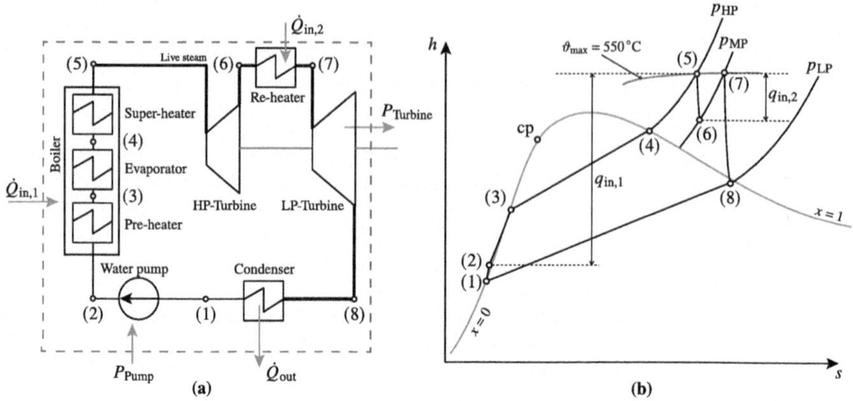

Fig. 7.7 Power plant with intermediate superheating: **a** Layout **b** h, s-diagram

the more energy is dissipated and the efficiency of the overall system decreases. Answer (c) is therefore not applicable.

7.4 **(b)**

The first law of thermodynamics in steady state for an adiabatic throttle with neglect of the outer energies is

$$h_1 + \underbrace{w_{t,12} + q_{12}}_{=0} = h_2 \tag{7.16}$$

that means, the throttle is isenthalp with $h_2 - h_2 = 0$. In contrast to the ideal gas, for which the specific enthalpy is only temperature-dependent, it also depends on the pressure for a real fluid, i.e.

$$dh = c_p\, dT + \underbrace{\left(\frac{\partial h}{\partial p}\right)_T}_{=\delta_T} dp. \tag{7.17}$$

The first law of thermodynamics has shown that the adiabatic throttle isenthalp, i.e. $dh = 0$. It follows that

$$0 = c_p\, dT + \delta_T dp. \tag{7.18}$$

The change of temperature with pressure can be determined with the gradient $\frac{dT}{dp}$ that follows

$$\boxed{\left(\frac{dT}{dp}\right)_h = -\frac{\delta_T}{c_p} = \delta_h} \tag{7.19}$$

According to this equation, a negative Joule–Thomson coefficient δ_h means that with adiabatic throttling, i.e. $\mathrm{d}p < 0$,[5] the temperature must rise, i.e. $\mathrm{d}T > 0$. Consequently, answer (b) is correct.

Problem 7.1 has shown that dissipation in an adiabatic throttle is necessary to reduce the pressure when the outer energies are neglected. Answer (c) is therefore incorrect.

7.5 **(c)**
The answer can be checked, for example, with Fig. 7.5. For superheated vapour at low pressures and high temperatures, the isotherms run horizontally. This is the range in which the strongly superheated vapour behaves like an ideal gas, i.e. the specific enthalpy is only a function of temperature. Because of the horizontal course, pressure changes in this region have no influence on the specific enthalpy.

Answer (a) applies to the wet steam region. There, the degree of freedom is $F = 1$, i.e. pressure and temperature depend on each other, only one variable can be freely chosen.

7.6 **(c)**
In Problem 7.1 we have already solved the task graphically and shown that the temperature decreases. Now we want to determine the solution for this problem mathematically. With the help of the steam table, cf. Appendix A, one obtains for state (1)

- Temperature: $\vartheta_1 = 200\,°\mathrm{C}$
- Specific enthalpy: $h_1 = h''\,(200\,°\mathrm{C}) = 2.7921 \times 10^3\,\frac{\mathrm{kJ}}{\mathrm{kg}}$
- Pressure: $p_1 = 15.5467\,\mathrm{bar}$

As has already been shown, cf. Problem 7.1, the throttling under the given circumstances is isenthalp, i.e.

$$h_2 = h_1 = 2.7921 \times 10^3\,\frac{\mathrm{kJ}}{\mathrm{kg}} \quad \text{and} \quad p_2 = 1\,\mathrm{bar}. \tag{7.20}$$

One can see that

$$h_2 > h''\,(1\,\mathrm{bar}) = 2.6749 \times 10^3\,\frac{\mathrm{kJ}}{\mathrm{kg}}. \tag{7.21}$$

Consequently, state (2) is in superheated state, the degree of freedom is $F = 2$, i.e. pressure and temperature are independent from each other. With the help of the steam table follows

$$\vartheta_2 = \vartheta\,(h_2,\ p_2) = 157.8020\,°\mathrm{C}. \tag{7.22}$$

[5] Hence, answer (a) is wrong. A throttle is utilised to *reduce* the pressure.

7.7 **(c)**
According to the task, the change of state should run from (′) to (″). If the water were vaporised isobarically in a *closed* system, the first law of thermodynamics would be

$$\delta q + \delta w = \delta u \tag{7.23}$$

with the partial energy equation

$$\delta w = \delta w_{\mathrm{V}} = -p\,\mathrm{d}v. \tag{7.24}$$

Hence, it yields for an isobaric, i.e. $\mathrm{d}p = 0$, change of state[6]

$$\boxed{\delta q = \delta u + p\,\mathrm{d}v = \mathrm{d}h} \tag{7.25}$$

In the case that isobaric vaporisation is carried out in an *open* system, this results in

$$\delta q + \delta w_{\mathrm{t}} = \delta h \tag{7.26}$$

with the partial energy equation

$$\delta w_{\mathrm{t}} = \delta y = v\,\mathrm{d}p = 0. \tag{7.27}$$

For an isobaric, i.e. $\mathrm{d}p = 0$, change of state one thus obtains the same result as for an open system, i.e.

$$\boxed{\delta q = \mathrm{d}h} \tag{7.28}$$

By integration, one obtains the total specific heat for the entire evaporation[7] process:

$$q_{12} = \int_{(')}^{('')} \delta q = \int_{(')}^{('')} \mathrm{d}h = h''\,(200\,^\circ\mathrm{C}) - h'\,(200\,^\circ\mathrm{C}) = \Delta h_{\mathrm{V}}\,(200\,^\circ\mathrm{C})\,. \tag{7.29}$$

From the steam table one gets

$$h''\,(200\,^\circ\mathrm{C}) = 2.7921 \times 10^3\,\frac{\mathrm{kJ}}{\mathrm{kg}} \tag{7.30}$$

and

$$h'\,(200\,^\circ\mathrm{C}) = 852.3931\,\frac{\mathrm{kJ}}{\mathrm{kg}} \tag{7.31}$$

[6] Note that $\mathrm{d}h = \mathrm{d}u + p\,\mathrm{d}v + v\,\mathrm{d}p$.

[7] Mind that the process is also isothermal.

Finally, the specific heat yields

$$q_{12} = h''\,(200\,°\text{C}) - h'\,(200\,°\text{C}) = 1.9397 \times 10^3\,\frac{\text{kJ}}{\text{kg}}. \tag{7.32}$$

Answer (c) is correct.

Alternatively, it applies

$$\text{d}s = \frac{\delta q}{T} + \underbrace{\frac{\delta \psi}{T}}_{=0}. \tag{7.33}$$

Since an isobaric vaporisation is also isothermal, the transformation and integration yields

$$q_{12} = \int_{(')}^{('')} T\,\text{d}s = T\left(s'' - s'\right) \tag{7.34}$$

with $T = 473.15\,\text{K}$ and

$$s'' = 6.4303\,\frac{\text{kJ}}{\text{kg K}} \quad \text{resp.} \quad s' = 2.3308\,\frac{\text{kJ}}{\text{kg K}}. \tag{7.35}$$

7.8 **(d)**
For a detailed explanation, please refer to the solution to Problems 7.1 and 7.6.

7.9 **(a)** During vaporisation, the temperature should remain constant so that the pressure also remains constant, since the degree of freedom in the wet steam region is $F = 1$. In general, the first law of thermodynamics is

$$\delta q + \delta w = \text{d}u + \text{d}e_\text{a} \tag{7.36}$$

with the partial energy equation

$$\delta w = -p\,\text{d}v + \delta \psi + \text{d}e_\text{a}. \tag{7.37}$$

Hence, it is under isobaric conditions

$$\delta q + \delta \psi = \text{d}u + p\,\text{d}v = \text{d}h. \tag{7.38}$$

It further applies

$$\text{d}s = \frac{\delta q}{T} + \frac{\delta \psi}{T} \tag{7.39}$$

so that

$$T\,\text{d}s = \delta q + \delta \psi. \tag{7.40}$$

If Eqs. 7.38 and 7.40 are combined, one obtains[8]

$$dh = T\,ds. \tag{7.41}$$

Integration leads to

$$\int_{(')}^{('')} dh = h'' - h' = \Delta h_V = \int_{(')}^{('')} T\,ds = T\left(s'' - s'\right). \tag{7.42}$$

Finally, the specific enthalpy of vaporisation is

$$\Delta h_V = T\left(s'' - s'\right) = 181.1081\,\frac{\text{kJ}}{\text{kg}}. \tag{7.43}$$

Answer (a) is correct.

7.10 **(b)**
In Problem 7.4 it has been shown in detail that the adiabatic throttle, i.e. $dp < 0$, is isenthalp, i.e. $dh = 0$. Hence, for a real fluid it yields

$$dh = 0 = c_p\,dT + \delta_T dp. \tag{7.44}$$

The change of temperature with pressure can be determined with the gradient $\frac{dT}{dp}$ that follows

$$\left(\frac{dT}{dp}\right)_h = -\frac{\delta_T}{c_p} = \delta_h \tag{7.45}$$

so that the direction of change of temperature depends on the Joule–Thomson coefficient δ_h:

- Case 1: Temperature rises, i.e. $dT > 0$, for $dp < 0$

$$\left(\frac{dT}{dp}\right)_h = \delta_h < 0. \tag{7.46}$$

- Case 2: Temperature drops, i.e. $dT < 0$, for $dp < 0$

$$\left(\frac{dT}{dp}\right)_h = \delta_h > 0. \tag{7.47}$$

- Case 3: Temperature remains constant, i.e. $dT = 0$, for $dp < 0$

$$\left(\frac{dT}{dp}\right)_h = \delta_h = 0. \tag{7.48}$$

[8] The same results from the fundamental equation of thermodynamics $dh = T\,ds + v\,dp$ for $dp = 0$.

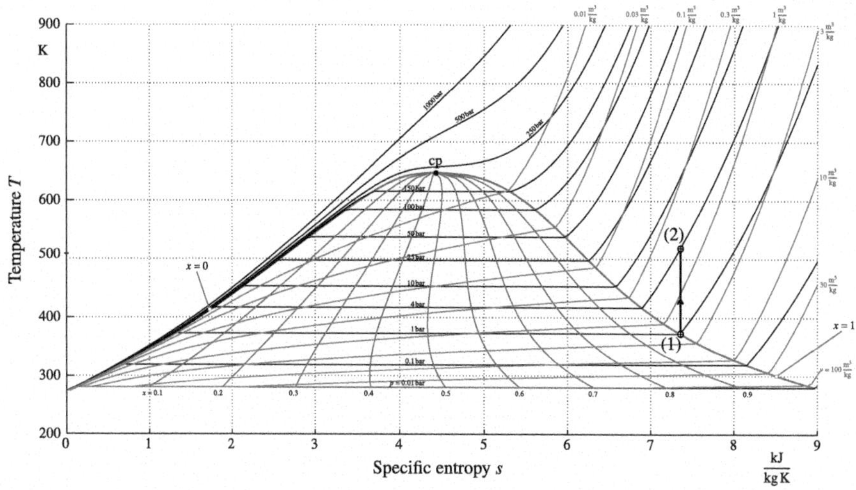

Fig. 7.8 Isentropic compression of water (saturated steam)

The Joule–Thomson coefficient is fluid-specific, so that answer (b) is correct. A throttle requires dissipation to reduce the pressure. Therefore, answer (a) does not make sense. Entropy has nothing to do with the first law of thermodynamics, yet entropy must increase because of the second law of thermodynamics. Answer (c) is not correct as well.

7.11 **(b)**

The change of state is shown in the T, s-diagram, see Fig. 7.8. State (1) lies on the intersection of the $x = 1$ isoline and the isobar $p = 1$ bar. State (2) must lie on a vertical line through state (1) because of the isentropic change of state.[9] In addition, state (2) lies on the $p = 4$ bar isobar. This means state (2) is in the superheated vapour region, i.e. answer (b) is correct. According to the T, s-diagram, the temperature has obviously risen, answer (a) is wrong. The temperature after compression can be determined with the help of the steam table. One obtains for state (1)

- Pressure: $p_1 = 1\,\text{bar}$
- Temperature: $\vartheta_1 = \vartheta_s\,(1\,\text{bar}) = 99.6059\,°\text{C}$
- Specific entropy: $s_1 = s''\,(1\,\text{bar}) = 7.3588\,\frac{\text{kJ}}{\text{kg K}}$

The change of state is isentropic, i.e.

$$s_2 = s_1 = 7.3588\,\frac{\text{kJ}}{\text{kg K}} \quad \text{and} \quad p_2 = 4\,\text{bar}. \tag{7.49}$$

[9] Due to the isentropic change of state, answer (c) does not make sense.

One can see that

$$s_2 > s''\,(4\,\text{bar}) = 6.8954\,\frac{\text{kJ}}{\text{kg K}}. \tag{7.50}$$

Consequently, state (2) is in superheated state, the degree of freedom is $F = 2$, i.e. pressure and temperature are independent from each other. With the help of the steam table follows

$$\vartheta_2 = \vartheta\,(s_2,\ p_2) = 244.5151\,°\text{C} \quad \text{resp.} \quad T_2 = 517.6651\,\text{K}. \tag{7.51}$$

7.12 **(c)**
It has been shown in Problem 7.9 that

$$\delta q + \delta\psi = \text{d}u + p\,\text{d}v = \text{d}h. \tag{7.52}$$

Since the change of state is reversible,[10] it yields

$$\delta q = \text{d}h \tag{7.53}$$

and by integration

$$q_{12} = h_2 - h_1. \tag{7.54}$$

It follows from the steam table that

$$h_1 = h''\,(150\,°\text{C}) = 2.7459 \times 10^3\,\frac{\text{kJ}}{\text{kg}} \quad \text{and} \quad h_2 = h'\,(150\,°\text{C}) = 632.2516\,\frac{\text{kJ}}{\text{kg}} \tag{7.55}$$

Finally, the specific heat reads as

$$q_{12} = h_2 - h_1 = -2.1137 \times 10^3\,\frac{\text{kJ}}{\text{kg}}. \tag{7.56}$$

Answer (c) is correct.

7.13 **(a)**
Figure 7.9 represents the $p,\ v,\ T$-space of a real fluid that includes the critical point (cp).

- Region 1 represents the solid phase.
- Liquid state is shown in region 2. However, liquid state and solid state are linked by region 6, i.e. the melting region.
- Region 3 shows the gas phase.
- Region 4 in the diagram represents the wet steam region. Liquid and vapour occur simultaneously in this region.
- Region 5 is the sublimation region, i.e. solid and vapour exist at the same time.

[10] The change of state takes place very slowly.

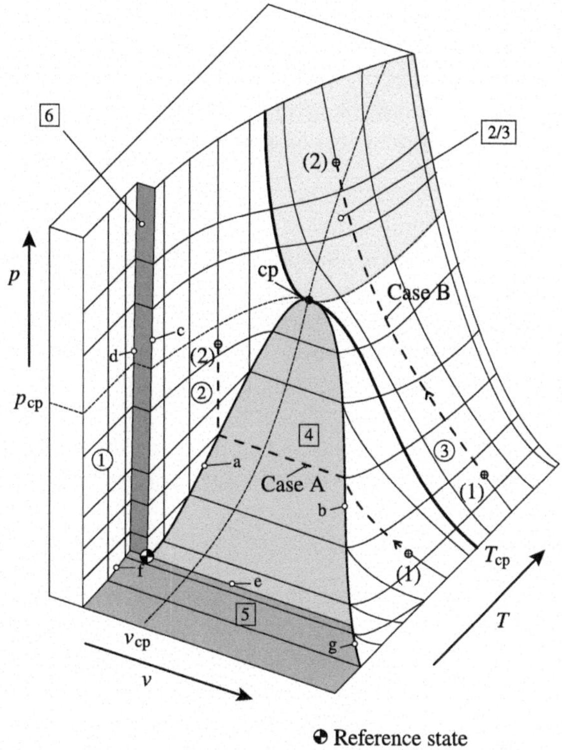

Fig. 7.9 p, v, T-space of a real fluid

- Region 2/3 represents the so-called supercritical gas/liquid region, i.e. $T > T_{cp}$ and $p > p_{cp}$. It is not possible to distinguish between liquid and gaseous states: the viscosity corresponds to that of a gas, the density to that of a liquid.

As can be seen from the Fig. 7.9, not all three states of aggregation exist at the critical point. Answer (b) is therefore wrong. The place where the three states of aggregation solid, liquid and gaseous exist is marked by curve e in Fig. 7.9. This curve represents the triple point line connecting sublimation and wet steam region.

The specific enthalpy of vaporisation, cf. Eq. 7.42 corresponds to

$$\Delta h_v = h'' - h' = T\left(s'' - s'\right) \quad (7.57)$$

where s'' is the specific entropy of the saturated steam, i.e. at $x = 1$, and s' is the specific entropy at boiling point, i.e. $x = 0$. As can be seen in the T, s-diagram, cf. Fig. 7.8, the horizontal distance between the $x = 0$ and $x = 1$ line, i.e. the specific enthalpy of vaporisation Δh_v, decreases towards the critical point. At the critical point it becomes minimal. Answer (c) is therefore wrong.

In the following, it will be investigated whether it is possible to liquefy a saturated steam at a constant temperature, cf. Fig. 7.9. State (1) is in the superheated vapour range. The change of state should be isothermal. Now the pressure is slowly increased, it can be seen how the specific volume decreases during the change of state due to the pressure increase at constant temperature. Two cases are examined.

- Case A: Temperature below critical temperature, i.e. $T < T_{cp}$
  The $x = 1$ boiling line is reached. Heat release causes the phase change to take place at constant pressure and constant temperature, respectively. The specific volume decreases until the boiling state is reached. By further isothermal pressure increase, finally state (2), i.e. supercooled liquid, is reached. Obviously, the liquefaction of the vapour has been achieved by increasing the pressure.
- Case B: Temperature above critical temperature, i.e. $T > T_{cp}$
  In this case, the wet steam region is not reached. Although the specific volume decreases with an increase in pressure, the fluid does not reach the liquid state. Liquefaction by pressure of a vapour above the critical temperature is not possible. Answer (a) is therefore correct.

7.14 **(b)**
The adiabatic throttle has already been dealt with several times, compare e.g. Problem 7.1. Basically, a throttle is used to reduce the pressure of a fluid. Therefore, answer (b) is correct. As already shown, the pressure reduction takes place by dissipation within the throttle. This is associated with the generation of entropy. However, since the throttle is adiabatic and thus no entropy can be dissipated with the heat to the environment, this entropy generation leads to an entropy increase of the fluid. Answer (a) is therefore incorrect. The throttle is a work-isolated, open system, so no technical work is exchanged with the environment. Answer (c) is therefore wrong as well.

7.15 **(a)**
According to Problem 7.10 it is

$$\left(\frac{\mathrm{d}T}{\mathrm{d}p}\right)_h = -\frac{\delta_T}{c_p} = \delta_h \tag{7.58}$$

It has been shown that a negative Joule–Thomson coefficient, i.e. Case 1, leads to a temperature rise when the fluid is throttled. Therefore, answer (a) is correct.

7.16 **(a)**
The partial energy equation[11] reads as

$$w_t = 0 = \int_1^2 v\,\mathrm{d}p + \psi_{12} + \underbrace{\Delta e_{a,12}}_{=0} \tag{7.59}$$

[11] Note that the heat exchanger is work-insulated.

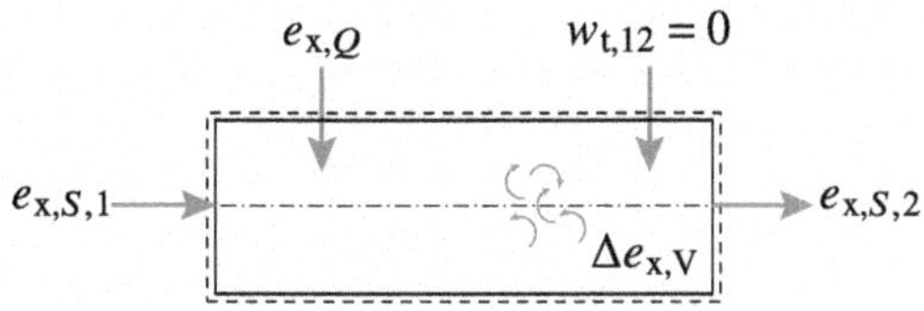

Fig. 7.10 Heat exchanger: exergy balance

respectively

$$\int_1^2 v\,dp = -\psi_{12} \leq 0. \tag{7.60}$$

As the pressure decreases, i.e. $dp < 0$, dissipation must occur in the heat exchanger. Answer (a) is correct.

The exergy balance for the heat exchanger follows according to Fig. 7.10. In steady state, the flux of exergy into the system must be balanced by the flux out of the system.

$$\underbrace{\dot{m}e_{x,S,1} + \dot{m}e_{x,Q}}_{\text{IN}} = \underbrace{\dot{m}e_{x,S,2} + \dot{m}\,\Delta e_{x,V}}_{\text{OUT}} \tag{7.61}$$

respectively

$$e_{x,S,2} - e_{x,S,1} = e_{x,Q} - \Delta e_{x,V}. \tag{7.62}$$

It further applies

$$\Delta e_{x,V} = T_{env} s_i = T_{env} \frac{\psi_{12}}{T_m} > 0 \tag{7.63}$$

and

$$e_{x,Q} = q \underbrace{\left(1 - \frac{T_{env}}{T_m}\right)}_{>0} < 0 \quad \text{with} \quad q < 0 \tag{7.64}$$

Hence,[12] the exergy decrease from (1) to (2) since

$$\Delta e_x = e_{x,S,2} - e_{x,S,1} = \underbrace{e_{x,Q}}_{<0} - \underbrace{\Delta e_{x,V}}_{>0} < 0. \tag{7.65}$$

The exergy decrease, so that answer (b) is wrong. The thermal equation of state can not be applied to a real fluid. According to the p, v, T-space, see Fig. 7.9, isobaric heat release of a vapour would be possible. Answer (c) does not make sense.

[12] Due to the condensation the fluid releases heat.

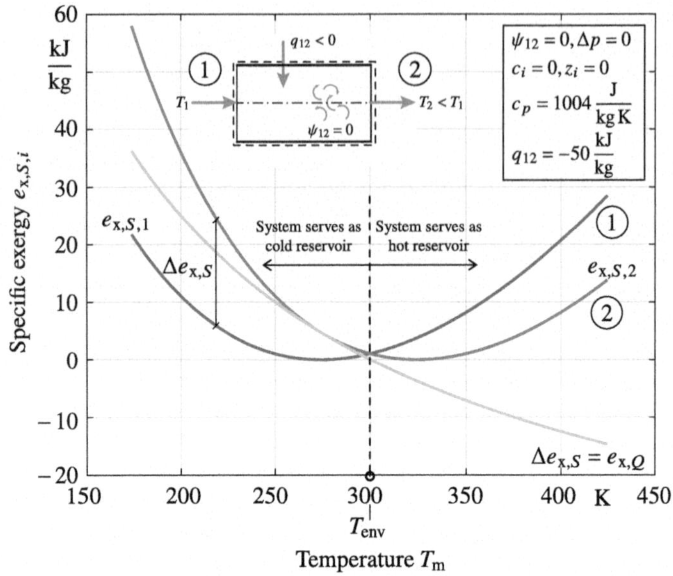

Fig. 7.11 Exergy of heat

Additional information:
In the following, the influence of the thermodynamic mean temperature on the exergetic balance is investigated. For the sake of simplicity, we will use an ideal gas that flows through the heat exchanger without pressure loss and cools down from T_1 to

$$T_2 = T_1 + \frac{q_{12}}{c_p} < T_1 \quad \text{with} \quad q_{12} < 0. \tag{7.66}$$

by releasing heat q_{12}, see Fig. 7.11. The thermodynamic mean temperature then obeys

$$T_m = \frac{T_2 - T_1}{\ln \frac{T_2}{T_1}} \tag{7.67}$$

Under the given conditions the exergy balance then is

$$e_{x,S,2} + e_{x,Q} = e_{x,S,1} \tag{7.68}$$

respectively

$$e_{x,S,2} - e_{x,S,1} = \Delta e_{x,S} = e_{x,Q} = \underbrace{q_{12}}_{<0} \left(1 - \frac{T_{env}}{T_m}\right) \tag{7.69}$$

Figure 7.11 shows the specific exergy of the entering flow (1) as well as of the leaving flow (2). According to Eq. 7.69, the difference between these two then represents the specific exergy of heat. The following cases can occur when the *heat*[13] is evaluated in terms of its working potential with respect to the environment, i.e. its exergy[14]:

- Case A: $T_m < T_{env}$
  In this case, the system is the cold reservoir while the environment serves as the hot reservoir, i.e. a reversible *heat pump* is used to shift the released heat of the system that has to be evaluated to the environment: The greater the temperature potential $T_{env} - T_m$ the heat pump has to overcome, the greater the power *supply* of the reversible heat pump. This positive power[15] equals the positive exergy $e_{x,Q}$. This is also expressed by Eq. 7.69:

$$e_{x,S,2} - e_{x,S,1} = \underbrace{q_{12}}_{<0} \underbrace{\left(1 - \frac{T_{env}}{T_m}\right)}_{<0} > 0 \tag{7.70}$$

  Heat and exergy have opposite signs. This case is represented by Fig. 7.12b, see also [1].
- Case B: $T_m = T_{env}$
  In this case, the heat has no exergy, so that the exergy of the inflow (1) is equal to the exergy of the outflow (2).
- Case C: $T_m > T_{env}$
  In this case the system is the hot reservoir, whereas the environment serves as cold reservoir, i.e. in order to determine the exergy $e_{x,Q}$ a reversible *thermal engine* is applied: The greater the temperature potential $T_m - T_{env}$ available to the thermal engine, the greater the power that can be *released*. This negative power[16] equals the exergy $e_{x,Q}$. This is also expressed by Eq. 7.69:

$$e_{x,S,2} - e_{x,S,1} = \underbrace{q_{12}}_{<0} \underbrace{\left(1 - \frac{T_{env}}{T_m}\right)}_{>0} < 0 \tag{7.71}$$

  Heat and exergy have the same signs. This case is represented by Fig. 7.12d, see also [1].

7.17 **(b)**
Graphs A and D show isothermal changes of state, i.e. these two solutions are wrong. An isochoric change of state is only shown in answer B, which means that graph C is also wrong. Since the fluid expands, it must be $p_2 < p_1$. The isobars in the wet steam region coincide with the horizontal isotherms in a T, s-diagram. Hence, graph B applies, as state (1) lies above state (2) within the wet steam region. Obviously, such an isochoric, reversible change of state can be achieved by cooling the system,

[13] And not the system!

[14] The exergy is evaluated by a reversible heat pump/thermal engine.

[15] Positive, because power is supplied.

[16] Negative, because power is released.

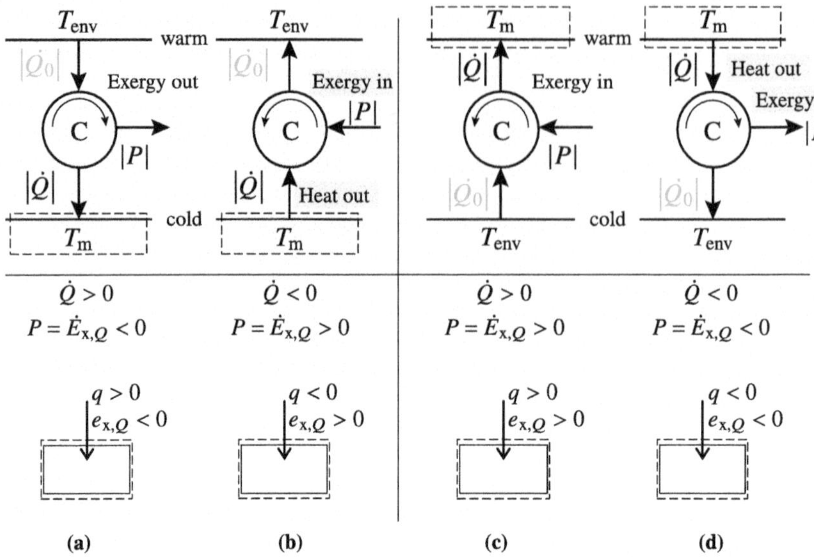

Fig. 7.12 Sign of exergy of heat

so that the specific entropy decreases. The specific heat required for this can be seen in diagram B as the area under the change of state. Answer (b) is correct.

7.18 **(c)**
The physical effect of efficiency increase by feedwater preheating is based on the so-called *carnotisation*, which was already described in Problem 7.3. The preheating is shown in Fig. 7.13. In a steam power process, raising the temperature of the feed water is thermodynamically unfavourable, since a lot of heat is supplied from outside at a low temperature level. According to Carnot, it makes more sense to supply the heat from the outside at high temperature. Therefore, the necessary feedwater preheating is done regeneratively, i.e. from the inside of the process: A partial steam flow is taken from the turbine and used thermally to preheat the feedwater. This increases the average temperature of the supplied heat from outside and, according to Carnot, improves the thermal efficiency, see Fig. 7.14. Hence, answer (c) is correct. Answers (a) and (b) do not make sense.

7.19 **(a)**
At a pressure of $p_{TP} = 0.006\,117$ bar and a temperature of $T_{TP} = 273.17$ K, water can exist in all three states of aggregation. Answer (a) is correct. One therefore speaks of the so-called triple point. If one marks the value pair p_{TP}, T_{TP} in the p, v, T-state space, see Fig. 7.9, the so-called triple point line results. On the left end of this line, the water is present as a solid, followed by liquid and finally, on the right end of the line, as vapour. Although pressure and temperature are the same along this line, the specific volume and the specific enthalpy/entropy differ, so answers (b) and (c) are wrong. For example, in order to vaporise the liquid, thermal energy must be

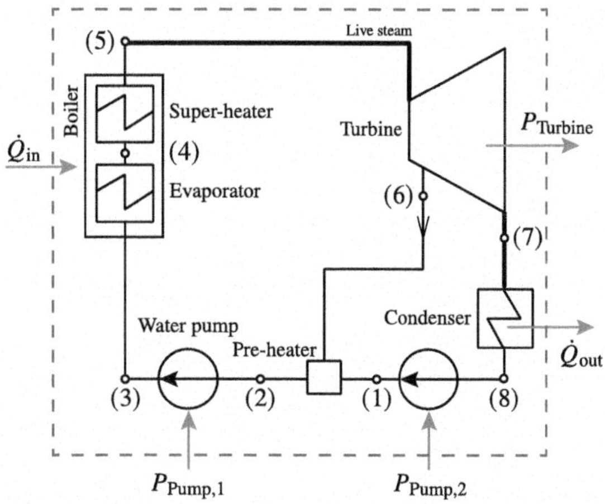

Fig. 7.13 Feedwater preheating

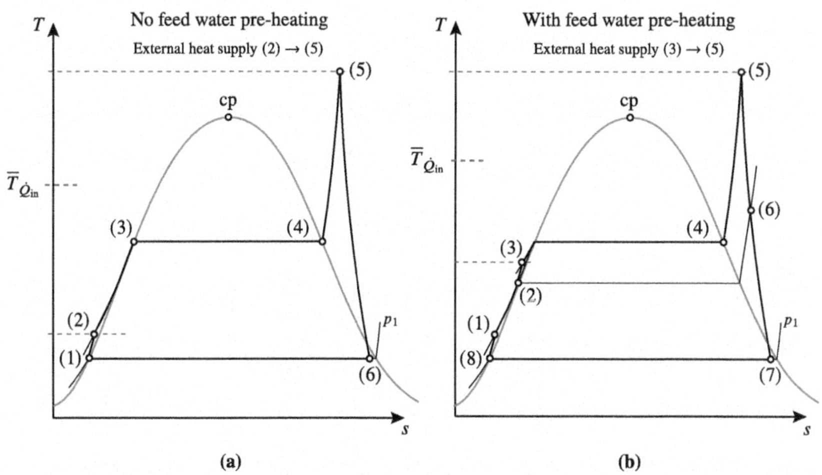

Fig. 7.14 Feedwater preheating in a T, s-diagram

supplied, so that both the specific enthalpy and the specific entropy increase. For the determination of the energetic reference, usually the boiling state of the liquid, i.e. (′) state, at p_{TP} is chosen.

7.20 The thermodynamic balances for the nozzle are sketched in Fig. 7.15. We start by determining the state values for the vapour. The thermal equation of state only applied to ideal gases, so the steam table for water, cf. Appendix A, is applied to solve this task. The specific volume yields

$$v_1 = v(p_1, \vartheta_1) = 0.1757 \, \frac{\text{m}^3}{\text{kg}}. \tag{7.72}$$

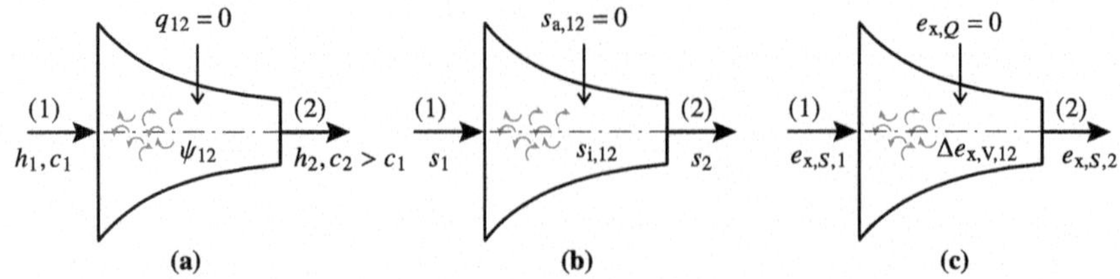

Fig. 7.15 Adiabatic nozzle: **a** Energy balance, **b** Entropy balance, **c** Exergy balance

To determine state (2), the polytropic correlation

$$pv^n = \text{const.} \rightarrow p_1 v_1^n = p_2 v_2^n = pv^n \tag{7.73}$$

is applied. This equation is universally valid regardless of the fluid. However, the equation

$$\frac{T_2}{T_1} = \left(\frac{p_2}{p_1}\right)^{\frac{n-1}{n}} \tag{7.74}$$

only applies to ideal gases, since Eq. 7.73 was transformed with the help of the thermal equation of state for ideal gases, i.e.

$$v_1 = \frac{RT_1}{p_1} \quad \text{and} \quad p_2 v_2 = \frac{RT_2}{p_2}. \tag{7.75}$$

Consequently, the specific volume in state (2) follows

$$v_2 = v_1 \left(\frac{p_1}{p_2}\right)^{\frac{1}{n}} = 0.2233\,\frac{\text{m}^3}{\text{kg}}. \tag{7.76}$$

To calculate the specific dissipation in the nozzle, the partial energy equation for open systems is applied, i.e.

$$w_t = 0 = \underbrace{\int_1^2 v\,\mathrm{d}p}_{=y_{12}} + \psi_{12} + \frac{1}{2}\left(c_2^2 - c_1^2\right) \tag{7.77}$$

Note that the nozzle is a work-insulated system, so no work is exchanged across the system boundary. The pressure work y_{12} can be determined if v is replaced as a function of p, i.e.

$$y_{12} = \int_1^2 v\,\mathrm{d}p \quad \text{with} \quad v = \left(\frac{p_1}{p}\right)^{\frac{1}{n}} v_1. \tag{7.78}$$

The[17] solution of the integral then is, cf. also Appendix E,

$$y_{12} = \int_1^2 \left(\frac{p_1}{p}\right)^{\frac{1}{n}} v_1 \, dp = \cdots = p_1 v_1 \frac{n}{n-1} \left[\left(\frac{v_1}{v_2}\right)^{n-1} - 1\right] = -98.6952 \, \frac{\mathrm{kJ}}{\mathrm{kg}} \tag{7.79}$$

Finally, the specific dissipation follows from Eq. 7.77 and obeys

$$\psi_{12} = \frac{1}{2}\left[c_1^2 - c_2^2\right] - y_{12} = 23.00 \, \frac{\mathrm{kJ}}{\mathrm{kg}}. \tag{7.80}$$

7.21 As already shown, compare Problem 7.1, the first law of thermodynamics is under the given boundary conditions

$$\underbrace{h_1 + \overbrace{q_{12} + w_{t,12} + \frac{1}{2}c_1^2 + g z_1}^{=0}}_{\text{IN}} = \underbrace{h_2 + \overbrace{\frac{1}{2}c_2^2 + g z_2}^{=0}}_{\text{OUT}}. \tag{7.81}$$

Hence, the change of state is isenthalp, i.e.

$$h_1 = h_2. \tag{7.82}$$

State (2) is fully thermodynamically given by temperature ϑ_2 and pressure p_2 and can be drawn in an h, s-diagram, see Fig. 7.16. It can be seen that the fluid is superheated vapour after leaving the throttle, since state (2) lies to the right of the $x = 1$ isoline. The specific enthalpy of state (2) can be taken from the diagram or the steam table, respectively, so that

$$h_2 = h\left(p_2, \vartheta_2\right) = 2726.7 \, \frac{\mathrm{kJ}}{\mathrm{kg}} = h_1. \tag{7.83}$$

The pressure p_1 and the specific enthalpy h_1 of state (1) are now known, i.e. the state is also thermodynamically determined. With these two indications, state (1) can also be drawn into the h, s-diagram, see Fig. 7.16. It can be clearly seen that the state is in the wet steam region, i.e. vapour and boiling liquid occur. Due to the entropy balance

$$\underbrace{s_1 + \overbrace{s_{a,12}}^{=0} + s_{i,12}}_{\text{IN}} = \underbrace{s_2}_{\text{OUT}} \tag{7.84}$$

the specific entropy must increase from (1) to (2), i.e.

$$s_2 - s_1 = s_{i,12} = \frac{\psi_{12}}{T_m} > 0. \tag{7.85}$$

[17] See Eq. 7.73.

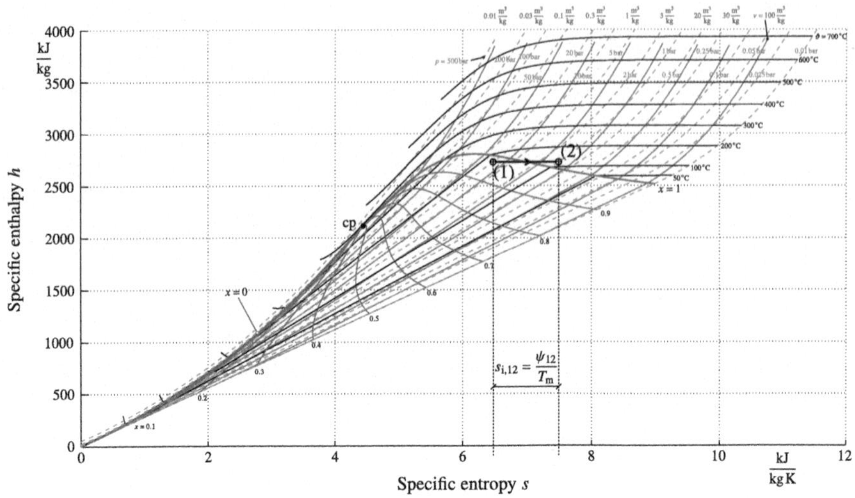

Fig. 7.16 Adiabatic throttling of water (saturated steam)

This is due to the entropy generation triggered by the required dissipation, which causes the pressure decrease. The specific enthalpy in state (1), i.e. in the wet steam region, is composed of the specific enthalpy of the boiling liquid (′) and the specific enthalpy of the saturated steam (″). In order to be able to specify the proportions, one introduces the vapour content x, i.e.

$$x = \frac{m_\mathrm{V}}{m_\mathrm{V} + m_\mathrm{L}}. \tag{7.86}$$

Since the enthalpy is an extensive state value and is composed additively of vapour and liquid, the following applies for state (1)

$$\left(m_{\mathrm{V},1} + m_{\mathrm{L},1}\right) h_1 = m_{\mathrm{V},1} h_1'' + m_{\mathrm{L},1} h_1' \tag{7.87}$$

so that

$$h_1 = x_1 h_1'' + (1 - x_1)\, h_1'. \tag{7.88}$$

From the steam table one obtains the specific enthalpies for (′) and (″) state, i.e.

$$h_1'(p_1) = 763.1814\,\frac{\mathrm{kJ}}{\mathrm{kg}} \quad \text{and} \quad h_1''(p_1) = 2.7772 \times 10^3\,\frac{\mathrm{kJ}}{\mathrm{kg}}. \tag{7.89}$$

Finally, the vapour content obeys

$$x_1 = \frac{h_1 - h_1'}{h_1'' - h_1'} = 0.9749. \tag{7.90}$$

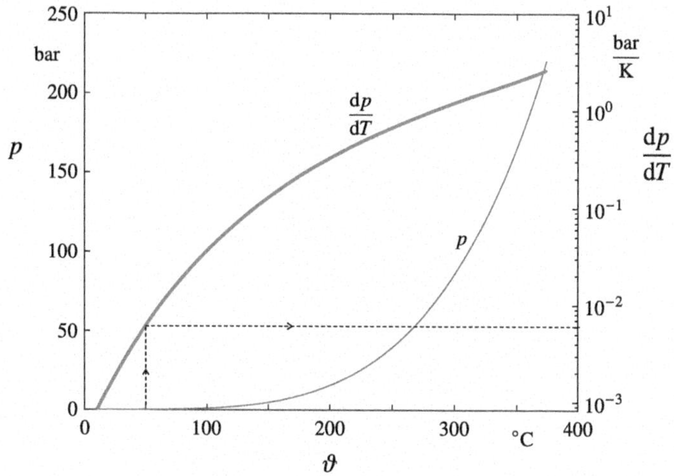

Fig. 7.17 Boiling point curve (water)

7.22 To solve this task, the Clausius-Clapeyron relation can be referred to, which is derived for example in [1]. It reads

$$\frac{\mathrm{d}p}{\mathrm{d}T} = \frac{s'' - s'}{v'' - v'} = \frac{\Delta h_{\mathrm{v}}}{T\,(v'' - v')} = \frac{h'' - h'}{T\,(v'' - v')}. \tag{7.91}$$

The thermodynamic state values in the (′) and (″) states respectively can be taken from the steam table:

$$h'\,(50\,°\mathrm{C}) = 209.3362\,\frac{\mathrm{kJ}}{\mathrm{kg}} \quad \text{and} \quad h''\,(50\,°\mathrm{C}) = 2.5913\times 10^3\,\frac{\mathrm{kJ}}{\mathrm{kg}} \tag{7.92}$$

and

$$v'\,(50\,°\mathrm{C}) = 0.0010\,\frac{\mathrm{m}^3}{\mathrm{kg}} \quad \text{and} \quad v''\,(50\,°\mathrm{C}) = 12.0279\,\frac{\mathrm{m}^3}{\mathrm{kg}} \tag{7.93}$$

Finally, the gradient of the boiling point curve at $T = 323.15\,\mathrm{K}$ is[18]

$$\frac{\mathrm{d}p}{\mathrm{d}T} = \frac{h'' - h'}{T\,(v'' - v')} = 612.8877\,\frac{\mathrm{Pa}}{\mathrm{K}} = 0.0061\,\frac{\mathrm{bar}}{\mathrm{K}}. \tag{7.94}$$

The result can also be determined graphically from Fig. 7.17.

[18] Note that the unit for the specific entaplie must be converted to $\frac{\mathrm{J}}{\mathrm{kg}}$.

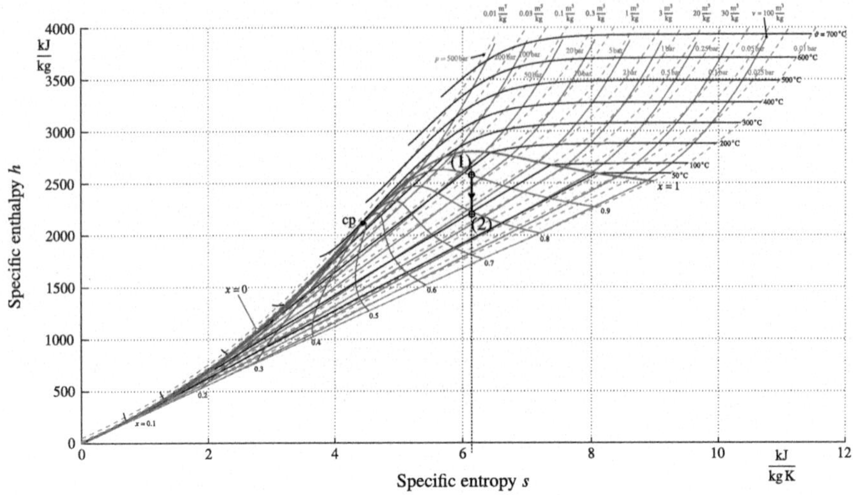

Fig. 7.18 Isentropic expansion of wet steam

7.23 For state (1), which is in the wet steam region, the following state values are specified

$$h_1 = 2575.6\,\frac{\text{kJ}}{\text{kg}} \quad \text{and} \quad s_1 = 6.1404\,\frac{\text{kJ}}{\text{kg K}}. \tag{7.95}$$

Obviously, this state can be clearly drawn in the h, s-diagram, see Fig. 7.18. State (2) has the same specific entropy as state (1) because of the isentropic change of state, i.e.

$$s_2 = s_1 = 6.1404\,\frac{\text{kJ}}{\text{kg K}}. \tag{7.96}$$

With the specific entropy s_2 and the pressure p_2, state (2) is unambiguously fixed, see Fig. 7.18. Obviously, state (2) is also in the wet steam region. The vapour content in state (2) yields

$$x_2 = \frac{s_2 - s_2'}{s_2'' - s_2'} = 0.7932 \tag{7.97}$$

with

$$s_2' = s'(p_2) = 1.2502\,\frac{\text{kJ}}{\text{kg K}} \quad \text{and} \quad s_2'' = s''(p_2) = 7.4150\,\frac{\text{kJ}}{\text{kg K}}. \tag{7.98}$$

Thus, the specific enthalpy in state (2) is as follows

$$h_2 = h_2' \cdot (1 - x_2) + h_2'' \cdot x_2 = 2198.4\,\frac{\text{kJ}}{\text{kg}} \tag{7.99}$$

with

$$h_2' = h'(p_2) = 398.0186\,\frac{\text{kJ}}{\text{kg}} \quad \text{and} \quad h_2'' = h''(p_2) = 2.6676 \times 10^3\,\frac{\text{kJ}}{\text{kg}}. \tag{7.100}$$

Finally, according to the first law of thermodynamics, it is

$$w_t + \underbrace{q_{12}}_{=0} = h_2 - h_1 = -377.24\,\frac{\text{kJ}}{\text{kg}} \tag{7.101}$$

The vapour content in state (1) obeys

$$x_1 = \frac{s_1 - s_1'(p_1)}{s_1''(p_1) - s_1'(p_1)} = \frac{h_1 - h_1'(p_1)}{h_1''(p_1) - h_1'(p_1)}. \tag{7.102}$$

In this equation, the pressure p_1 is the only unknown that can be determined in an interactive, numerical procedure using the steam table. The result is

$$p_1 = 9.9937\,\text{bar}. \tag{7.103}$$

7.24 The following state values result for state (1)

$$p_1 = 20\,\text{bar} \quad \text{and} \quad \vartheta_1 = 500\,^\circ\text{C}. \tag{7.104}$$

It is obviously superheated vapour, since according to the vapour table

$$\vartheta_1 > \vartheta_s(p_1) = 212.3845\,^\circ\text{C}. \tag{7.105}$$

State (1) can be unambiguously fixed in a h, s-diagram, see Fig. 7.19. From the steam table, respectively from the h, s-diagram in a somewhat less precise way, the specific volume of state (1) obeys

$$v_1 = v(p_1, T_1) = 0.1757\,\frac{\text{m}^3}{\text{kg}}. \tag{7.106}$$

From the polytropic relation

$$p_1 v_1^n = p_2 v_2^n = p v^n \tag{7.107}$$

one gets

$$v_2 = v_1 \left(\frac{p_1}{p_2}\right)^{\frac{1}{n}} = 0.2233\,\frac{\text{m}^3}{\text{kg}}. \tag{7.108}$$

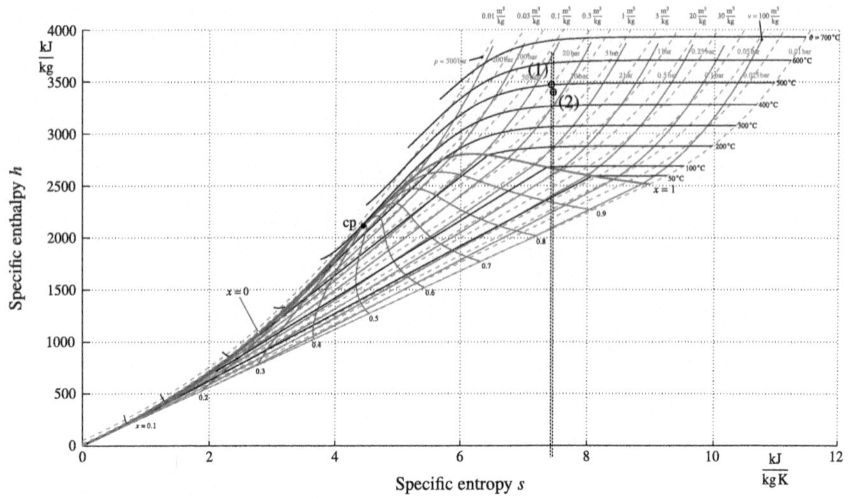

Fig. 7.19 Polytropic expansion of water vapour

With the pressure and the specific volume, state (2) can also be drawn in the h, s-diagram. The specific pressure work follows, cf. Appendeix E,

$$y_{12} = \int_1^2 v \, \mathrm{d}p = p_1 v_1 \frac{n}{n-1} \left[\left(\frac{v_1}{v_2} \right)^{n-1} - 1 \right] = -98.6952 \, \frac{\mathrm{kJ}}{\mathrm{kg}}. \tag{7.109}$$

The partial energy equation obeys

$$w_\mathrm{t} = 0 = y_{12} + \psi_{12} + \frac{1}{2} \left(c_2^2 - c_1^2 \right) \tag{7.110}$$

so that the specific dissipation follows

$$\psi_{12} = -y_{12} - \frac{1}{2} \left(c_2^2 - c_1^2 \right) = 23.0030 \, \frac{\mathrm{kJ}}{\mathrm{kg}}. \tag{7.111}$$

The thermodynamic mean temperature follows from a balance of entropy

$$\mathrm{d}s = \frac{\delta \psi}{T} + \underbrace{\frac{\delta q}{T}}_{=0} \tag{7.112}$$

so that

$$s_2 - s_1 = \frac{\psi_{12}}{T_\mathrm{m}}. \tag{7.113}$$

Finally, it applies

$$T_{\mathrm{m}} = \frac{\psi_{12}}{s_2 - s_1} = 754.7113\,\mathrm{K} \tag{7.114}$$

with

$$s_1 = s\,(p_1,\,T_1) = 7.4335\,\frac{\mathrm{kJ}}{\mathrm{kg\,K}} \quad \text{and} \quad s_2 = s\,(p_2,\,v_2) = 7.4640\,\frac{\mathrm{kJ}}{\mathrm{kg\,K}}. \tag{7.115}$$

Alternatively, the thermodynamic mean temperature can be calculated according to

$$T_{\mathrm{m}} = \frac{T_2 - T_1}{\ln \frac{T_2}{T_1}} = 754.3815\,\mathrm{K} \tag{7.116}$$

with the temperature T_2 taken from the steam table, i.e.

$$T_2 = T_2\,(p_2,\,v_2) = 735.9193\,\mathrm{K}. \tag{7.117}$$

The deviation between the two calculation methods, cf. Eqs. 7.114 and 7.116, for determining T_{m} is approx. 0.044 % and is due to deviations within the steam table.

7.25 The specific volume of the saturated steam in state (1) can be determined with the help of the steam table, i.e.

$$v_1 = v''\,(p_1) = 0.4624\,\frac{\mathrm{m}^3}{\mathrm{kg}}. \tag{7.118}$$

From the polytropic relation

$$p_1 v_1^n = p_2 v_2^n = p v^n \tag{7.119}$$

one gets

$$v_2 = v_1 \left(\frac{p_1}{p_2}\right)^{\frac{1}{n}} = 0.2867\,\frac{\mathrm{m}^3}{\mathrm{kg}}. \tag{7.120}$$

According to the Appendix E, the specific pressure work is

$$y_{12} = \int\limits_1^2 v\,\mathrm{d}p = n \cdot \frac{p_1 v_1}{n-1}\left[\left(\frac{v_1}{v_2}\right)^{n-1} - 1\right] = 1.4304 \times 10^5\,\frac{\mathrm{J}}{\mathrm{kg}}. \tag{7.121}$$

7.26 The entropy balance for the system shown in Fig. 7.20 is

$$\underbrace{\dot{m}s_1 + \dot{S}_{\mathrm{i}}}_{\mathrm{IN}} = \underbrace{\dot{m}s_4 + \dot{S}_{\mathrm{a}}}_{\mathrm{OUT}} \tag{7.122}$$

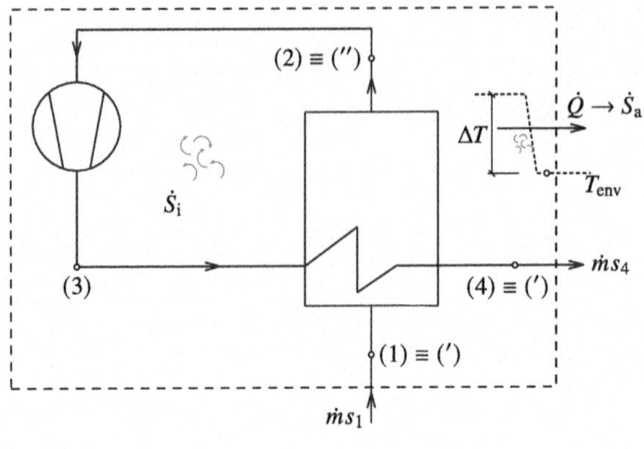

Fig. 7.20 Entropy balance

The system boundary is sufficiently large so that the heat is released at T_{env}.[19] The entropy flux released with the heat flux is therefore

$$\dot{S}_{\text{a}} = \frac{\dot{Q}}{T_{\text{env}}}. \tag{7.123}$$

The specific entropies of states (1) and (4) are determined with the steam table, i.e.

$$s_1 = s'(p_1) = 1.7766 \, \frac{\text{kJ}}{\text{kg K}} \tag{7.124}$$

and

$$s_4 = s'(p_4) = 2.0460 \, \frac{\text{kJ}}{\text{kg K}} \tag{7.125}$$

Finally, the loss of exergy obeys

$$\Delta \dot{E}_{\text{x,V}} = T_{\text{env}} \dot{S}_{\text{i}} \tag{7.126}$$

respectively

$$\Delta \dot{E}_{\text{x,V}} = \dot{m} T_{\text{env}} (s_4 - s_1) + \dot{Q} = 153.7143 \, \text{kW}. \tag{7.127}$$

7.27 The change of state (1) to (2) in the turbine is shown in Fig. 7.21. States (1) and (2) are uniquely determined by pressure and temperature and are in the region of superheated vapour. With the steam table one determines the specific enthalpies to

$$h_1 = h(p_1, \vartheta_1) = 3.4009 \times 10^3 \, \frac{\text{kJ}}{\text{kg}} \tag{7.128}$$

---

[19] I.e., the entropy generated within the system boundary also includes the part of the heat transport within the system with $\Delta T > 0$.

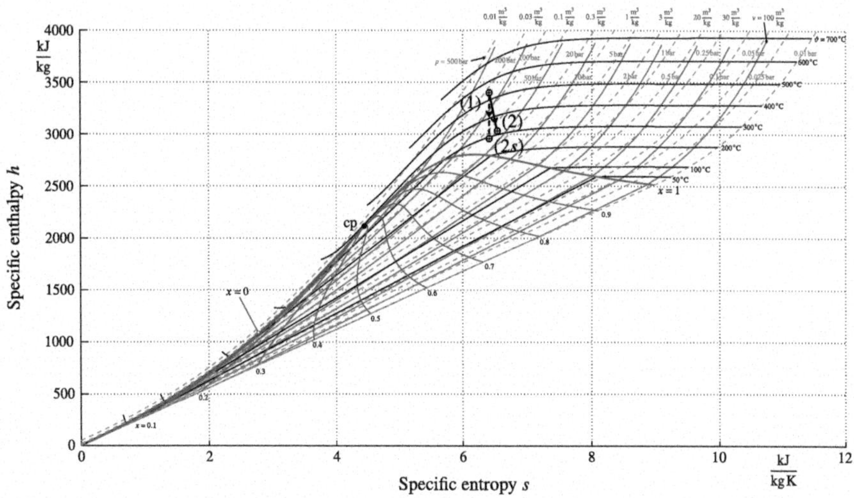

Fig. 7.21 Polytropic, adiabatic turbine

and

$$h_2 = h(p_2, \vartheta_2) = 3.0305 \times 10^3 \, \frac{\text{kJ}}{\text{kg}}. \tag{7.129}$$

The specific entropies follow analogously, i.e.

$$s_1 = s(p_1, \vartheta_1) = 6.4106 \, \frac{\text{kJ}}{\text{kg K}} \tag{7.130}$$

and

$$s_2 = s(p_2, \vartheta_2) = 6.5377 \, \frac{\text{kJ}}{\text{kg K}}. \tag{7.131}$$

An adiabatic, frictionless, i.e. isentropic, reference turbine is used to determine the isentropic turbine efficiency. This turbine reaches the same final pressure p_2 as the turbine under investigation. For the isentropic turbine (1) to (2s), the specific enthalpy results accordingly with

$$h_{2s} = h(p_2, s_1) = 2.9569 \times 10^3 \, \frac{\text{kJ}}{\text{kg}} \tag{7.132}$$

from the steam table. State (2s) is also sketched in Fig. 7.21. Finally, the isentropic efficiency obeys

$$\eta_{s,\text{T}} = \frac{h_2 - h_1}{h_{2s} - h_1} = 0.8343. \tag{7.133}$$

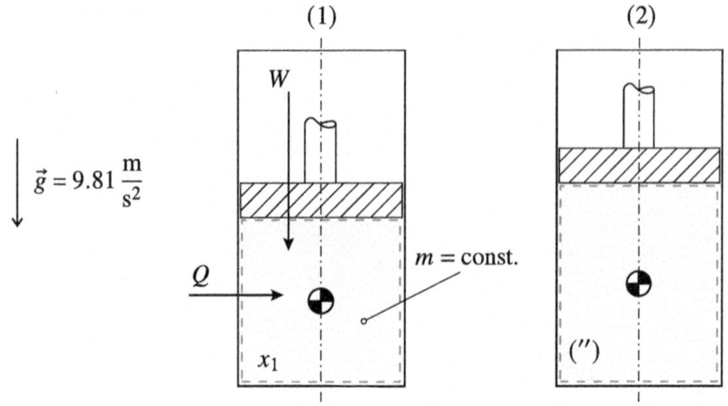

Fig. 7.22 Closed system

7.28 The problem is outlined in Fig. 7.22. In state (1), the fluid is present as wet steam. The supply of heat increases the volume, since vaporisation turns liquid water in the boiling state into saturated steam. The change of state is isobaric, since the environment and the mass of the piston imprint the pressure in the system and the change of state proceeds very slowly, i.e. reversibly without friction. Therefore, during vaporisation, the pressure and thus the temperature remain constant. As soon as the temperature rises, vaporisation is complete and the supply of heat ends. State (2) is therefore saturated steam in (″)-state. The first law of thermodynamics applies to the given system boundary,[20] i.e.

$$Q + W = U_2 - U_1 + \Delta E_{pot}. \tag{7.134}$$

The partial energy equation for the work W is composed of the volume work to increase the volume of the enclosed fluid and the mechanical work to raise the centre of gravity of the trapped vapour, i.e.

$$W = W_V + W_{mech} + \underbrace{\Psi}_{=0} = -\int_1^2 p\,dV + \Delta E_{pot}. \tag{7.135}$$

Hence, the first law of thermodynamics is

$$Q - \int_1^2 p\,dV + \Delta E_{pot} = U_2 - U_1 + \Delta E_{pot}. \tag{7.136}$$

[20] It is assumed that the centre of gravity is at rest in state (1) as well as in state (2), i.e. kinetic energy is not relevant.

The integral can be easily solved because the pressure is constant, i.e.

$$Q - p\,(V_2 - V_1) = U_2 - U_1 \tag{7.137}$$

respectively

$$Q = U_2 + pV_2 - U_1 - pV_1 = H_2 - H_1 = m\,(h_2 - h_1)\,. \tag{7.138}$$

The specific enthalpy in state (2) can be determined as follows

$$h_2 = h''\,(p = 4\,\text{bar}) = 2.7381 \times 10^3\,\frac{\text{kJ}}{\text{kg}}. \tag{7.139}$$

Thus, solving for h_1 reads as

$$h_1 = h_2 - \frac{Q}{m} = 1.9381 \times 10^3\,\frac{\text{kJ}}{\text{kg}} \tag{7.140}$$

Finally, the vapour content in state (1) obeys

$$x_1 = \frac{h_1 - h'}{h'' - h'} = 0.6250 \tag{7.141}$$

with

$$h' = h'\,(p = 4\,\text{bar}) = 604.7235\,\frac{\text{kJ}}{\text{kg}} \quad \text{and} \quad h'' = h_2 = 2.7381 \times 10^3\,\frac{\text{kJ}}{\text{kg}}. \tag{7.142}$$

7.29

(a) The following applies to the vapour content in state (4)

$$x_4 = \frac{h_4 - h_4'}{h_4'' - h_4'} \tag{7.143}$$

with the specific enthalpies taken from the steam table

$$h_4'' = h''\,(1\,\text{bar}) = 2.6749 \times 10^3\,\frac{\text{kJ}}{\text{kg}} \tag{7.144}$$

and

$$h_4' = h'\,(1\,\text{bar}) = 417.4365\,\frac{\text{kJ}}{\text{kg}} \tag{7.145}$$

Hence, the specific enthalpy of state (4) follows

$$h_4 = h_4' + x_4\left(h_4'' - h_4'\right) = 2.6524 \times 10^3\,\frac{\text{kJ}}{\text{kg}}. \tag{7.146}$$

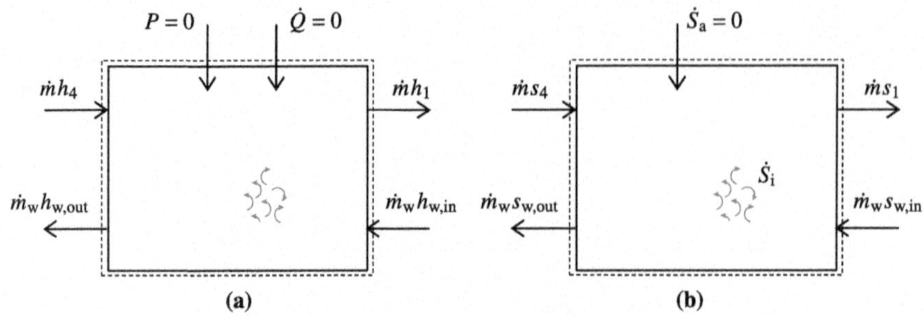

Fig. 7.23 Condenser: **a** Energy balance **b** Entropy balance

The first law of thermodynamics for the adiabatic turbine obeys

$$P_t = \dot{m}\,(h_4 - h_3) \quad \text{with} \quad P_t = -200 \times 10^3\,\text{kW}. \tag{7.147}$$

The specific enthalpy in state (3) then is

$$h_3 = h_4 - \frac{P_t}{\dot{m}} = 3.6524 \times 10^3\,\frac{\text{kJ}}{\text{kg}} \quad \text{with} \quad \dot{m} = 200\,\frac{\text{kg}}{\text{s}}. \tag{7.148}$$

With the steam table one finally obtains the temperature of the super-heated state (3), i.e.

$$\vartheta_3 = \vartheta\,(h_3,\ p_3) = 600.7345\,°\text{C}. \tag{7.149}$$

(b) The first law of thermodynamics for the condenser obeys, cf. Fig. 7.23a,

$$\underbrace{\dot{m}h_4 + \dot{m}_w h_{w,in} + \overbrace{\dot{Q} + P}^{=0}}_{\text{IN}} = \underbrace{\dot{m}h_1 + \dot{m}_w h_{w,out}}_{\text{OUT}} \tag{7.150}$$

so that

$$\begin{aligned} h_1 &= h_4 - \frac{\dot{m}_w}{\dot{m}}\left(h_{w,out} - h_{w,in}\right) \\ &= h_4 - \frac{\dot{m}_w}{\dot{m}} c_w \left(\vartheta_{w,out} - \vartheta_{w,in}\right) \\ &= 169.6233\,\frac{\text{kJ}}{\text{kg}}. \end{aligned} \tag{7.151}$$

With the steam table one finally obtains the temperature of the subcooled state (1), i.e.

$$\vartheta_1 = \vartheta\,(h_1,\ p_1) = 40.4943\,°\text{C}. \tag{7.152}$$

The specific enthalpy of state (2) follows from the isentropic efficiency of the feedwater pump, i.e.

$$\eta_{sP} = \frac{h_{2s} - h_1}{h_2 - h_1} \tag{7.153}$$

so that

$$h_2 = h_1 + \frac{h_{2s} - h_1}{\eta_{sP}}. \tag{7.154}$$

It further applies

$$h_{2s} = h\,(s_1, p_2) = 176.6201\,\frac{\text{kJ}}{\text{kg}} \quad \text{with} \quad s_1 = s\,(p_1, h_1) = 0.5790\,\frac{\text{kJ}}{\text{kg K}}. \tag{7.155}$$

The specific enthalpy in state (2) finally is

$$h_2 = h_1 + \frac{h_{2s} - h_1}{\eta_{sP}} = 179.6187\,\frac{\text{kJ}}{\text{kg}}. \tag{7.156}$$

(c) The supplied heat in the combustion chamber is

$$\dot{Q} = \dot{m}\,(h_3 - h_2) = 6.9455 \times 10^5\,\text{kW} \tag{7.157}$$

and the released heat in the condenser

$$\dot{Q}_{\text{w}} = \dot{m}\,(h_1 - h_4) = -4.9655 \times 10^5\,\text{kW}. \tag{7.158}$$

Finally, the thermal efficiency is

$$\eta_{\text{th}} = 1 - \frac{|\dot{Q}_{\text{w}}|}{|\dot{Q}|} = 28.5078\,\%. \tag{7.159}$$

The thermal efficiency can be increased by the following measures:

- Intermediate superheating
- Feed water preheating
- Increase in superheating temperature.

(d) The entropy balance for the condenser is, cf. Fig. 7.23b,

$$\underbrace{\dot{m}s_4 + \dot{m}_{\text{w}}s_{\text{w,in}} + \dot{S}_{\text{i}} + \overbrace{\dot{S}_{\text{a}}}^{=0}}_{\text{IN}} = \underbrace{\dot{m}s_1 + \dot{m}_{\text{w}}s_{\text{w,out}}}_{\text{OUT}} \tag{7.160}$$

so that

$$\dot{S}_{\text{i}} = \dot{m}\,(s_1 - s_4) + \dot{m}_{\text{w}}\left(s_{\text{w,out}} - s_{\text{w,in}}\right) = \dot{m}\,(s_1 - s_4) + \dot{m}_{\text{w}} c_{\text{w}} \ln \frac{T_{\text{w,out}}}{T_{\text{w,in}}} \tag{7.161}$$

With

$$s_1 = s\,(p_1,\, h_1) = 0.5790\,\frac{\text{kJ}}{\text{kg K}} \tag{7.162}$$

and

$$s_4 = s\,(p_4 = p_1,\, h_4) = 7.2982\,\frac{\text{kJ}}{\text{kg K}} \tag{7.163}$$

it finally is

$$\dot{S}_{\text{i}} = \dot{m}\,(s_1 - s_4) + \dot{m}_{\text{w}} c_{\text{w}} \ln \frac{T_{\text{w,out}}}{T_{\text{w,in}}} = 379.5562\,\frac{\text{kW}}{\text{K}} \tag{7.164}$$

respectively

$$\Delta \dot{E}_{\text{x,V}} = \dot{S}_{\text{i}} T_{\text{env}} = 1.0747 \times 10^5\,\text{kW}. \tag{7.165}$$

Chapter 8
Mixtures of Ideal Gases and Humid Air

In addition to pure substances, mixtures of substances play a major role in thermodynamics. For example, our ambient air is a mixture of many components, such as nitrogen and oxygen, which make up the largest proportion of the air. In the exercises presented here, the topic is how thermodynamic state variables can be determined for such mixtures. If the state variables are accessible, the first and second laws of thermodynamics can be applied as usual. To stay with the example of ambient air, it is known that air can also carry water vapour. The higher the temperature of the air, the greater the absorption capacity for water vapour. The tasks presented here deal with this so-called humid air and examine in particular what happens when the water vapour condenses out. Tasks are presented that address the conditioning of indoor air and according to which criteria air conditioning systems are designed. In particular, graphical solution methods are applied.

8.1 Problems

8.1 An ideal gas mixture consists of equal volume fractions of O_2 and N_2. What is the gas constant of the mixture?

(a) $287.14\,\frac{\mathrm{J}}{\mathrm{kg\,K}}$
(b) $277.14\,\frac{\mathrm{J}}{\mathrm{kg\,K}}$
(c) $267.14\,\frac{\mathrm{J}}{\mathrm{kg\,K}}$
(d) $257.14\,\frac{\mathrm{J}}{\mathrm{kg\,K}}$

8.2 Two humid, unsaturated air flows are mixed adiabatically, isobarically. Which statement applies?

(a) There is a decrease in entropy.
(b) The mixed air must also be unsaturated.
(c) The mixing air may be in the fog area when one air flow is significantly colder than the other.
(d) None of the above statements applies.

A. Schmidt, *Technical Thermodynamics Workbook for Engineers*,
https://doi.org/10.1007/978-3-031-50172-2_8

8.3 Heat is supplied isobarically to humid, unsaturated air. Which statement applies?

(a) The entropy of the air decreases.
(b) The relative humidity decreases.
(c) The water content increases.
(d) None of the above statements applies.

8.4 Two non-identical ideal gases mix adiabatically and isothermally. Which statement applies?

(a) The total entropy remains constant.
(b) Water condenses out.
(c) Exergy rises.
(d) None of the above statements applies.

8.5 Which statement is correct?

(a) Humid, unsaturated air is heavier than dry air.
(b) Humid, unsaturated air is lighter than dry air.
(c) Humid, unsaturated and dry air are equally heavy.
(d) None of the above statements applies.

8.6 To what temperature can a 30 °C warm bottle of beer be cooled in the outside air in a light wind if a cloth soaked in water is wrapped around the bottle and it is ensured that the cloth is always sufficiently moist? The outside air has a temperature of 25 °C with a relative humidity of 30 %. The air pressure is $p = 1$ bar.

(a) 10.3 °C
(b) 14.4 °C
(c) 17.4 °C
(d) Cooling is not possible.

8.7 Humid air (25 °C, 30 % relative humidity, 1 bar constant total pressure) is sprayed with so much liquid water from 0 °C in a steady state flow process that the water content increases by 3 $\frac{\text{g}}{\text{kg}}$. What is the temperature of the humid air in thermodynamic equilibrium when it leaves the humidifier?

(a) 26.72 °C
(b) 17.52 °C
(c) 14.12 °C
(d) 6.77 °C

8.8 A 30 °C warm bottle of beer with a cloth soaked in water wrapped around it is cooled in the outside air in a light wind. This ensures that the cloth is always sufficiently moistened. The resulting temperature of the beer is 12 °C. The outside air has a temperature of 20 °C. The air pressure is $p = 1$ bar. What is the relative humidity of the outside air?

(a) 37.9 %
(b) 45.5 %

(c) 67.1 %
(d) Cooling of the beer is not possible under these circumstances.

8.9 Two different quantities of an identical ideal gas with the same temperature and pressure are in an adiabatic vessel and are initially separated by a wall. Now the wall is removed. Let the work to remove the wall be negligible. Which statement applies?

(a) The temperature is increasing.
(b) Due to the mixing, entropy is generated.
(c) The pressure is increasing.
(d) None of the above statements applies.

8.10 An air flow of humid air (1 bar, $\varphi = 50\,\%$, 25 °C) is isobarically and adiabatically injected with liquid water of 25 °C. What is the maximum mass flux of water that can be injected until the air flow is saturated in thermodynamic equilibrium? The mass flux of dry air is $1\,\frac{\text{kg}}{\text{s}}$.

(a) Approx. $3.0\,\frac{\text{g}}{\text{s}}$
(b) Approx. $6.8\,\frac{\text{g}}{\text{s}}$
(c) Approx. $12.8\,\frac{\text{g}}{\text{s}}$
(d) The air cannot reach saturated state.

8.11 What is the specific enthalpy h_{1+x} of humid air with a temperature of 20 °C, a relative humidity of 80 % at a total pressure of 2 bar?

(a) Approx. $20\,\frac{\text{kJ}}{\text{kg}}$
(b) Approx. $35\,\frac{\text{kJ}}{\text{kg}}$
(c) Approx. $42\,\frac{\text{kJ}}{\text{kg}}$
(d) Approx. $50\,\frac{\text{kJ}}{\text{kg}}$

8.12 A closed vessel initially contains dry air (ideal gas) and liquid water at 0.96 bar and 20 °C. Now the vessel is heated to 100 °C. What is the pressure at thermodynamic equilibrium where liquid water is still present? The change in volume of the liquid water is to be neglected compared to the volume of air.

8.13 Humid air with $\varphi_1 = 0.225$, $\vartheta_1 = 49.45\,°\text{C}$, $p_1 = 0.9\,\text{bar}$ passes a cooled diffuser and leaves with $p_2 = 1\,\text{bar}$, $\vartheta_2 = 21\,°\text{C}$. The velocity of the humid air when entering the diffuser is $c_1 = 200\,\frac{\text{m}}{\text{s}}$, when leaving it is negligibly small. The cross section at the inlet shall be $A_1 = 0.1\,\text{m}^2$.

a) Determine the mass flow rate of dry air.
b) Calculate the mass flow rate of liquid water (fog) at the outlet.
c) What is the heat flux $\dot{Q}_{12}$ exchanged with the environment?
d) Calculate the flux of loss of exergy $\dot{E}_{\text{x,V}}$. Ambient temperature is $\vartheta_{\text{env}} = 0\,°\text{C}$.

8.14 Two air flows are mixed adiabatically, isobarically at $p = 1\,\text{bar}$ in a mixing chamber. Air flow (1) has a temperature of $\vartheta_1 = 15\,°\text{C}$ and a water content of $x_1 = 0.004$. Air flow (2) has a temperature of $\vartheta_2 = 35\,°\text{C}$ and a water content of $x_2 = 0.012$. The mass fluxes of the dry air are identical for both air flows. What is the mixing temperature?

8.15 An air flow (1) with $\vartheta_1 = 25\,°\mathrm{C}$, $p_1 = 1\,\mathrm{bar}$, $x_1 = 5\,\frac{\mathrm{g}}{\mathrm{kg}}$ and $\dot{m}_{a,1} = 250\,\frac{\mathrm{kg}}{\mathrm{h}}$ is isobarically mixed with $\dot{m}_v = 2.5\,\frac{\mathrm{kg}}{\mathrm{h}}$ vapour of 100 °C in an adiabatic chamber. What is the temperature of the air (2) when it leaves the chamber in steady state?

8.16 In an adiabatic tank ($V = 1\,\mathrm{m}^3$) there is nitrogen N_2 and oxygen O_2. The gases are to be considered ideal and are initially separated by a wall. Both gases have a pressure of $p_1 = 1\,\mathrm{bar}$, a temperature of 300 K and each occupy a volume of $0.5\,\mathrm{m}^3$. The separating wall is now removed and the gases mix. How much energy is dissipated until thermodynamic equilibrium is reached?

8.17 Two air flows are mixed adiabatically, isobarically at $p = 1\,\mathrm{bar}$ in a mixing chamber. Air flow (1) has a temperature of $\vartheta_1 = 15\,°\mathrm{C}$ and a water content of $x_1 = 0.004$. Air flow (2) has a temperature of $\vartheta_2 = 35\,°\mathrm{C}$ and a water content of $x_2 = 0.016$. The mass flux of dry air from state (1) is twice that of state (2). What is the mixing temperature?

8.18 Humid air ($c_1 = 200\,\frac{\mathrm{m}}{\mathrm{s}}$, $x_1 = 0.0193$, $\vartheta_1 = 50\,°\mathrm{C}$, $p_1 = 0.9\,\mathrm{bar}$) passes a cooled diffusor into the environment ($\vartheta_{env} = 0\,°\mathrm{C}$, $p_{env} = 1.0\,\mathrm{bar}$). At outlet (2), the velocity is negligible and $p_2 = p_{env}$, $\vartheta_2 = 21\,°\mathrm{C}$. The mass flow rate of the dry air is $18.85\,\frac{\mathrm{kg}}{\mathrm{s}}$. What is the heat flux exchanged with the environment?

8.19 By isobaric injection of air, the CO_2-concentration of an exhaust gas flow in an adiabatic mixing chamber is to be reduced to $x_{3,CO_2} = 0.02$, see Fig. 8.1. The static pressure of all flows is 1.5 bar. All gases are to be regarded as ideal gases with constant specific heat capacities ($c_{p,N_2} = 1.04\,\frac{\mathrm{kJ}}{\mathrm{kg\,K}}$, $c_{p,O_2} = 0.91\,\frac{\mathrm{kJ}}{\mathrm{kg\,K}}$, $c_{p,CO_2} = 0.89\,\frac{\mathrm{kJ}}{\mathrm{kg\,K}}$). It further is

- Air (1)

$$x_{1,O_2} = 0.21,\; x_{2,N_2} = 0.79 \tag{8.1}$$

$$\vartheta_1 = 23\,°\mathrm{C} \tag{8.2}$$

- Exhaust gas (2)

$$x_{2,CO_2} = 0.1,\; x_{2,N_2} = 0.8,\; x_{2,O_2} = 0.1 \tag{8.3}$$

$$\dot{n}_2 = 50\,\frac{\mathrm{kmol}}{\mathrm{h}},\; \vartheta_2 = 150\,°\mathrm{C} \tag{8.4}$$

Fig. 8.1 Exhaust gas

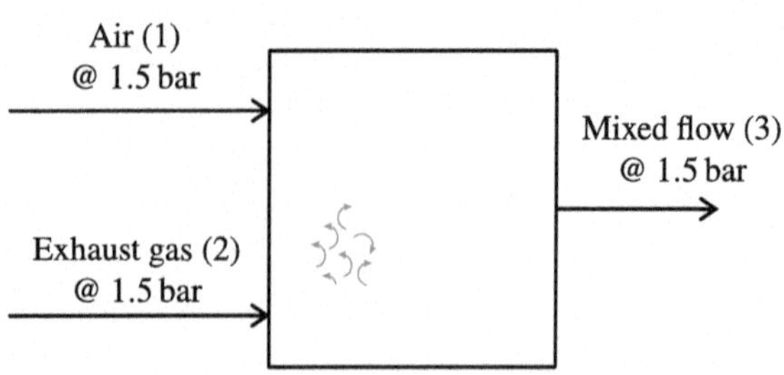

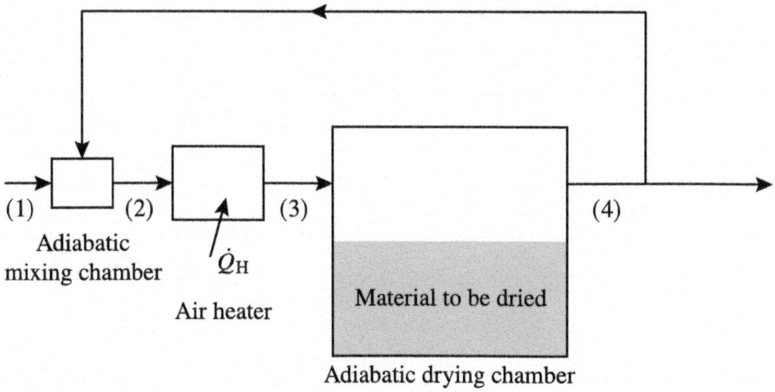

Fig. 8.2 Dehumidification

- Mixed flow (3)

$$x_{3,CO_2} = 0.02. \tag{8.5}$$

(a) Calcuate the molar flux $\dot{n}_1$ of air.
(b) Determine the partial pressures of the exhaust gas before and after the injection of air.
(c) What is the temperature ϑ_3?
(d) What is the flux of generated entropy?

8.20 In a drying unit, moisture is removed from a material to be dried and supplied to the air. Ambient air ($\vartheta_1 = 20\,°C$, $\varphi_1 = 70\,\%$) is used for this purpose and mixed with exhaust air from the drying chamber. The mixed air in state (2) is then heated ($\vartheta_3 = 50\,°C$, $\varphi_3 = 20\,\%$) and supplied to the drying chamber. There, it absorbs moisture and leaves the chamber at a temperature of $\vartheta_4 = 29.5\,°C$, see Fig. 8.2.

- The total pressure is 1 bar throughout. Mixing/drying chamber and piping are adiabatic.
- Kinetic, potential energies and heat losses can be neglected.
- The water released by the drying material is absorbed by the air with ϑ_4.

(a) Determine the water content x and the specific enthalpies h_{1+x} of states (1) and (3).
(b) What is the water content x and the enthalpy h_{1+x} at state (4)?
(c) Determine the enthalpy h_{1+x} and the temperature of state (2). What is the ratio of the dry air mixed in the mixing chamber?
(d) What mass flux of mixed air (3) is required to absorb 180 kg of water per hour from the material to be dried?
(e) What heat flux $\dot{Q}_H$ is required?

8.21 With the foehn, warm and humid air masses move northwards from the Mediterranean. These air masses rise on the southern side of the Alps. At altitude z_2 the dew point is reached. The condensing water forms clouds which continue to rise with the air masses up to $z_3 = 2000\,\text{m}$. At the main ridge of the Alps, the condensed water is completely separated, state (4). After crossing the main Alpine ridge, the air masses fall to a height of $z_5 = 0\,\text{m}$, see Fig. 8.3. The following assumptions apply:

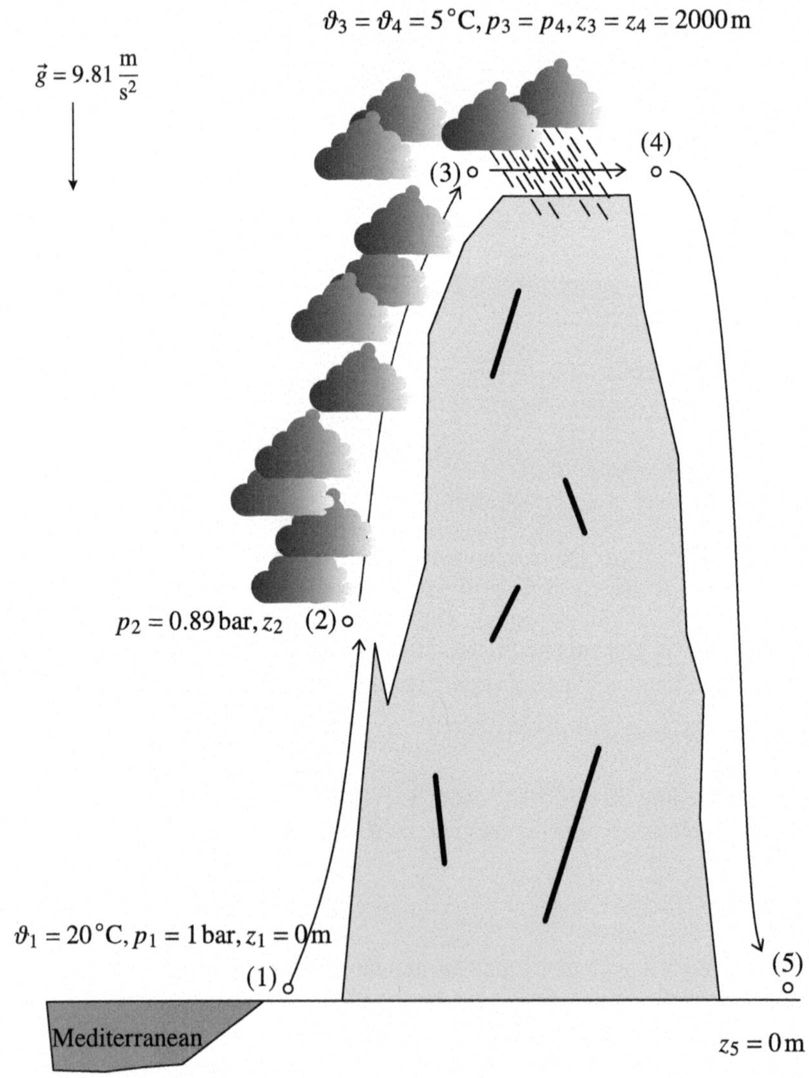

Fig. 8.3 Cloud formation in the Alps

- The usual assumptions for humid air apply.
- The isentropic exponent of the gaseous humid air is $\kappa = 1.4$.
- The changes of state occur adiabatically and frictionless.
- Changes in kinetic energy are negligible.

(a) At what height z_2 do the first clouds form?
(b) What is the pressure p_3 on the main ridge of the Alps?
(c) What is the temperature in state (5)?

8.2 Solutions

8.1 **(b)**
The following applies to the gas constant of an ideal gas mixture

$$R = \sum_i \xi_i R_i . \tag{8.6}$$

However, in this problem the mass fractions ξ_i are not known, but the volume fractions σ_i, so that Eq. 8.6 cannot be applied, i.e.

$$\xi_i \neq \sigma_i . \tag{8.7}$$

Nevertheless, volume fractions correspond to molar fractions and pressure fractions, respectively, i.e.

$$\sigma_i = x_i = \pi_i \quad \text{with} \quad \pi_i = \frac{p_i}{p} . \tag{8.8}$$

With the help of the molar fractions, the molar mass M of the mixture can now be determined, i.e.

$$M = \sum_i x_i M_i = x_{O_2} M_{O_2} + x_{N_2} M_{N_2} \quad \text{with} \quad x_{O_2} = x_{N_2} = 0.5 \tag{8.9}$$

and

$$M_{O_2} = 32 \, \frac{\text{kg}}{\text{kmol}} \quad \text{resp.} \quad M_{N_2} = 28 \, \frac{\text{kg}}{\text{kmol}} . \tag{8.10}$$

Hence, the molar mass of the mixture yields

$$M = \sum_i x_i M_i = x_{O_2} M_{O_2} + x_{N_2} M_{N_2} = 30 \, \frac{\text{kg}}{\text{kmol}} . \tag{8.11}$$

Finally, the gas constant of the mixture obeys

$$R = \frac{R_{\mathrm{M}}}{M} = 277.1433 \, \frac{\mathrm{J}}{\mathrm{kg\,K}} \quad \text{with} \quad R_{\mathrm{M}} = 8.3143 \, \frac{\mathrm{kJ}}{\mathrm{kmol\,K}}. \tag{8.12}$$

Answer (b) is correct.

8.2 **(c)**
Mixing is an irreversible process that is usually driven by chemical potential. Once mixing has occurred, the system is in thermodynamic equilibrium, the potentials are balanced and thus there is no driver back to the initial state. This means that irreversible mixing processes are accompanied by the generation of entropy. Since in the present case the mixing is adiabatic, no entropy is released from the system by a heat flux. The entropy must therefore increase due to the irreversibility. Answer (a) is wrong.

If two mass fluxes of humid air (1) and (2) are mixed isobarically, adiabatically, the mass balances follow:

- Dry air:

$$\underbrace{\dot{m}_{\mathrm{a},1} + \dot{m}_{\mathrm{a},2}}_{\mathrm{IN}} = \underbrace{\dot{m}_{\mathrm{a},3}}_{\mathrm{OUT}} \tag{8.13}$$

- Water:

$$\underbrace{\dot{m}_{\mathrm{w},1} + \dot{m}_{\mathrm{w},2}}_{\mathrm{IN}} = \underbrace{\dot{m}_{\mathrm{w},3}}_{\mathrm{OUT}} \tag{8.14}$$

respectively

$$\dot{m}_{\mathrm{a},1} x_1 + \dot{m}_{\mathrm{a},2} x_2 = \left(\dot{m}_{\mathrm{a},1} + \dot{m}_{\mathrm{a},2}\right) x_3 \tag{8.15}$$

so that

$$x_3 = \frac{\dot{m}_{\mathrm{a},1} x_1 + \dot{m}_{\mathrm{a},2} x_2}{\dot{m}_{\mathrm{a},1} + \dot{m}_{\mathrm{a},2}} \tag{8.16}$$

The energy balance yields

$$\underbrace{\dot{m}_{\mathrm{a},1} h_{1+x,1} + \dot{m}_{\mathrm{a},2} h_{1+x,2}}_{\mathrm{IN}} = \underbrace{\dot{m}_{\mathrm{a},3} h_{1+x,3}}_{\mathrm{OUT}}. \tag{8.17}$$

so that

$$h_{1+x,3} = \frac{\dot{m}_{\mathrm{a},1} h_{1+x,1} + \dot{m}_{\mathrm{a},2} h_{1+x,2}}{\dot{m}_{\mathrm{a},1} + \dot{m}_{\mathrm{a},2}} \tag{8.18}$$

It has been shown, see e.g. [1], that the mixing point (3) lies on the so-called mixing straight in the h_{1+x}, x-diagram, i.e. the connecting line between the states (1) and (2). How far away state (3) is on the straight line from states (1) and (2), respectively, depends on the ratio of the dry air mass flows, i.e.

$$\frac{\dot{m}_{\text{a},2}}{\dot{m}_{\text{a},1}} = \frac{\overline{13}}{\overline{23}}. \tag{8.19}$$

One can easily show with the h_{1+x}, x-diagram that it is possible to mix two unsaturated air flows so that the mixed air is saturated, respectively liquid water additionally appears in addition, see Fig. 8.4. This happens when one air flow, in this case state

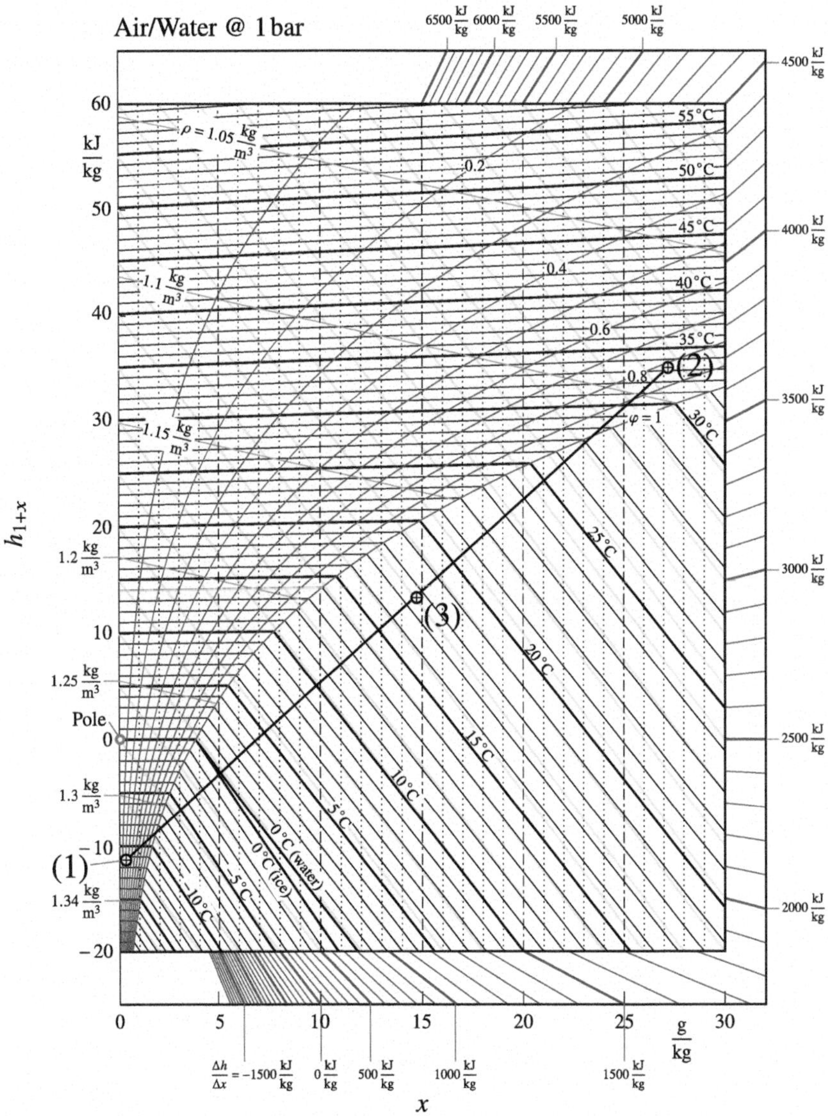

Fig. 8.4 Mixing of two flows of unsaturated humid air

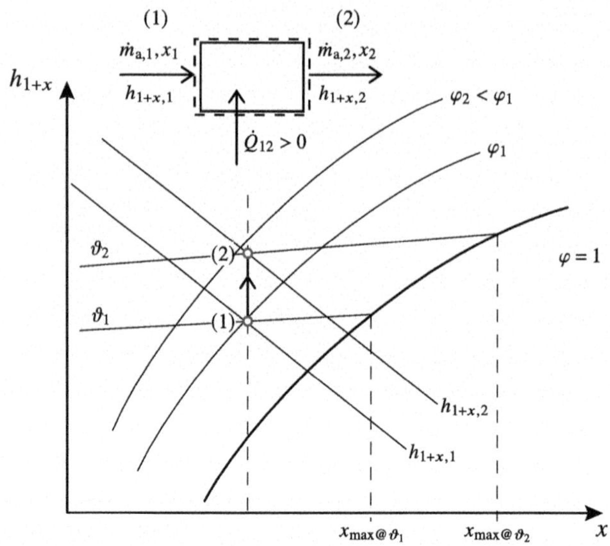

Fig. 8.5 Pure heating of unsaturated humid air

(1), is very cold and therefore very dry and the second air flow (2) is very warm and has a high water content $x = \frac{\dot{m}_w}{\dot{m}_a}$. This phenomenon can be observed, for example, on cold winter days when the air breathed forms mist when exhaled. Answer (b) is wrong and answer (c) applies.

8.3 **(b)**
In Fig. 8.5, the process is sketched schematically as well as the change of state (1) to (2) in a h_{1+x}, x-diagram. The mass balances for the change of state as shown in Fig. 8.5 are:

- Dry air:

$$\underbrace{\dot{m}_{a,1}}_{\text{IN}} = \underbrace{\dot{m}_{a,2}}_{\text{OUT}} \tag{8.20}$$

- Water:

$$\underbrace{\dot{m}_{w,1}}_{\text{IN}} = \underbrace{\dot{m}_{w,2}}_{\text{OUT}} \tag{8.21}$$

respectively

$$\dot{m}_{a,1} x_1 = \dot{m}_{a,2} x_2 \tag{8.22}$$

so that with the mass balance for dry air it yields

$$\boxed{x_1 = x_2} \tag{8.23}$$

The water content x of the air is unaffected by the heat supply. The change of state in a h_{1+x}, x-diagram must follow a vertical line. Answer (c) is wrong.

The first law of thermodynamics obeys

$$\underbrace{\dot{H}_1 + \dot{Q}_{12}}_{\text{IN}} = \underbrace{\dot{H}_2}_{\text{OUT}} \tag{8.24}$$

respectively

$$\dot{m}_{a,1} h_{1+x,1} + \dot{Q}_{12} = \dot{m}_{a,2} h_{1+x,2} = \dot{m}_{a,1} h_{1+x,2}. \tag{8.25}$$

Hence, it is

$$\boxed{h_{1+x,2} - h_{1+x,1} = \frac{\dot{Q}_{12}}{\dot{m}_{a,1}} > 0} \tag{8.26}$$

The specific enthalpy from (1) to (2) must therefore increase due to the supply of heat. The change of state in the h_{1+x}, x-diagram thus runs vertically upwards, see Fig. 8.5. The entropy balance yields[1]

$$\underbrace{\dot{S}_1 + \overbrace{\dot{S}_{a,12}}^{>0} + \dot{S}_{i,12}}_{\text{IN}} = \underbrace{\dot{S}_2}_{\text{OUT}} \tag{8.27}$$

so that

$$\dot{S}_2 - \dot{S}_1 = \dot{S}_{a,12} + \dot{S}_{i,12} > 0. \tag{8.28}$$

Consequently, answer (a) does not apply.

According to h_{1+x}, x-diagram, the air in state (1) is unsaturated, i.e. $\varphi_1 < 1$. Unsaturated because the vapour content x_1 is smaller than the amount of vapour that air with a temperature ϑ_1 could maximally absorb. This maximum vapour content $x_{\max@\vartheta_1}$ can be found in the diagram by extending the isobar of state (1) until the saturation line is reached.

For state (2), which is vertically above (1), one can proceed similarly. Here, one finds a maximum vapour content $x_{\max@\vartheta_2}$, which, due to $\vartheta_2 > \vartheta_1$, is greater than $x_{\max@\vartheta_1}$. Obviously, the warmer the air, the more vapour it can hold. Anyhow, the relative humidity expresses how close the vapour content is to the maximum possible vapour content. In case (2), the air could absorb much more vapour than would be possible in state (1) due to the higher temperature. Therefore, the relative humidity in state (2) is smaller than in state (1), i.e. $\varphi_2 < \varphi_1$. This can also be seen in the φ-isolines in Fig. 8.5. Answer (b) is therefore correct.

[1] In this case, the change of state is isobaric, so that there is no dissipation respectively entropy generation if the outer energies are neglected.

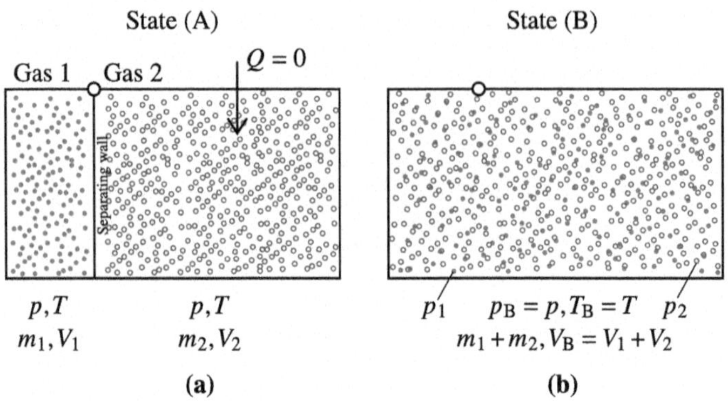

Fig. 8.6 Adiabatic, isothermal mixing

8.4 (**d**)
For the mixing process, see Fig. 8.6, the entropy balance is as follows

$$S_{\mathrm{B}} = S_{\mathrm{A}} + S_{\mathrm{i,AB}} + \underbrace{S_{\mathrm{a,AB}}}_{=0} \tag{8.29}$$

respectively

$$S_{\mathrm{i,AB}} = S_{\mathrm{B}} - S_{\mathrm{A}}. \tag{8.30}$$

Entropy as an extensive state value is composed of the parts for gas 1 and gas 2, i.e.

$$S_{\mathrm{i,AB}} = S_{\mathrm{B}} - S_{\mathrm{A}} = m_1 \left(s_{\mathrm{B}} - s_{\mathrm{A}}\right)_1 + m_2 \left(s_{\mathrm{B}} - s_{\mathrm{A}}\right)_2. \tag{8.31}$$

It follows with the caloric equation of state

$$\begin{aligned} S_{\mathrm{i,AB}} &= m_1 \left(c_{p,1} \ln \frac{T}{T} - R_1 \ln \frac{p_1}{p}\right)_1 + m_2 \left(c_{p,2} \ln \frac{T}{T} - R_2 \ln \frac{p_2}{p}\right)_2 \\ &= -m_1 R_1 \ln \frac{p_1}{p} - m_2 R_2 \ln \frac{p_2}{p} > 0. \end{aligned} \tag{8.32}$$

Since the partial pressures p_1 and p_2 are smaller than p, entropy is generated during the mixing.[2] In other words, the mixing process is irreversible. The driver for the transient process is the concentration differences. When the concentrations are balanced in the state of equilibrium, the system can no longer move back to the initial state on its own. Since the process is adiabatic, the entropy of the system must therefore increase. Answer (a) is wrong. Entropy generation goes hand in hand with the loss of exergy

[2] However, jf the described mixing is carried out with two identical gases, no entropy is generated. This is the Gibbs paradox.

$$\Delta E_{\mathrm{x,V}} = T_{\mathrm{env}} S_{\mathrm{i,12}}. \tag{8.33}$$

As no further exergy is supplied to the adiabatic system from outside, the exergy in the system must decrease. Answer (c) does not apply. However, answer (b) does not make sense because it was not mentioned that there should be water in the system. Consequently, answer (d) applies.

8.5 **(b)**
The density of unsaturated humid air, depending on its vapour content x_{v}, yields, cf. [1],

$$\rho = (1 + x_{\mathrm{v}}) \cdot \rho_{1+x} = \frac{p}{R_{\mathrm{a}} T} \cdot \left[\frac{1 + x_{\mathrm{v}}}{1 + 1.608 \cdot x_{\mathrm{v}}}\right]. \tag{8.34}$$

The following cases shall now be examined:

- density of dry air ($x_{\mathrm{v}} = 0$):

$$\rho\,(x_{\mathrm{v}} = 0) = \frac{p}{R_{\mathrm{a}} T} \tag{8.35}$$

- density of humid air ($x_{\mathrm{v}} > 0$):

$$\rho\,(x_{\mathrm{v}} > 0) = \underbrace{\frac{p}{R_{\mathrm{a}} T}}_{\rho(x_{\mathrm{v}}=0)} \cdot \underbrace{\left[\frac{1 + x_{\mathrm{v}}}{1 + 1.608 \cdot x_{\mathrm{v}}}\right]}_{<1} \tag{8.36}$$

Obviously, the density of unsaturated humid air is smaller than the density of dry air, i.e.

$$\rho\,(x_{\mathrm{v}} > 0) < \rho\,(x_{\mathrm{v}} = 0)\,. \tag{8.37}$$

Thus, answer (b) is correct.

8.6 **(b)**
This problem is about the so-called wet-bulb temperature, which is based on evaporative cooling. This effect is based on the fact that unsaturated air flows over a surface wetted with water. Directly at the water surface, individual water molecules can escape the liquid and form a liquid/vapour interface. The gas phase directly at the liquid surface is in a saturated state, i.e. it carries its maximum vapour load. A thermodynamic equilibrium is formed, i.e. temperature potential and chemical potential are balanced. The phase transition now takes place until the entire system, see Fig. 8.7, has reached a thermodynamic equilibrium. This is when the air at the outlet of the sketched system boundary (2) is also in a saturated state. Finally, at equilibrium, the cloth, beer and exiting saturated air have the same temperature ϑ_2. For the liquid/gaseous phase transition at the water surface, enthalpy of vaporisation is required, which is taken from the air/substance/beer environment. Consequently, the environment cools down as a result. Obviously, this effect does not work if the

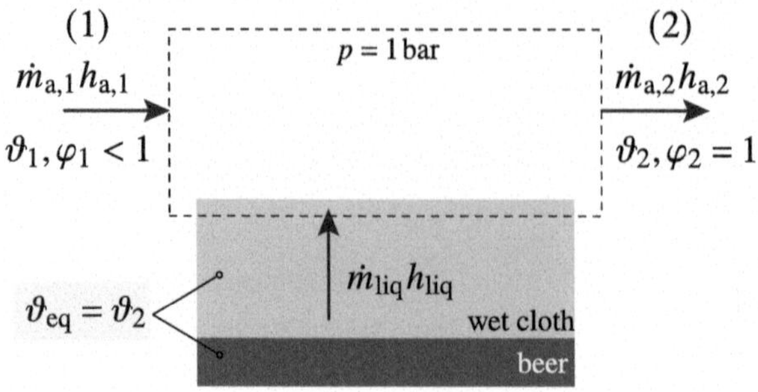

Fig. 8.7 Evaporative cooling

entering air (1) is already saturated. In the first part we want to solve the problem mathematically. The following is known from state (1)

$$\varphi_1 = 0.3 \quad \text{and} \quad \vartheta_1 = 25\,°\text{C}. \tag{8.38}$$

The vapour pressure in state (1) yields

$$p_{v,1} = \varphi_1 p_s(\vartheta_1) = 0.0095\,\text{bar} \quad \text{with} \quad p_s(\vartheta_1) = 0.0317\,\text{bar}. \tag{8.39}$$

Hence, the vapour content in state (1) obeys

$$x_1 = 0.622 \frac{p_{v,1}}{p - p_{v,1}} = 0.0060 \tag{8.40}$$

and the specific enthalpy is

$$h_1 = c_{p,a}\vartheta_1 + x_1 \left[\Delta h_{v,0} + c_{p,v}\vartheta_1\right] = 40.3065\,\frac{\text{kJ}}{\text{kg}}. \tag{8.41}$$

In state (2) the air is saturated in thermodynamic equilibrium, i.e. $\varphi_2 = 1$. Its temperature ϑ_2 is unknown. However, it applies

$$p_{v,2} = p_s(\vartheta_2) \tag{8.42}$$

and

$$x_2 = 0.622 \frac{p_s(\vartheta_2)}{p - p_s(\vartheta_2)}. \tag{8.43}$$

The specific enthalpy in state (2) obeys

$$h_2 = c_{p,a}\vartheta_2 + x_2 \left[\Delta h_{v,0} + c_{p,v}\vartheta_2\right]. \tag{8.44}$$

The mass balance for the dry air is

$$\dot{m}_{\mathrm{a},1} = \dot{m}_{\mathrm{a},2} \tag{8.45}$$

and the mass balance for the water yields

$$\underbrace{\dot{m}_{\mathrm{a},1} x_1 + \dot{m}_{\mathrm{liq}}}_{\mathrm{IN}} = \underbrace{\dot{m}_{\mathrm{a},2} x_2}_{\mathrm{OUT}} \tag{8.46}$$

so that

$$\dot{m}_{\mathrm{liq}} = \dot{m}_{\mathrm{a},1} \left(x_2 - x_1 \right). \tag{8.47}$$

The energy balance[3] for the steady state, open system is as follows

$$\underbrace{\dot{m}_{\mathrm{a},1} h_{1+x,1} + \dot{m}_{\mathrm{liq}} h_{\mathrm{liq}}}_{\mathrm{IN}} = \underbrace{\dot{m}_{\mathrm{a},2} h_{1+x,2}}_{\mathrm{OUT}} \quad \text{with} \quad h_{\mathrm{liq}} = c_{\mathrm{liq}} \vartheta_2. \tag{8.48}$$

Hence, it is

$$h_{1+x,1} + \left(x_2 - x_1 \right) c_{\mathrm{liq}} \vartheta_2 = h_{1+x,2} \tag{8.49}$$

respectively

$$\boxed{h_{1+x,1} + \left(0.622 \frac{p_{\mathrm{s}}(\vartheta_2)}{p - p_{\mathrm{s}}(\vartheta_2)} - x_1 \right) c_{\mathrm{liq}} \vartheta_2 = c_{p,\mathrm{a}} \vartheta_2 + 0.622 \frac{p_{\mathrm{s}}(\vartheta_2)}{p - p_{\mathrm{s}}(\vartheta_2)} \left[\Delta h_{\mathrm{v},0} + c_{p,\mathrm{v}} \vartheta_2 \right]} \tag{8.50}$$

This equation contains the equilibrium temperature ϑ_2 as an unknown and can be solved numerically. The solution is $\vartheta_2 = 14.3607\,°\mathrm{C}$.

In the second part we solve the problem graphically with the h_{1+x}, x-diagram, see Fig. 8.8. State (1) can be easily depicted in such a diagram, since temperature and relative humidity are given at $p = 1$ bar. State (2) is now reached by adiabatic humidification with liquid water. For such a change of state, one determines the slope of the change of state

$$\frac{\Delta h}{\Delta x} = h_{\mathrm{liq}} = c_{\mathrm{liq}} \vartheta_2. \tag{8.51}$$

and marks it with the help of the pole construction. Unfortunately, due to the missing temperature the slope can not be calculated explicitly. However, the slope of such a humidification with liquid water corresponds exactly to the slope of the ϑ_2-isotherm in the fog area, cf. [1]. This circumstance is now utilised: The determination of (2) is carried out with the help of a ruler, which is placed at point (1) and rotated in it until the straight line coincides exactly with an isotherm in the fog area, see Fig. 8.8.

---

[3] The initial temperature of the beer is irrelevant, as we are only investigating the steady state. However, the initial temperature has an influence on how long it takes to reach this state. But this is not the question here.

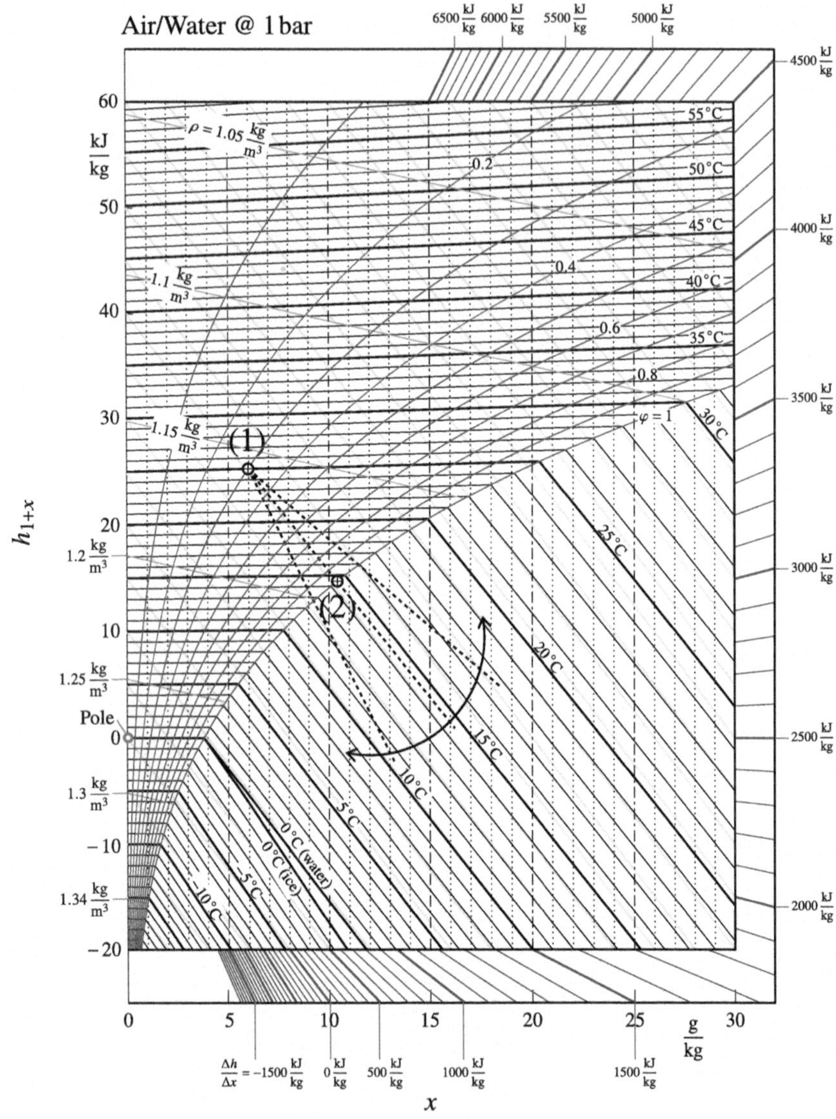

Fig. 8.8 Evaporative cooling

The intersection of this straight line with the $\varphi = 1$-isoline results in state (2) that is in saturated state. The change of state now has the slope of the isotherm ϑ_2 in the fog region, which corresponds to the slope of a humidification with liquid water, cf. Eq. 8.51. A temperature ϑ_2 of approx. 14.4 °C is determined in the h_{1+x}, x-diagram. Answer (b) is correct.

8.7 **(b)**
The following is known from state (1)

$$\varphi_1 = 0.3 \quad \text{and} \quad \vartheta_1 = 25\,^{\circ}\text{C}. \tag{8.52}$$

The vapour pressure in state (1) yields

$$p_{\text{v},1} = \varphi_1 p_{\text{s}}(\vartheta_1) = 0.0095\,\text{bar} \quad \text{with} \quad p_{\text{s}}(\vartheta_1) = 0.0317\,\text{bar}. \tag{8.53}$$

Hence, the vapour content in state (1) obeys

$$x_1 = 0.622 \frac{p_{\text{v},1}}{p - p_{\text{v},1}} = 0.0060 \tag{8.54}$$

and the specific enthalpy is

$$h_{1+x,1} = c_{p,\text{a}}\vartheta_1 + x_1\left[\Delta h_{\text{v},0} + c_{p,\text{v}}\vartheta_1\right] = 40.3065\,\frac{\text{kJ}}{\text{kg}}. \tag{8.55}$$

The water content increases by $\Delta x = 3\,\frac{\text{g}}{\text{kg}}$, so that

$$x_2 = x_1 + \Delta x = 0.0090. \tag{8.56}$$

The mass balance, see Fig. 8.9, for the dry air is

$$\dot{m}_{\text{a},1} = \dot{m}_{\text{a},2} \tag{8.57}$$

and the mass balance for the water yields

$$\underbrace{\dot{m}_{\text{a},1}x_1 + \dot{m}_{\text{liq}}}_{\text{IN}} = \underbrace{\dot{m}_{\text{a},2}x_2}_{\text{OUT}} \tag{8.58}$$

so that

$$\dot{m}_{\text{liq}} = \dot{m}_{\text{a},1}\Delta x. \tag{8.59}$$

The energy balance, see Fig. 8.9, for the steady state, open system is as follows

$$\underbrace{\dot{m}_{\text{a},1}h_{1+x,1} + \dot{m}_{\text{liq}}h_{\text{liq}}}_{\text{IN}} = \underbrace{\dot{m}_{\text{a},2}h_{1+x,2}}_{\text{OUT}} \quad \text{with} \quad h_{\text{liq}} = c_{\text{liq}}\vartheta_{\text{liq}}. \tag{8.60}$$

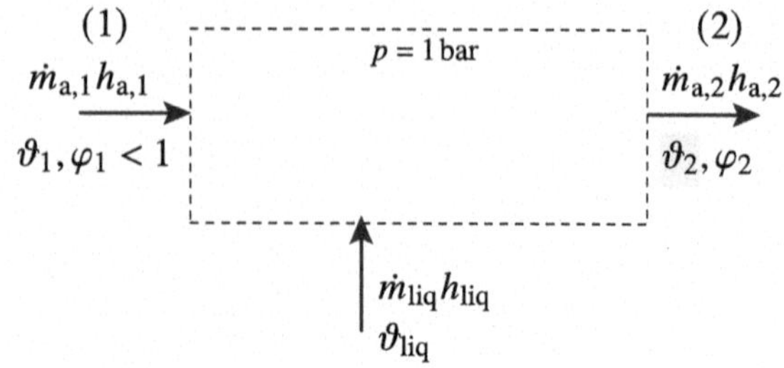

Fig. 8.9 Humidification

Hence, it is

$$h_{1+x,2} = h_{1+x,1} + \Delta x\, c_{liq}\vartheta_{liq} = h_{1+x,1} = 40.3065\,\frac{kJ}{kg} \quad \text{with} \quad \vartheta_{liq} = 0\,°C. \quad (8.61)$$

Finally, the temperature ϑ_2 at the outlet needs to fulfill

$$h_{1+x,2} = c_{p,a}\vartheta_2 + x_2\left[\Delta h_{v,0} + c_{p,v}\vartheta_2\right] \quad (8.62)$$

so that

$$\vartheta_2 = \frac{h_{1+x,2} - x_2\Delta h_{v,0}}{c_{p,a} + x_2 c_{p,v}} = 17.5153\,°C. \quad (8.63)$$

Note that has been assumed that state (2) is unsaturated. Since

$$x_2 = 0.0090 < 0.622\frac{p_s(\vartheta_2)}{p - p_s(\vartheta_2)} = 0.0127 \quad (8.64)$$

the assumption has been correct.

Alternatively, a graphical solution is possible, see Fig. 8.10. State (1) is unambiguously fixed due to the specification of ϑ_1, φ_1 and p and can be drawn into the h_{1+x}, x-diagram, see Fig. 8.10. Because of the specification of $\Delta x = 3\,\frac{g}{kg}$ the water content x_2 of state (2) is known and can be drawn as a vertical line in the diagram. However, we only know that (2) lies on this vertical line. State (2) cannot yet be clearly assigned so that a second information is required. The change of state (1) to (2) corresponds to an adiabatic humidification with liquid water. The gradient of such a change of state yields, cf. [1],

$$\frac{\Delta h}{\Delta x} = h_{liq} = c_{liq}\vartheta_{liq} = 0\,\frac{kJ}{kg} \quad \text{with} \quad \vartheta_{liq} = 0\,°C. \quad (8.65)$$

This slope is drawn in with the border scale and the pole of the h_{1+x}, x-diagram, see Fig. 8.10. This slope is now shifted parallel through state (1). State (2) is where

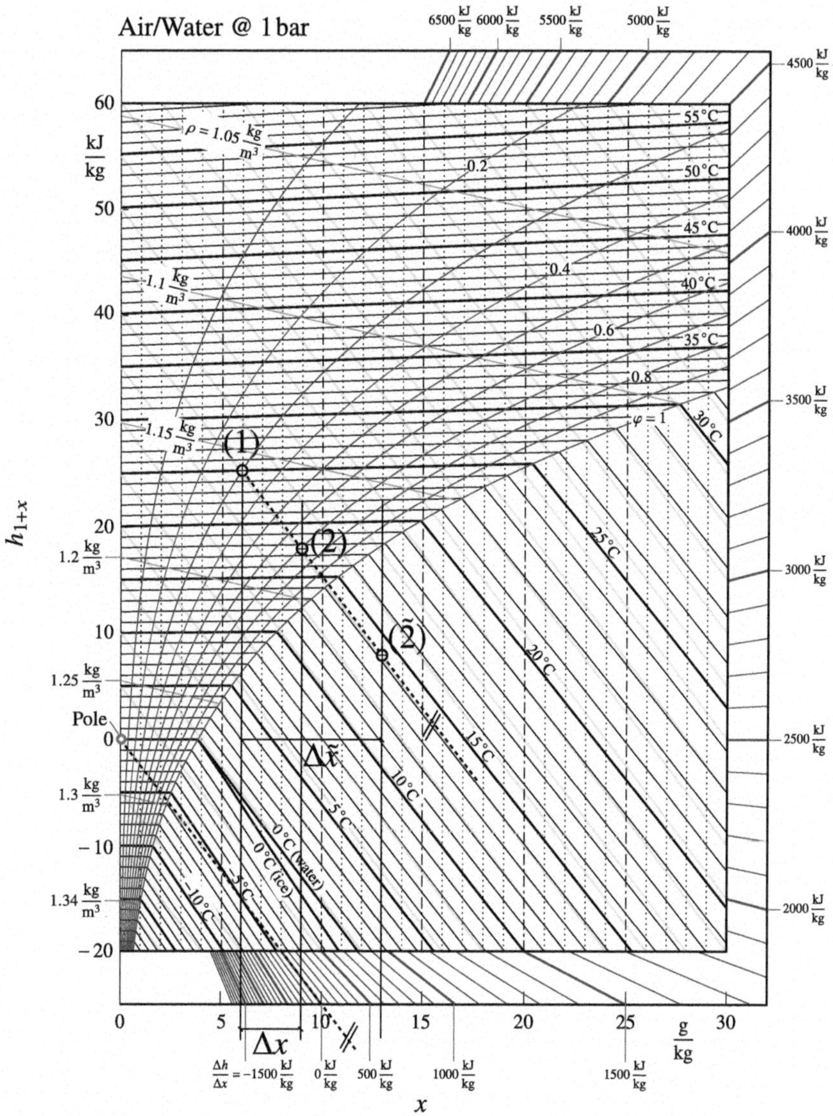

Fig. 8.10 Humidification

the straight line intersects the x_2-vertical. A temperature ϑ_2 of approx. 17.5 °C is determined in the h_{1+x}, x-diagram. State (2) is unsaturated. Answer (b) is correct.

What would happen if more water was injected under the same boundary conditions, e.g. $\Delta\tilde{x} = 7\,\frac{\text{g}}{\text{kg}}$, i.e. $x_{\tilde{2}} = x_1 + \Delta x = 0.0130$?

The graphical solution (1) to ($\tilde{2}$) would be carried out analogously to the first part, see Fig. 8.10. It can be seen, however, that state ($\tilde{2}$) then ends up in the fog area. A temperature of approx. 14.2 °C can be determined. Mathematically, we derive the same balances as in the first part, i.e.

$$h_{1+x,\tilde{2}} = h_{1+x,1} + \Delta x\, c_{\text{liq}}\vartheta_{\text{liq}} = h_{1+x,1} = 40.3065\,\frac{\text{kJ}}{\text{kg}} \quad \text{with} \quad \vartheta_{\text{liq}} = 0\,^\circ\text{C}. \tag{8.66}$$

However, the determination of the specific enthalpy of ($\tilde{2}$) is now somewhat more complex, since the air is saturated with vapour and additional liquid water occurs, i.e.

$$h_{1+x,\tilde{2}} = c_{p,\text{a}}\vartheta_{\tilde{2}} + x_{\text{s}}\left[\Delta h_{\text{v},0} + c_{p,\text{v}}\vartheta_{\tilde{2}}\right] + \left(x_{\tilde{2}} - x_{\text{s}}\right) c_{\text{liq}}\vartheta_{\tilde{2}} \tag{8.67}$$

with

$$x_{\text{s}} = 0.622\frac{p_{\text{s}}\left(\vartheta_{\tilde{2}}\right)}{p - p_{\text{s}}\left(\vartheta_{\tilde{2}}\right)}. \tag{8.68}$$

Hence, the following equation

$$\begin{aligned} &c_{p,\text{a}}\vartheta_{\tilde{2}} + 0.622\frac{p_{\text{s}}\left(\vartheta_{\tilde{2}}\right)}{p - p_{\text{s}}\left(\vartheta_{\tilde{2}}\right)}\left[\Delta h_{\text{v},0} + c_{p,\text{v}}\vartheta_{\tilde{2}}\right] + \\ &+ \left(x_{\tilde{2}} - 0.622\frac{p_{\text{s}}\left(\vartheta_{\tilde{2}}\right)}{p - p_{\text{s}}\left(\vartheta_{\tilde{2}}\right)}\right) c_{\text{liq}}\vartheta_{\tilde{2}} = h_{1+x,1} \end{aligned} \tag{8.69}$$

needs to be solved. A numerical solution is found with $\vartheta_{\tilde{2}} = 14.2051$ °C.

8.8 **(a)**
The basics of evaporative cooling have already been discussed in Problem 8.6, so we jump straight into the solution, see also Fig. 8.11. In the first part we want to solve the problem mathematically. The following is known from state (2)

$$\varphi_2 = 1 \quad \text{and} \quad \vartheta_2 = 12\,^\circ\text{C}. \tag{8.70}$$

The vapour pressure in state (2) yields

$$p_{\text{v},2} = p_{\text{s}}\left(\vartheta_2\right) = 0.0140\,\text{bar}. \tag{8.71}$$

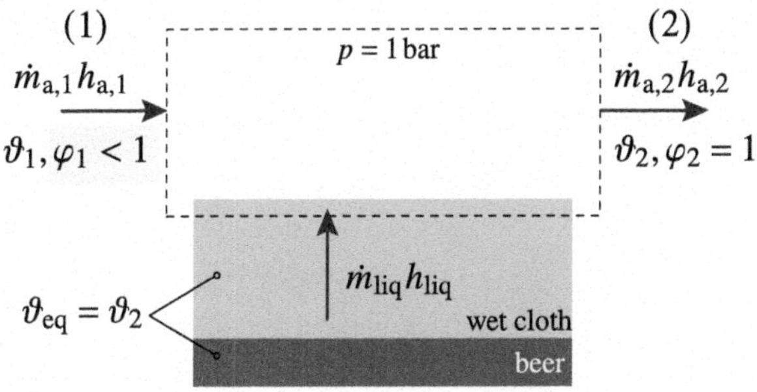

Fig. 8.11 Evaporative cooling

Hence, the vapour content in state (2) obeys

$$x_2 = 0.622 \frac{p_{v,2}}{p - p_{v,2}} = 0.0088 \tag{8.72}$$

and the specific enthalpy is

$$h_{1+x,2} = c_{p,a}\vartheta_2 + x_2 \left[\Delta h_{v,0} + c_{p,v}\vartheta_2\right] = 34.3698 \, \frac{\text{kJ}}{\text{kg}}. \tag{8.73}$$

In state (1) the air is unsaturated and its temperature is $\vartheta_1 = 20\,°\text{C}$. It applies

$$p_{v,1} = \varphi_1 p_s(\vartheta_1) \quad \text{with} \quad p_s(\vartheta_1) = 0.0234\,\text{bar} \tag{8.74}$$

and

$$x_1 = 0.622 \frac{\varphi_1 p_s(\vartheta_1)}{p - \varphi_1 p_s(\vartheta_1)}. \tag{8.75}$$

The specific enthalpy in state (1) obeys

$$h_{1+x,1} = c_{p,a}\vartheta_1 + x_1 \left[\Delta h_{v,0} + c_{p,v}\vartheta_1\right]. \tag{8.76}$$

The mass balance for the dry air is

$$\dot{m}_{a,1} = \dot{m}_{a,2} \tag{8.77}$$

and the mass balance for the water yields

$$\underbrace{\dot{m}_{a,1}x_1 + \dot{m}_{liq}}_{\text{IN}} = \underbrace{\dot{m}_{a,2}x_2}_{\text{OUT}} \tag{8.78}$$

so that

$$\dot{m}_{liq} = \dot{m}_{a,1}\left(x_2 - x_1\right). \tag{8.79}$$

The energy balance[4] for the steady state, open system is as follows

$$\underbrace{\dot{m}_{a,1}h_{1+x,1} + \dot{m}_{liq}h_{liq}}_{IN} = \underbrace{\dot{m}_{a,2}h_{1+x,2}}_{OUT} \quad \text{with} \quad h_{liq} = c_{liq}\vartheta_2 = 50.2800\,\frac{kJ}{kg}. \tag{8.80}$$

Hence, it is

$$h_{1+x,1} + (x_2 - x_1)\,c_{liq}\vartheta_2 = h_{1+x,2} \tag{8.81}$$

respectively

$$\boxed{c_{p,a}\vartheta_1 + 0.622\frac{\varphi_1 p_s(\vartheta_1)}{p - \varphi_1 p_s(\vartheta_1)}\left[\Delta h_{v,0} + c_{p,v}\vartheta_1\right] + \left(x_2 - 0.622\frac{\varphi_1 p_s(\vartheta_1)}{p - \varphi_1 p_s(\vartheta_1)}\right)h_{liq} = h_{1+x,2}} \tag{8.82}$$

This equation contains the relative humidity φ_1 as an unknown and can be solved numerically. The solution is $\varphi_1 = 0.3792$.

In the second part we solve the problem graphically with the h_{1+x}, x-diagram, see Fig. 8.12. State (2) can be easily depicted in such a diagram, since temperature and relative humidity $\varphi_2 = 1$ are given at $p = 1$ bar. Starting from state (1), which has not yet been defined in the diagram, adiabatic humidification with liquid water is carried out. For such a change of state, one determines the slope of the change of state

$$\frac{\Delta h}{\Delta x} = h_{liq} = c_{liq}\vartheta_2 = 50.2800\,\frac{kJ}{kg}. \tag{8.83}$$

This slope is drawn in with the border scale and the pole of the h_{1+x}, x-diagram, see Fig. 8.12. This slope is now shifted parallel through state (2). Where the straight line then intersects the ϑ_1-isotherm is state (1). Alternatively, since the slope of the ϑ_2-isotherm in the fog region corresponds to the slope of the change of state, see Eq. 8.83, one can extend the fog isotherm of (2) into the region of unsaturated air. The intersection with the ϑ_1-isotherm yields state (1). A relative humidity φ_1 of approx. 38 % is determined in the h_{1+x}, x-diagram. Answer (a) is correct.

8.9 **(d)**
The adiabatic isothermal/isobaric mixing process for two different ideal gases has already been discussed in Problem 8.4 and will now only be briefly summarised, cf. Fig. 8.13. The first law of thermodynamics obeys

$$Q_{AB} + W_{AB} = 0 = \Delta U = \Delta U_1 + \Delta U_2 \tag{8.84}$$

---

[4] The initial temperature of the beer is irrelevant, as we are only investigating the steady state. However, the initial temperature has an influence on how long it takes to reach this state. But this is not the question here.

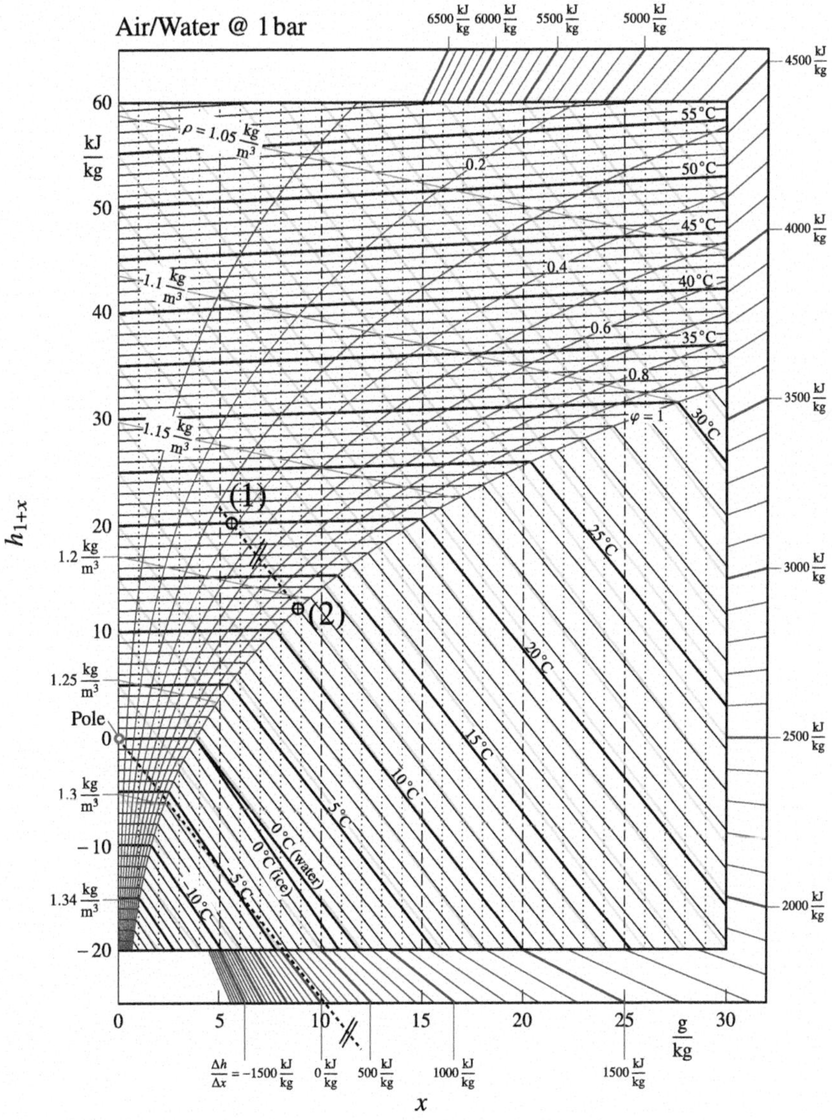

Fig. 8.12 Evaporative cooling

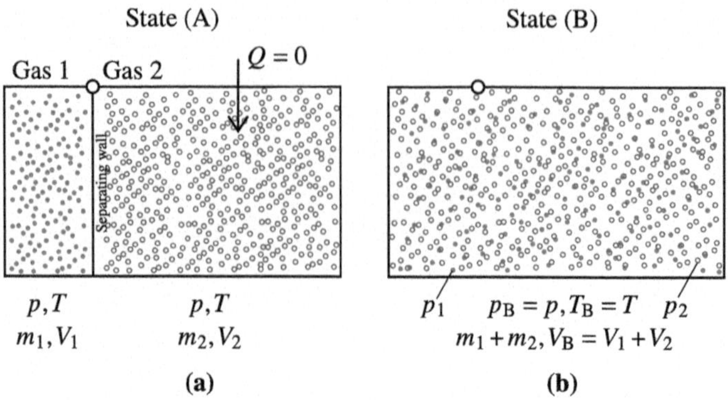

Fig. 8.13 Adiabatic, isothermal/isobaric mixing

respectively with the caloric equation of state

$$m_1 c_v (T_B - T) + m_2 c_v (T_B - T) = 0. \tag{8.85}$$

Therefore, $T_B = T$, so that the temperature before and after mixing is identical. Answer (a) does not apply. The thermal equation of state for gas 1 obeys

- State A

$$pV_1 = m_1 R_1 T \tag{8.86}$$

- State B

$$p_1 (V_1 + V_2) = m_1 R_1 T = pV_1 \tag{8.87}$$

and for gas 2

- State A

$$pV_2 = m_2 R_2 T \tag{8.88}$$

- State B

$$p_2 (V_1 + V_2) = m_2 R_2 T = pV_2. \tag{8.89}$$

Consequently, it is

$$p_1 + p_2 = p. \tag{8.90}$$

The sum of the partial pressures p_1 and p_2 gives the total pressure p. Answer (c) does not apply. The entropy balance follows

$$S_B = S_A + S_{i,AB} + \underbrace{S_{a,AB}}_{=0} \tag{8.91}$$

respectively

$$S_{\mathrm{i,AB}} = S_\mathrm{B} - S_\mathrm{A}. \tag{8.92}$$

Entropy as an extensive state value is composed of the parts for gas 1 and gas 2, i.e.

$$S_{\mathrm{i,AB}} = S_\mathrm{B} - S_\mathrm{A} = m_1\left(s_\mathrm{B} - s_\mathrm{A}\right)_1 + m_2\left(s_\mathrm{B} - s_\mathrm{A}\right)_2. \tag{8.93}$$

It follows with the caloric equation of state

$$\begin{aligned} S_{\mathrm{i,AB}} &= m_1\left(c_{p,1}\ln\frac{T}{T} - R_1\ln\frac{p_1}{p}\right)_1 + m_2\left(c_{p,2}\ln\frac{T}{T} - R_2\ln\frac{p_2}{p}\right)_2 \\ &= -m_1R_1\ln\frac{p_1}{p} - m_2R_2\ln\frac{p_2}{p} > 0. \end{aligned} \tag{8.94}$$

Since the partial pressures p_1 and p_2 are smaller than p, cf. Eq. 8.90, entropy is generated during the mixing. In other words, the mixing process is irreversible. The driver for the transient process is the concentration differences. When the concentrations are balanced in the state of equilibrium, the system can no longer move back to the initial state on its own. Since the process is adiabatic, the entropy of the system must therefore increase. One would therefore tend to choose answer (b). However, the above equation only applies if the fluids are *different* ideal gases 1 and 2. If the gases are identical, as in this case, no entropy is generated. This is called Gibbs paradox. Answer (b) is therefore wrong and answer (d) is correct.

8.10 **(a)**
The following is known from state (1)

$$\varphi_1 = 0.5 \quad \text{and} \quad \vartheta_1 = 25\,°\mathrm{C}. \tag{8.95}$$

The vapour pressure in state (1) yields

$$p_{\mathrm{v},1} = \varphi_1 p_\mathrm{s}\left(\vartheta_1\right) = 0.0158\,\mathrm{bar} \quad \text{with} \quad p_\mathrm{s}\left(\vartheta_1\right) = 0.0317\,\mathrm{bar}. \tag{8.96}$$

Hence, the vapour content in state (1) obeys

$$x_1 = 0.622\frac{p_{\mathrm{v},1}}{p - p_{\mathrm{v},1}} = 0.0100 \tag{8.97}$$

and the specific enthalpy is

$$h_{1+x,1} = c_{p,\mathrm{a}}\vartheta_1 + x_1\left[\Delta h_{\mathrm{v},0} + c_{p,\mathrm{v}}\vartheta_1\right] = 50.6074\,\frac{\mathrm{kJ}}{\mathrm{kg}}. \tag{8.98}$$

The water content increases by Δx, until state (2) is saturated, i.e.

$$\varphi_2 = 1 \quad \text{and} \quad p_{\mathrm{v},2} = p_\mathrm{s}\left(\vartheta_2\right) \tag{8.99}$$

The vapour content then is

$$x_2 = x_1 + \Delta x = 0.622 \frac{p_s(\vartheta_2)}{p - p_s(\vartheta_2)}. \tag{8.100}$$

The mass balance, see Fig. 8.14, for the dry air is

$$\dot{m}_{a,1} = \dot{m}_{a,2} \tag{8.101}$$

and the mass balance for the water yields

$$\underbrace{\dot{m}_{a,1} x_1 + \dot{m}_{liq}}_{\text{IN}} = \underbrace{\dot{m}_{a,2} x_2}_{\text{OUT}} \tag{8.102}$$

so that

$$\dot{m}_{liq} = \dot{m}_{a,1} \Delta x. \tag{8.103}$$

The energy balance, see Fig. 8.14, for the steady state, open system is as follows

$$\underbrace{\dot{m}_{a,1} h_{1+x,1} + \dot{m}_{liq} h_{liq}}_{\text{IN}} = \underbrace{\dot{m}_{a,2} h_{1+x,2}}_{\text{OUT}} \quad \text{with} \quad h_{liq} = c_{liq} \vartheta_{liq}. \tag{8.104}$$

Hence, it is

$$h_{1+x,2} = h_{1+x,1} + \Delta x \, c_{liq} \vartheta_{liq} \quad \text{with} \quad \vartheta_{liq} = 25\,°\text{C}. \tag{8.105}$$

with the specific enthalpy in the saturated state (2)

$$h_{1+x,2} = c_{p,a} \vartheta_2 + x_2 \left[\Delta h_{v,0} + c_{p,v} \vartheta_2 \right]. \tag{8.106}$$

Finally, the following equation needs to be solved

$$\boxed{c_{p,a} \vartheta_2 + 0.622 \frac{p_s(\vartheta_2)}{p - p_s(\vartheta_2)} \left[\Delta h_{v,0} + c_{p,v} \vartheta_2 \right] = h_{1+x,1} + \left(0.622 \frac{p_s(\vartheta_2)}{p - p_s(\vartheta_2)} - x_1 \right) c_{liq} \vartheta_{liq}} \tag{8.107}$$

The only unknown in this equation is temperature ϑ_2 of the saturated state (2). A numerical solution leads to $\vartheta_2 = 17.8821\,°\text{C}$. The vapor content then is

$$x_2 = 0.622 \frac{p_s(\vartheta_2)}{p - p_s(\vartheta_2)} = 0.0130 \quad \text{with} \quad p_s(\vartheta_2) = 0.0205\,\text{bar}. \tag{8.108}$$

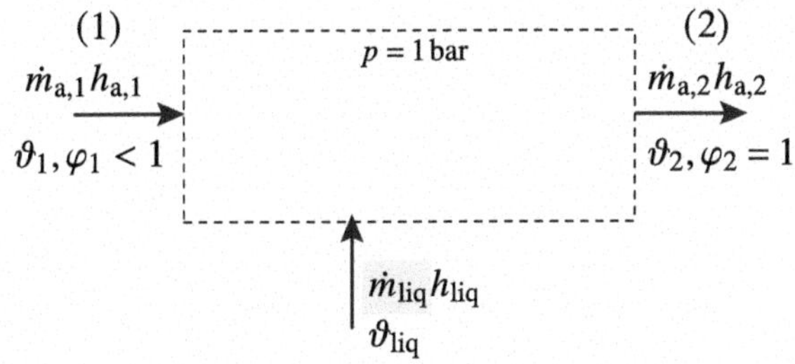

Fig. 8.14 Humidification

Finally, the mass flow rate of the injected water obeys

$$\dot{m}_{liq} = \dot{m}_{a,1} \Delta x = \dot{m}_{a,1} (x_2 - x_1) = 2.9973 \,\frac{\text{g}}{\text{s}}. \tag{8.109}$$

Alternatively, a graphical solution is possible, see Fig. 8.15. State (1) is unambiguously fixed due to the specification of ϑ_1, φ_1 and p and can be drawn into the h_{1+x}, x-diagram, see Fig. 8.15. We further know that (2) lies on the $\varphi = 1$-saturation curve. State (2) cannot yet be clearly assigned so that a second information is required. The change of state (1) to (2) corresponds to an adiabatic humidification with liquid water. The gradient of such a change of state yields, cf. [1],

$$\frac{\Delta h}{\Delta x} = h_{liq} = c_{liq} \vartheta_{liq} = 104.75 \,\frac{\text{kJ}}{\text{kg}} \quad \text{with} \quad \vartheta_{liq} = 25\,°\text{C}. \tag{8.110}$$

This slope is drawn in with the border scale and the pole of the h_{1+x}, x-diagram, see Fig. 8.15. This slope is now shifted parallel through state (1). State (2) is where the straight line intersects the $\varphi = 1$-saturation curve. A temperature ϑ_2 of approx. 17.9 °C is determined in the h_{1+x}, x-diagram. Furthermore, the vapour content is approx. $x_2 = 13 \,\frac{\text{g}}{\text{kg}} = 0.013$. Hence, the mass flow rate of the injected water obeys

$$\dot{m}_{liq} = \dot{m}_{a,1} \Delta x = \dot{m}_{a,1} (x_2 - x_1) = 3.0 \,\frac{\text{g}}{\text{s}}. \tag{8.111}$$

Therefore, answer (a) is correct.

8.11 **(b)**
The following information on state (1) is known

$$\vartheta_1 = 20\,°\text{C} \quad \text{with} \quad \varphi_1 = 0.8 \quad \text{at} \quad p_1 = 2\,\text{bar}. \tag{8.112}$$

In the first part, the problem is solved mathematically. The vapour pressure of state (1) obeys

$$p_{v,1} = \varphi_1 p_s(\vartheta_1) = 0.0187\,\text{bar} \quad \text{with} \quad p_s(\vartheta_1) = 0.0234\,\text{bar}. \tag{8.113}$$

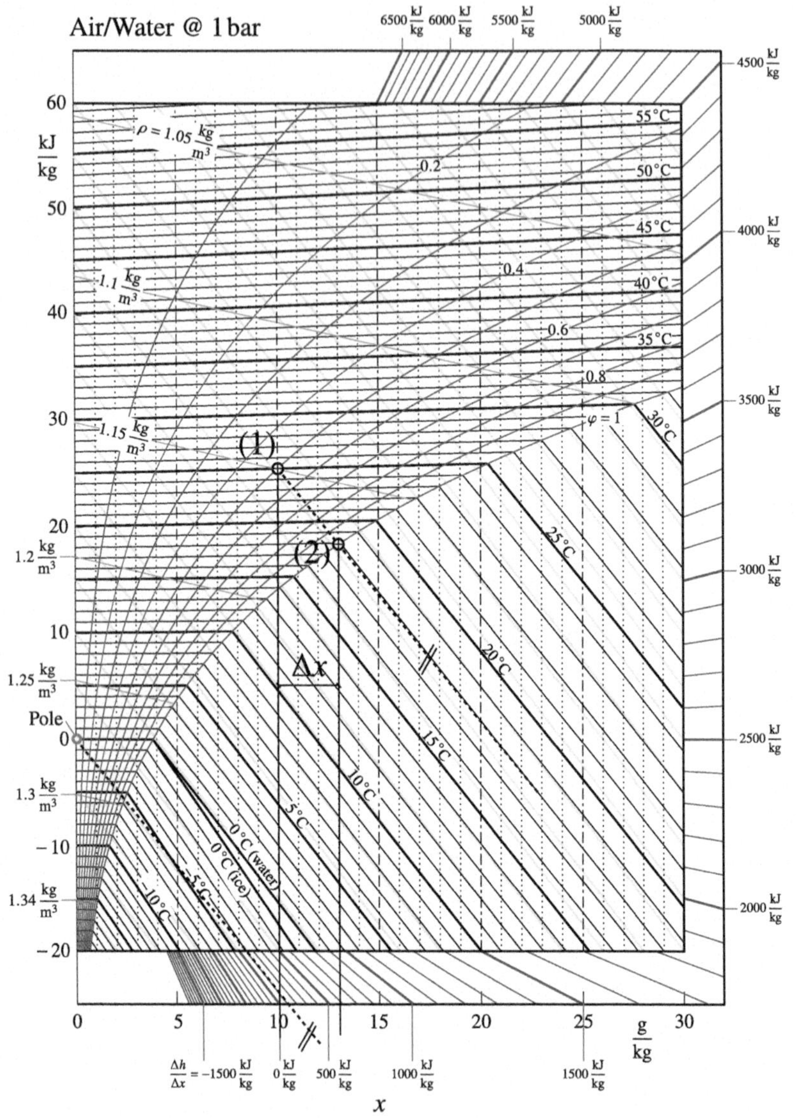

Fig. 8.15 Humidification

The vapour content reads as

$$x_1 = 0.622 \frac{p_{v,1}}{p_1 - p_{v,1}} = 0.0059. \tag{8.114}$$

This results in a specific enthalpy of

$$h_{1+x,1} = c_{p,a}\vartheta_1 + x_1 \left[\Delta h_{v,0} + c_{p,v}\vartheta_1\right] = 34.9859 \, \frac{\text{kJ}}{\text{kg}}. \tag{8.115}$$

The second part deals with a graphical solution in the h_{1+x}, x-diagram. This is not quite as simple as it looks at first glance, since state (1) has a pressure of 2 bar, but the h_{1+x}, x-diagram applies to 1 bar! It is

$$\varphi = \frac{p_v}{p_s(\vartheta)} \tag{8.116}$$

and

$$x = 0.622 \frac{p_v}{p - p_v} \rightarrow p_v = \frac{xp}{0.622 + x} \tag{8.117}$$

Combing these two equations leads to

$$\frac{\varphi}{p} = \frac{x}{0.622 + x} \cdot \frac{1}{p_s(\vartheta)}. \tag{8.118}$$

Now we transfer state (1) to the reference diagram (Index: ref) for 1 bar. Temperature and vapour content remain constant when transferred to the reference diagram, as they are independent of pressure, so that

$$\frac{\varphi}{p} = \frac{x}{0.622 + x} \cdot \frac{1}{p_s(\vartheta)} = \text{const.} \tag{8.119}$$

Hence it is

$$\frac{\varphi}{p} = \frac{\varphi_1}{p_1} = \frac{\varphi_{\text{ref}}}{p_{\text{ref}}} = \text{const.} \tag{8.120}$$

respectively

$$\varphi_{\text{ref}} = p_{\text{ref}} \frac{\varphi_1}{p_1} = 0.4 \quad \text{with} \quad p_{\text{ref}} = 1\,\text{bar}. \tag{8.121}$$

Now state (1) can be transferred to a standard h_{1+x}, x-diagram for 1 bar. To do this, fix the state with

$$(\vartheta_1, \varphi_1) \rightarrow \boxed{(\vartheta_1, \varphi_{\text{ref}})} \tag{8.122}$$

The saturation isolines must therefore be converted, see Fig. 8.16. The other state values can be read off as usual. One finds for the specific enthalpy

$$h_{1+x,1} = 35\,\frac{\text{kJ}}{\text{kg}}. \tag{8.123}$$

Answer (d) is correct.

8.12 The problem is outlined in Fig. 8.16. In state (1), the pressure in the gas phase is only imposed by the dry air, i.e.

$$p_{a,1} = p_1 = 0.96\,\text{bar} \tag{8.124}$$

and the partial pressure of the vapour must be

$$p_{v,1} = p_1 - p_{a,1} = 0\,\text{bar}. \tag{8.125}$$

Since the volume change of the liquid water is to be neglected, i.e. $V_{a,1} = V_{a,2}$, it is an isochoric change of state[5] (1) to (2). The thermal equation of state for the dry air in state (1) is

$$p_{a,1} V_{a,1} = R_a T_1 \tag{8.126}$$

and for state (2)

$$p_{a,2} V_{a,2} = R_a T_2. \tag{8.127}$$

This results in

$$p_{a,2} = p_{a,1}\frac{T_2}{T_1} = 1.222\,\text{bar}. \tag{8.128}$$

Thermodynamic equilibrium also means that the chemical potentials, in this case between the liquid water and the vapour in the gas phase, are balanced (Fig. 8.17). This is the case when the vapour in the air has reached saturation state, i.e.

$$p_{v,2} = p_s\,(100\,°\text{C}) = 1.0142\,\text{bar}. \tag{8.129}$$

The total pressure in state (2) is then the sum of the partial pressures of vapour and dry air

$$p_2 = p_{a,2} + p_{v,2} = 2.2362\,\text{bar}. \tag{8.130}$$

8.13

(a) The mass flow rate obeys the thermal equation of state, i.e.

$$\dot{m}_a R_a T_1 = p_{a,1} \dot{V} \tag{8.131}$$

[5] Note that in thermal equilibrium the gas phase and the liquid have a uniform temperature.

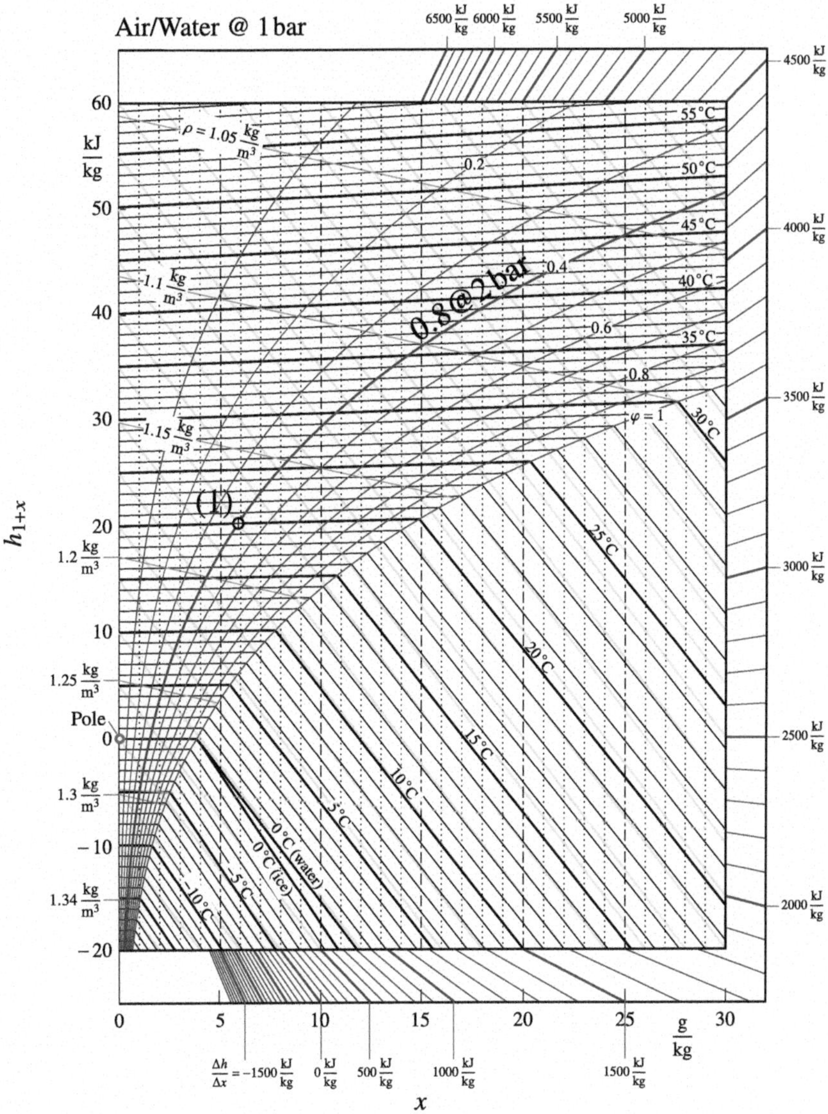

Fig. 8.16 h_{1+x}, x-diagram @2 bar

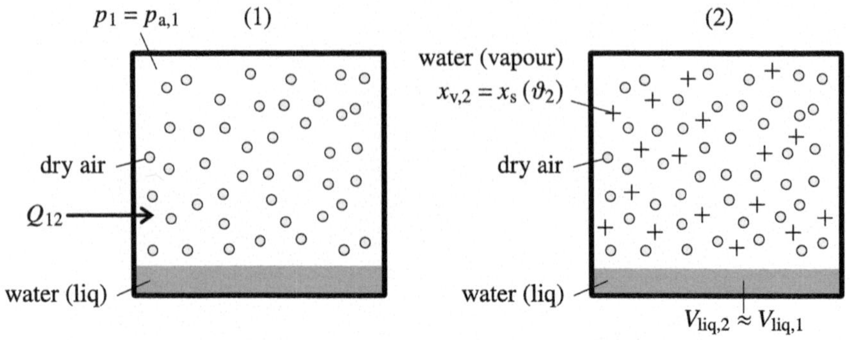

Fig. 8.17 Closed vessel

with

$$\dot{V} = A_1 c_1 = 20\,\frac{\mathrm{m}^3}{\mathrm{s}}. \tag{8.132}$$

The partial pressures yield

$$p_{v,1} = \varphi_1 p_s(\vartheta_1) = 0.0270\,\mathrm{bar} \tag{8.133}$$

and

$$p_{a,1} = p_1 - p_{v,1} = 0.8730\,\mathrm{bar}. \tag{8.134}$$

The mass flow rate of dry air then is

$$\dot{m}_a = \frac{p_{a,1}\dot{V}}{R_a T_1} = 18.8507\,\frac{\mathrm{kg}}{\mathrm{s}}. \tag{8.135}$$

(b) The water content in state (1) is

$$x_1 = x_{v,1} = 0.622\frac{p_{v,1}}{p_1 - p_{v,1}} = 0.0193. \tag{8.136}$$

The flow rate of water in state (1) then reads as

$$\dot{m}_{w,1} = \dot{m}_{v,1} = x_{v,1}\dot{m}_a = 0.3632\,\frac{\mathrm{kg}}{\mathrm{s}}. \tag{8.137}$$

According to its temperature, the maximum load of vapour in state (2) follows

$$x'_{v,2} = 0.622\frac{p_s(\vartheta_2)}{p_2 - p_s(\vartheta_2)} = 0.0159. \tag{8.138}$$

The mass flow rate of liquid water yields

$$\dot{m}_{\text{liq},2} = \dot{m}_{\text{w},1} - \dot{m}_{\text{v},2} = \dot{m}_{\text{w},1} - \dot{m}_{\text{a}} x'_{\text{v},2} = 0.0640\,\frac{\text{kg}}{\text{s}}. \tag{8.139}$$

(c) The first law of thermodynamics obeys[6]

$$\dot{Q}_{12} + \underbrace{P_{12}}_{=0} = \dot{H}_2 - \dot{H}_1 + \underbrace{\dot{E}_{\text{kin},2}}_{=0} - \dot{E}_{\text{kin},1} \tag{8.140}$$

respectively

$$\dot{Q}_{12} = \dot{m}_{\text{a}}\left[h_{1+x,2} - h_{1+x,1}\right] - \left(\dot{m}_{\text{a}} + \dot{m}_{\text{v},1}\right)\frac{1}{2}c_1^2. \tag{8.141}$$

The specific enthalpies in state (1) and state (2) follow

$$h_{1+x,1} = c_{p,\text{a}}\vartheta_1 + x_1\left[\Delta h_{\text{v},0} + c_{p,\text{v}}\vartheta_1\right] = 99.5874\,\frac{\text{kJ}}{\text{kg}} \tag{8.142}$$

and

$$h_{1+x,2} = c_{p,\text{a}}\vartheta_2 + x'_{\text{v},2}\left[\Delta h_{\text{v},0} + c_{p,\text{v}}\vartheta_2\right] + \left(x_1 - x'_{\text{v},2}\right)c_{\text{liq}}\vartheta_2 = 61.6799\,\frac{\text{kJ}}{\text{kg}}. \tag{8.143}$$

Finally, the heat flux is

$$\dot{Q}_{12} = \dot{m}_{\text{a}}\left[h_{1+x,2} - h_{1+x,1} - (1 + x_1)\frac{c_1^2}{2}\right] = -1.0989 \times 10^3\,\text{kW}. \tag{8.144}$$

(d) In order to calculate the flux of loss of exergy

$$\dot{E}_{\text{x,V}} = T_{\text{env}}\dot{S}_{\text{i}} \tag{8.145}$$

an entropy balance needs to be carried out, see Fig. 8.18. Such a balance obeys

$$\dot{m}_{\text{a}} s_{1+x,1} + \dot{S}_{\text{i}} + \dot{S}_{\text{a}} = \dot{m}_{\text{a}} s_{1+x,2} \tag{8.146}$$

with[7]

$$\dot{S}_{\text{a}} = \frac{\dot{Q}_{12}}{T_{\text{env}}}. \tag{8.147}$$

[6] Potential energies are disregarded.

[7] According to the selected system boundary, heat is exchanged at ambient temperature.

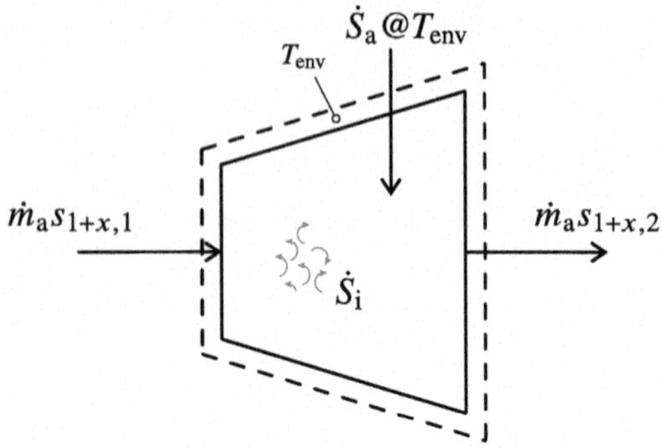

Fig. 8.18 Entropy balance diffusor

The flux of entropy generation is therefore

$$\dot{S}_i = \dot{m}_a \left[s_{1+x,2} - s_{1+x,1}\right] - \frac{\dot{Q}_{12}}{T_{env}}. \tag{8.148}$$

With the following reference states

$$T_{0,a} = 273.15\,\mathrm{K} \quad p_{0,a} = 1\,\mathrm{bar} \tag{8.149}$$

$$T_{0,w} = 273.16\,\mathrm{K} \quad p_{0,w} = 0.006\,117\,\mathrm{bar} \tag{8.150}$$

the specific entropies are

- State (1)—Unsaturated

$$\begin{aligned} s_{1+x,1} =& c_{p,a} \ln \frac{T_1}{T_{0,a}} - R_a \ln \frac{p_1 - p_{v,1}}{p_{0,a}} + \\ &+ x_{v,1} \left[c_{p,v} \ln \frac{T_1}{T_{0,w}} - R_v \ln \frac{p_{v,1}}{p_{0,w}} + \frac{\Delta h_{v,0}}{T_{0,a}} \right] \\ &= 0.3752\,\frac{\mathrm{kJ}}{\mathrm{kg\,K}}. \end{aligned} \tag{8.151}$$

- State (2)—Fog region

$$\begin{aligned} s'_{1+x,2} =& c_{p,a} \ln \frac{T_2}{T_{0,a}} - R_a \ln \frac{p_2 - p_s(\vartheta_2)}{p_{0,a}} + \\ &+ x'_{v,2} \left[c_{p,v} \ln \frac{T_2}{T_{0,w}} - R_v \ln \frac{p_s(\vartheta_2)}{p_{0,w}} + \frac{\Delta h_{v,0}}{T_{0,a}} \right] \\ &= 0.2188\,\frac{\mathrm{kJ}}{\mathrm{kg\,K}} \end{aligned} \tag{8.152}$$

so that

$$s_{1+x,2} = s'_{1+x,2} + \left(x_1 - x'_{v,2}\right) \cdot c_{liq} \ln \frac{T_2}{T_{0,w}} = 0.2198 \, \frac{kJ}{kg\,K}. \tag{8.153}$$

The flux of entropy generation thus yields

$$\dot{S}_i = \dot{m}_a \left[s_{1+x,2} - s_{1+x,1}\right] - \frac{\dot{Q}_{12}}{T_{env}} = 1.0948 \, \frac{kW}{K}. \tag{8.154}$$

Finally, the flux of loss of exergy follows

$$\dot{E}_{x,V} = T_{env}\dot{S}_i = 299.0411 \, kW. \tag{8.155}$$

8.14 The mixing process is shown schematically in Fig. 8.19. In a h_{1+x}, x-diagram, see Fig. 8.20, one can verify that both state (1) and state (2) are unsaturated. In the first part, the problem is solved mathematically. The mass balance for dry air obeys

$$\dot{m}_{a,1} + \dot{m}_{a,2} = \dot{m}_{a,3} \quad \text{with} \quad \dot{m}_{a,1} = \dot{m}_{a,2} = \dot{m}_a \tag{8.156}$$

so that

$$\dot{m}_{a,3} = 2\dot{m}_a \tag{8.157}$$

The balance for the water reads as

$$\underbrace{\dot{m}_a x_1}_{=\dot{m}_{w,1}} + \underbrace{\dot{m}_a x_2}_{=\dot{m}_{w,2}} = \underbrace{2\dot{m}_a x_3}_{=\dot{m}_{w,3}} \tag{8.158}$$

so that

$$x_3 = \frac{1}{2}\left(x_1 + x_2\right) = 0.008. \tag{8.159}$$

The energy balance obeys

$$\underbrace{\dot{m}_a h_{1+x,1} + \dot{m}_a h_{1+x,2}}_{\text{IN}} = \underbrace{2\dot{m}_a h_{1+x,3}}_{\text{OUT}} \tag{8.160}$$

with the specific enthalpies[8] of state (1)

$$h_{1+x,1} = c_{p,a}\vartheta_1 + x_1\left(\Delta h_{v,0} + c_{p,v}\vartheta_1\right) = 25.1716 \, \frac{kJ}{kg} \tag{8.161}$$

[8] Mind that both states are unsaturated.

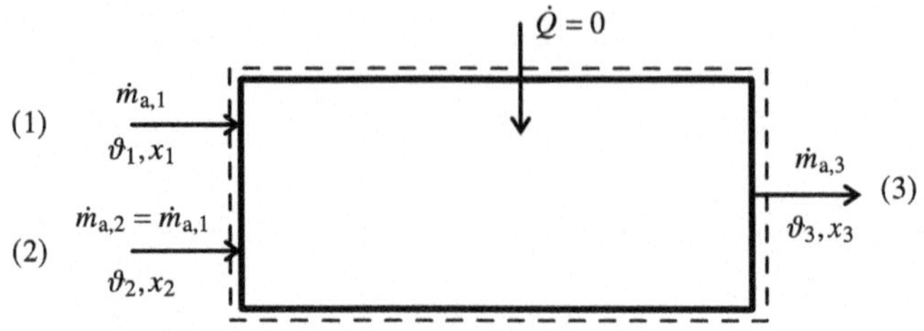

Fig. 8.19 Mixing of two air flows

and state (2)

$$h_{1+x,2} = c_{p,\mathrm{a}}\vartheta_2 + x_2\left(\Delta h_{\mathrm{v},0} + c_{p,\mathrm{v}}\vartheta_2\right) = 65.9212\,\frac{\mathrm{kJ}}{\mathrm{kg}}. \tag{8.162}$$

Hence, the specific enthalpy of state (3) is

$$h_{1+x,3} = \frac{1}{2}\left(h_{1+x,1} + h_{1+x,2}\right) = 45.5464\,\frac{\mathrm{kJ}}{\mathrm{kg}}. \tag{8.163}$$

In the h_{1+x}, x diagram, see Fig. 8.20, it can be identified with $h_{1+x,3}$ and x_3 that state (3) is also unsaturated, i.e.

$$h_{1+x,3} = c_{p,\mathrm{a}}\vartheta_3 + x_3\left(\Delta h_{\mathrm{v},0} + c_{p,\mathrm{v}}\vartheta_3\right). \tag{8.164}$$

Finally, the temperature in state (3) is

$$\vartheta_3 = \frac{h_{1+x,3} - x_3\Delta h_{\mathrm{v},0}}{c_{p,\mathrm{a}} + x_3 c_{p,\mathrm{v}}} = 25.073\,^{\circ}\mathrm{C} \tag{8.165}$$

Alternatively, in the second part, a graphical solution is given, see Fig. 8.20. States (1) and (2) can be clearly drawn in the h_{1+x}, x-diagram. Since this is adiabatic, isobaric mixing, state (3) lies on a connecting straight between (1) and (2). As the mass fluxes of the dry air in state (1) and (2) are identical, according to [1] it yields

$$\frac{\dot{m}_{\mathrm{a},2}}{\dot{m}_{\mathrm{a},1}} = 1 = \frac{a}{b} \tag{8.166}$$

This means that state (3) is exactly in the centre of the connecting line $\overline{12}$. In the diagram one can determine that $\vartheta_3 = 25\,^{\circ}\mathrm{C}$.

8.15 The problem is illustrated in Fig. 8.21. In a h_{1+x}, x-diagram, see Fig. 8.22, one can verify that state (1) is unsaturated. In the first part, the problem is solved mathematically. The mass balance for dry air obeys

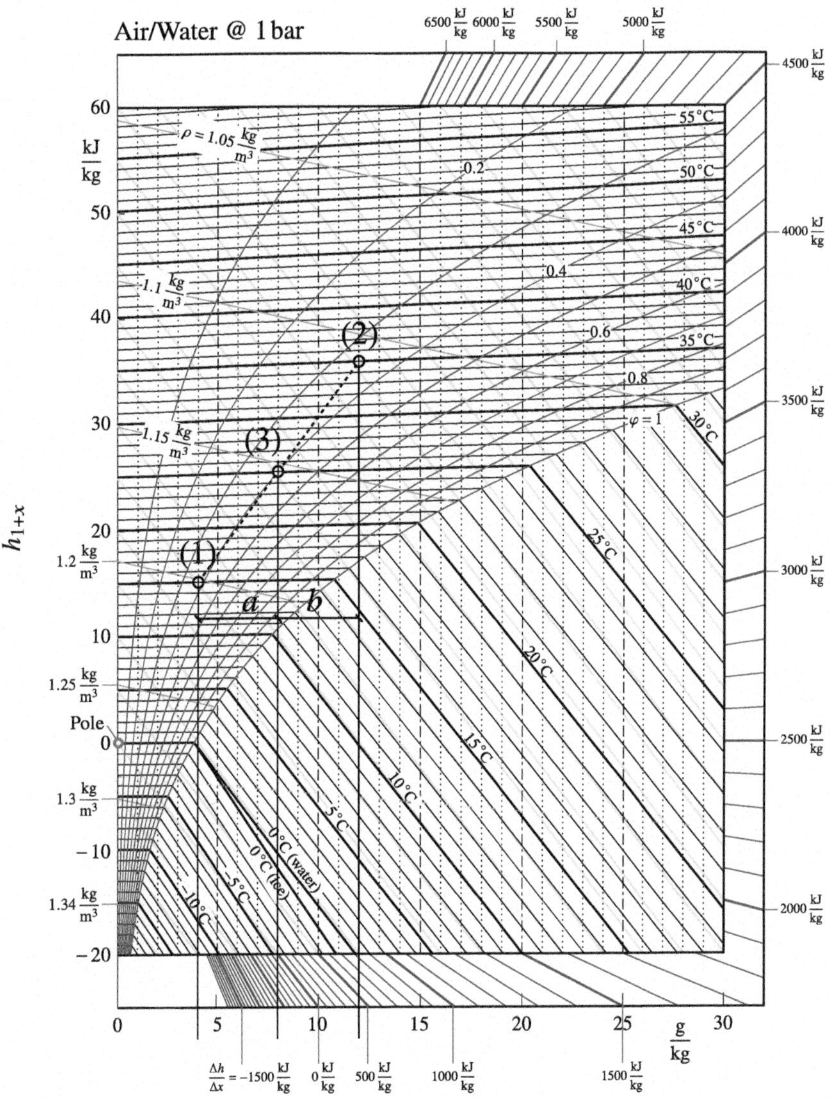

Fig. 8.20 Adiabatic mixing of two air flows

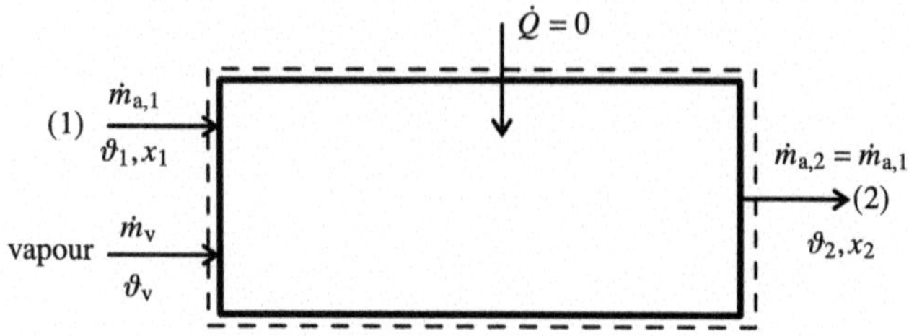

Fig. 8.21 Injection of vapour

$$\dot{m}_{a,1} = \dot{m}_{a,2} = \dot{m}_a = 250\,\frac{\text{kg}}{\text{h}}. \tag{8.167}$$

The balance for the water reads as

$$\dot{m}_a x_1 + \dot{m}_v = \dot{m}_a x_2 \tag{8.168}$$

so that

$$x_2 = x_1 + \frac{\dot{m}_v}{\dot{m}_a} = 0.015. \tag{8.169}$$

The energy balance states

$$\underbrace{\dot{m}_a h_{1+x,1} + \dot{m}_v h_v}_{\text{IN}} = \underbrace{\dot{m}_a h_{1+x,2}}_{\text{OUT}} \quad \text{with} \quad h_v = \Delta h_{v,0} + c_{p,v}\vartheta_v \tag{8.170}$$

with

$$h_{1+x,1} = c_{p,a}\vartheta_1 + x_1\left(\Delta h_{v,0} + c_{p,v}\vartheta_1\right) = 37.8325\,\frac{\text{kJ}}{\text{kg}}. \tag{8.171}$$

Hence, the specific enthalpy in state (2) is

$$h_{1+x,2} = h_{1+x,1} + \frac{\dot{m}_v}{\dot{m}_a}\left(\Delta h_{v,0} + c_{p,v}\vartheta_v\right) = 64.6925\,\frac{\text{kJ}}{\text{kg}} \tag{8.172}$$

It is assumed that state (2) is unsaturated, so that

$$h_{1+x,2} = c_{p,a}\vartheta_2 + x_2\left(\Delta h_{v,0} + c_{p,v}\vartheta_2\right). \tag{8.173}$$

Finally, the temperature in state (2) is

$$\vartheta_2 = \frac{h_{1+x,2} - x_2\Delta h_{v,0}}{c_{p,a} + x_2 c_{p,v}} = 26.35\,^\circ\text{C} \tag{8.174}$$

Lastly, it must be checked whether the assumption that state (2) was unsaturated is justified. Since

$$x_2 < x_s(\vartheta_2) = 0.622 \frac{p_s(\vartheta_2)}{p - p_s(\vartheta_2)} = 0.0221 \quad \text{with} \quad p_s(\vartheta_2) = 0.0343\,\text{bar} \tag{8.175}$$

the assumption was correct.

Alternatively, in the second part, a graphical solution is given, see Fig. 8.22. State (1) can be clearly drawn in the h_{1+x}, x-diagram. It is further known that state (2) must lie on the $x_2 = 0.015$-vertical. In addition, the direction of the change of state is known due to the injection of the vapour. This follows the gradient

$$\frac{\Delta h}{\Delta x} = h_v = \Delta h_{v,0} + c_{p,v}\vartheta_v = 2686\,\frac{\text{kJ}}{\text{kg}}. \tag{8.176}$$

This slope is drawn in with the border scale and the pole of the h_{1+x}, x-diagram, see Fig. 8.22. This slope is now shifted parallel through state (1). State (2) is where this straight intersects the x_2-vertical. A temperature ϑ_2 of approx. 26.3 °C is determined in the h_{1+x}, x-diagram.

8.16 To determine the temperature in thermodynamic equilibrium, the first law of thermodynamics is applied, cf. Fig. 8.22.[9] It reads

$$Q_{12} + W_{12} = 0 = \Delta U = \Delta U_{N_2} + \Delta U_{O_2} \tag{8.177}$$

respectively with the caloric equation of state

$$0 = m_{N_2} c_{v,N_2} (T_2 - T_1) + m_{O_2} c_{v,O_2} (T_2 - T_1)\,. \tag{8.178}$$

Hence, it follows

$$T_2 = T_1 = T. \tag{8.179}$$

The thermal equation of state for nitrogen obeys

- State (1)

$$p_1 V_{N_2} = m_{N_2} R_{N_2} T \tag{8.180}$$

- State (2)

$$p_{N_2} \underbrace{\left(V_{N_2} + V_{O_2}\right)}_{=V} = m_{N_2} R_{N_2} T = p_1 V_{N_2} \tag{8.181}$$

and for oxygen

[9] According to Avogadro's law, equal volumes of ideal gases at the same temperature and pressure have the same number of molecules.

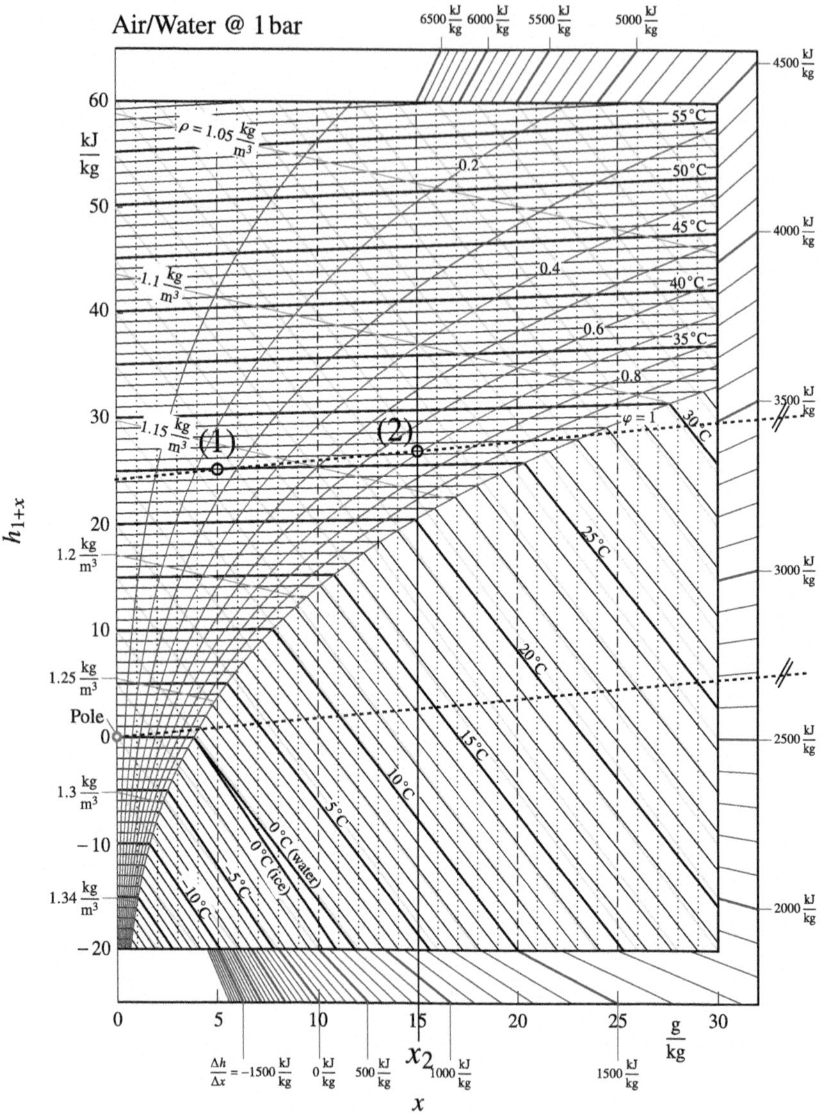

Fig. 8.22 Injection of vapour

- State (1)

$$p_1 V_{O_2} = m_{O_2} R_{O_2} T \tag{8.182}$$

- State (2)

$$p_{O_2} \underbrace{\left(V_{N_2} + V_{O_2}\right)}_{=V} = m_{O_2} R_{O_2} T = p_1 V_{O_2} \tag{8.183}$$

Consequently, it is

$$\underbrace{p_{N_2+O_2}}_{=p_2} = p_1 . \tag{8.184}$$

Anyhow, it follows

$$p_{N_2} = p_1 \frac{V_{N_2}}{V} = 0.5\,\text{bar} \tag{8.185}$$

and

$$p_{O_2} = p_1 \frac{V_{O_2}}{V} = 0.5\,\text{bar}. \tag{8.186}$$

The balance of entropy obeys

$$S_2 = S_1 + \underbrace{S_{a,12}}_{=0} + S_{i,12} \tag{8.187}$$

so that

$$S_{i,12} = S_2 - S_1 = m_{N_2} \left(s_2 - s_1\right)_{N_2} + m_{O_2} \left(s_2 - s_1\right)_{O_2} . \tag{8.188}$$

The caloric equations of state follow

$$\left(s_2 - s_1\right)_{N_2} = c_{p,N_2} \underbrace{\ln \frac{T_2}{T_1}}_{=0} - R_{N_2} \ln \frac{p_{N_2}}{p_1} = R_{N_2} \ln \frac{p_1}{p_{N_2}} \tag{8.189}$$

and

$$\left(s_2 - s_1\right)_{O_2} = c_{p,O_2} \underbrace{\ln \frac{T_2}{T_1}}_{=0} - R_{O_2} \ln \frac{p_{O_2}}{p_1} = R_{O_2} \ln \frac{p_1}{p_{O_2}} . \tag{8.190}$$

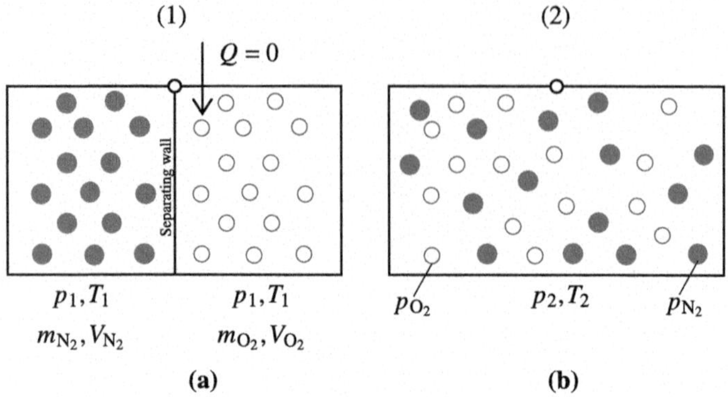

Fig. 8.23 Adiabatic mixing

The entropy generated, which indicates an irreversible change of state, thus reads as

$$S_{i,12} = m_{N_2} R_{N_2} \ln \frac{p_1}{p_{N_2}} + m_{O_2} R_{O_2} \ln \frac{p_1}{p_{O_2}}$$
$$= \frac{p_1 V_{N_2}}{T} \ln \frac{p_1}{p_{N_2}} + \frac{p_1 V_{O_2}}{T} \ln \frac{p_1}{p_{O_2}} \tag{8.191}$$
$$= 231.0491 \, \frac{\text{J}}{\text{K}}.$$

Finally, the dissipated energy follows from

$$S_{i,12} = \int_1^2 \frac{\delta \Psi}{T} \tag{8.192}$$

Since the change of state is isothermal, cf. [1], it is

$$\Psi_{12} = T S_{i,12} = 6.9315 \times 10^4 \, \text{J}. \tag{8.193}$$

8.17 The mixing process is shown schematically in Fig. 8.24. In a h_{1+x}, x-diagram, see Fig. 8.25, one can verify that both state (1) and state (2) are unsaturated. In the first part, the problem is solved mathematically. The mass balance for dry air obeys

$$\dot{m}_{a,1} + \dot{m}_{a,2} = \dot{m}_{a,3} \quad \text{with} \quad \dot{m}_{a,1} = 2\dot{m}_{a,2} \tag{8.194}$$

so that

$$\dot{m}_{a,3} = 3\dot{m}_{a,2} \tag{8.195}$$

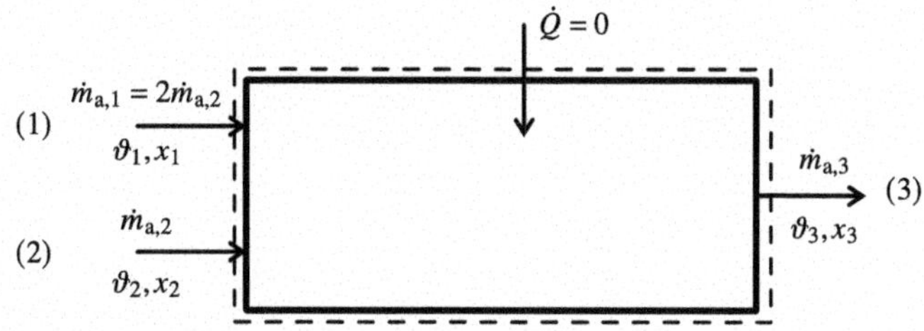

Fig. 8.24 Mixing of two air flows

The balance for the water reads as

$$\underbrace{2\dot{m}_{a,2}x_1}_{=\dot{m}_{w,1}} + \underbrace{\dot{m}_{a,2}x_2}_{=\dot{m}_{w,2}} = \underbrace{3\dot{m}_{a,2}x_3}_{=\dot{m}_{w,3}} \tag{8.196}$$

so that

$$x_3 = \frac{1}{3}\left(2x_1 + x_2\right) = 0.008. \tag{8.197}$$

The energy balance obeys

$$\underbrace{2\dot{m}_{a,2}h_{1+x,1} + \dot{m}_{a,2}h_{1+x,2}}_{\text{IN}} = \underbrace{3\dot{m}_{a,2}h_{1+x,3}}_{\text{OUT}} \tag{8.198}$$

with the specific enthalpies[10] of state (1)

$$h_{1+x,1} = c_{p,a}\vartheta_1 + x_1\left(\Delta h_{v,0} + c_{p,v}\vartheta_1\right) = 25.1716\,\frac{\text{kJ}}{\text{kg}} \tag{8.199}$$

and state (2)

$$h_{1+x,2} = c_{p,a}\vartheta_2 + x_2\left(\Delta h_{v,0} + c_{p,v}\vartheta_2\right) = 76.1816\,\frac{\text{kJ}}{\text{kg}} \tag{8.200}$$

Hence, the specific enthalpy of state (3) is

$$h_{1+x,3} = \frac{1}{3}\left(2h_{1+x,1} + h_{1+x,2}\right) = 42.1749\,\frac{\text{kJ}}{\text{kg}} \tag{8.201}$$

In the h_{1+x}, x diagram, see Fig. 8.25, it can be identified with $h_{1+x,3}$ and x_3 that state (3) is also unsaturated, i.e.

$$h_{1+x,3} = c_{p,a}\vartheta_3 + x_3\left(\Delta h_{v,0} + c_{p,v}\vartheta_3\right). \tag{8.202}$$

[10] Mind that both states are unsaturated.

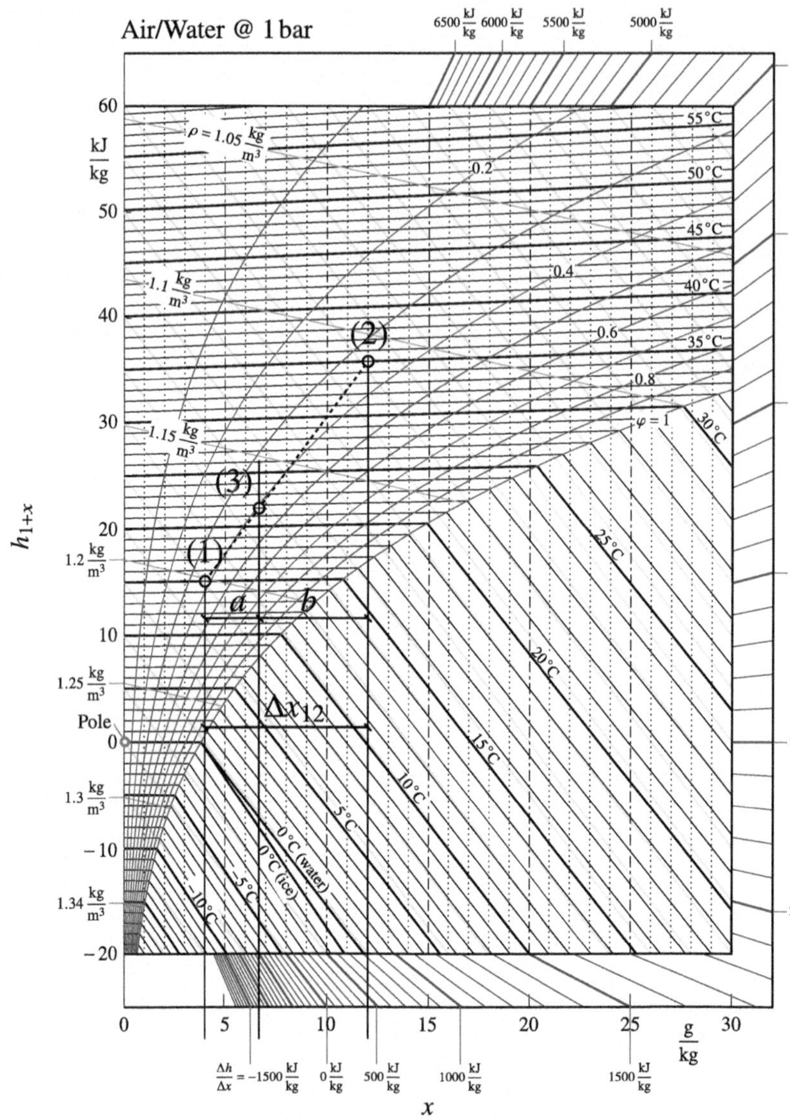

Fig. 8.25 Adiabatic mixing of two air flows

Finally, the temperature in state (3) is

$$\vartheta_3 = \frac{h_{1+x,3} - x_3 \Delta h_{\mathrm{v},0}}{c_{p,\mathrm{a}} + x_3 c_{p,\mathrm{v}}} = 21.764\,^\circ\mathrm{C} \tag{8.203}$$

Alternatively, in the second part, a graphical solution is given, see Fig. 8.25. States (1) and (2) can be clearly drawn in the h_{1+x}, x-diagram. Since this is adiabatic, isobaric mixing, state (3) lies on a connecting straight between (1) and (2). As the mass fluxes of the dry air in state (1) and (2) are identical, according to [1] it yields

$$\frac{\dot{m}_{\mathrm{a},2}}{\dot{m}_{\mathrm{a},1}} = \frac{1}{2} = \frac{a}{b} \tag{8.204}$$

According to Fig. 8.25 it is

$$a + b = \Delta x_{12} \tag{8.205}$$

Hence, with Eq. 8.204 it applies

$$b = \frac{2}{3} \Delta x_{12} \tag{8.206}$$

respectively

$$a = \frac{1}{3} \Delta x_{12} \tag{8.207}$$

Consequently, state (3) can be constructed in the diagram. A temperature of $\vartheta_3 = 21.7\,^\circ\mathrm{C}$ can be determined.

8.18 The problem is depicted in Fig. 8.26. The water balance obeys

$$\underbrace{\dot{m}_{\mathrm{a}} x_1}_{\dot{m}_{\mathrm{w},1}} = \underbrace{\dot{m}_{\mathrm{a}} x_2}_{\dot{m}_{\mathrm{w},2}} \tag{8.208}$$

so that

$$x_1 = x_2. \tag{8.209}$$

Since pressure and temperature of state (2) are known, one can check if state (2) is unsaturated. The saturation pressure for ϑ_2 can be determined with the steam table, i.e.

$$p_{\mathrm{s}}(\vartheta_2) = 0.0249\,\mathrm{bar}. \tag{8.210}$$

The maximum vapour content of state (2) then is

$$x_2' = 0.622 \frac{p_{\mathrm{s}}(\vartheta_2)}{p_2 - p_{\mathrm{s}}(\vartheta_2)} = 0.0159. \tag{8.211}$$

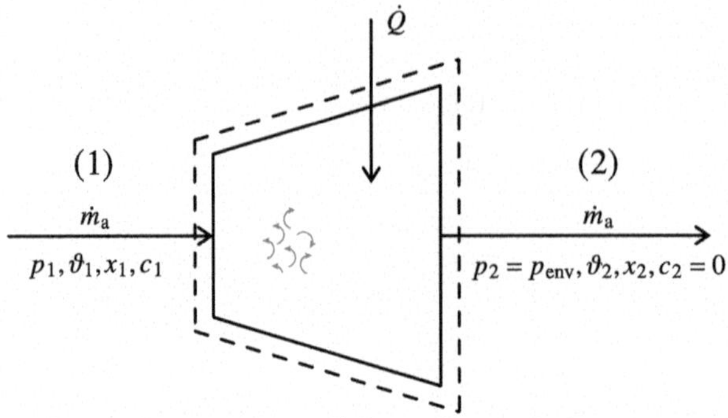

Fig. 8.26 Cooled diffuser

Consequently, since

$$x_2 > x_2' \tag{8.212}$$

state (2) is in the fog region, i.e. the air is saturated with vapour and additionally carries liquid water. The specific enthalpy of the unsaturated state (1) is

$$h_{1+x,1} = c_{p,a}\vartheta_1 + x_1 \cdot \left[\Delta h_{v,0} + c_{p,v}\vartheta_1\right] = 100.2449\,\frac{kJ}{kg} \tag{8.213}$$

and for state (2) which is in the fog region

$$h_{1+x,2} = c_{p,a}\vartheta_2 + x_2' \cdot \left[\Delta h_{v,0} + c_{p,v}\vartheta_2\right] + \left(x_1 - x_2'\right) \cdot c_{liq}\vartheta_2 = 61.6828\,\frac{kJ}{kg}. \tag{8.214}$$

The energy balance for the diffuser obeys

$$\underbrace{\dot{Q} + \dot{H}_1 + \dot{E}_{kin,1}}_{IN} = \underbrace{\dot{H}_2}_{OUT} \tag{8.215}$$

respectively

$$\dot{Q} = \dot{m}_a h_{1+x,2} - \dot{m}_a h_{1+x,1} - \dot{m}_{total} e_{kin,1} \quad \text{with} \quad \dot{m}_{total} = \dot{m}_a + \dot{m}_v = \dot{m}_a \cdot (1 + x_1). \tag{8.216}$$

Finally, it is

$$\dot{Q} = \dot{m}_a\left[\left(h_{1+x,2} - h_{1+x,1}\right) - (1 + x_1) \cdot \frac{c_1^2}{2}\right] = -1.1112 \times 10^3\,\text{kW}. \tag{8.217}$$

8.19

(a) The mass balance for carbon dioxide in steady state obeys

$$\dot{m}_{1,CO_2} + \dot{m}_{2,CO_2} = \dot{m}_{3,CO_2} \tag{8.218}$$

Dividing by M_{CO_2} leads to the following balance of the molar quantities

$$\dot{n}_{1,CO_2} + \dot{n}_{2,CO_2} = \dot{n}_{3,CO_2} \tag{8.219}$$

respectively

$$\underbrace{\dot{n}_1 x_{1,CO_2}}_{=0} + \dot{n}_2 x_{2,CO_2} = \dot{n}_3 x_{3,CO_2} \tag{8.220}$$

with

$$\dot{n}_1 + \dot{n}_2 = \dot{n}_3. \tag{8.221}$$

Thus, it is

$$\dot{n}_1 = \frac{\dot{n}_2 x_{2,CO_2} - \dot{n}_2 x_{3,CO_2}}{x_{3,CO_2}} = 200\,\frac{\text{kmol}}{\text{h}}. \tag{8.222}$$

(b) It applies

$$x_i = \frac{\dot{n}_i}{\dot{n}} = \frac{p_i}{p} \rightarrow p_i = x_i\,p. \tag{8.223}$$

- Exhaust gas before mixing (2)

$$p_{2,CO_2} = x_{2,CO_2}\,p_2 = 0.15\,\text{bar} \tag{8.224}$$

$$p_{2,N_2} = x_{2,N_2}\,p_2 = 1.2\,\text{bar} \tag{8.225}$$

$$p_{2,O_2} = x_{2,O_2}\,p_2 = 0.15\,\text{bar} \tag{8.226}$$

- Exhaust gas after mixing (3) It is

$$x_{3,N_2} = \frac{\dot{n}_{3,N_2}}{\dot{n}_3} = \frac{x_{1,N_2}\dot{n}_1 + x_{2,N_2}\dot{n}_2}{\dot{n}_1 + \dot{n}_2} = 0.7920 \tag{8.227}$$

and

$$x_{3,O_2} = \frac{\dot{n}_{3,O_2}}{\dot{n}_3} = \frac{x_{1,O_2}\dot{n}_1 + x_{2,O_2}\dot{n}_2}{\dot{n}_1 + \dot{n}_2} = 0.188 \tag{8.228}$$

so that

$$p_{3,CO_2} = x_{3,CO_2}\,p_3 = 0.03\,\text{bar} \tag{8.229}$$

$$p_{3,N_2} = x_{3,N_2} p_3 = 1.188\,\mathrm{bar} \tag{8.230}$$

$$p_{3,O_2} = x_{3,O_2} p_3 = 0.2820\,\mathrm{bar}. \tag{8.231}$$

(c) The mass fluxes for the components in states (1), (2) and (3) follow

- Air (1)

$$\dot{m}_{1,O_2} = \dot{n}_{1,O_2} M_{O_2} = \dot{n}_1 x_{1,O_2} M_{O_2} = 1344\,\frac{\mathrm{kg}}{\mathrm{h}} \tag{8.232}$$

$$\dot{m}_{1,N_2} = \dot{n}_{1,N_2} M_{N_2} = \dot{n}_1 x_{1,N_2} M_{N_2} = 4424\,\frac{\mathrm{kg}}{\mathrm{h}} \tag{8.233}$$

- Exhaust gas before mixing (2)

$$\dot{m}_{2,O_2} = \dot{n}_{2,O_2} M_{O_2} = \dot{n}_2 x_{2,O_2} M_{O_2} = 160\,\frac{\mathrm{kg}}{\mathrm{h}} \tag{8.234}$$

$$\dot{m}_{2,N_2} = \dot{n}_{2,N_2} M_{N_2} = \dot{n}_2 x_{2,N_2} M_{N_2} = 1120\,\frac{\mathrm{kg}}{\mathrm{h}} \tag{8.235}$$

$$\dot{m}_{2,CO_2} = \dot{n}_{2,CO_2} M_{CO_2} = \dot{n}_2 x_{2,CO_2} M_{CO_2} = 220\,\frac{\mathrm{kg}}{\mathrm{h}} \tag{8.236}$$

- Exhaust gas after mixing (3)

$$\dot{m}_{3,O_2} = \dot{n}_{3,O_2} M_{O_2} = \dot{n}_3 x_{3,O_2} M_{O_2} = 1504\,\frac{\mathrm{kg}}{\mathrm{h}} \tag{8.237}$$

$$\dot{m}_{3,N_2} = \dot{n}_{3,N_2} M_{N_2} = \dot{n}_3 x_{3,N_2} M_{N_2} = 5544\,\frac{\mathrm{kg}}{\mathrm{h}} \tag{8.238}$$

$$\dot{m}_{3,CO_2} = \dot{n}_{3,CO_2} M_{CO_2} = \dot{n}_3 x_{3,CO_2} M_{CO_2} = 220\,\frac{\mathrm{kg}}{\mathrm{h}}. \tag{8.239}$$

The first law of thermodynamics in steady state obeys

$$\begin{aligned}\overbrace{P + \dot{Q}}^{=0} &+ \dot{m}_{1,O_2} h_{O_2}(T_1) + \dot{m}_{1,N_2} h_{N_2}(T_1) + \dot{m}_{2,N_2} h_{N_2}(T_2) + \dot{m}_{2,O_2} h_{O_2}(T_2) + \\ &+ \dot{m}_{2,CO_2} h_{CO_2}(T_2) = \dot{m}_{3,N_2} h_{N_2}(T_3) + \dot{m}_{3,O_2} h_{O_2}(T_3) + \dot{m}_{3,CO_2} h_{CO_2}(T_3)\end{aligned} \tag{8.240}$$

with

$$\dot{m}_{3,N_2} = \dot{m}_{1,N_2} + \dot{m}_{2,N_2} \tag{8.241}$$

$$\dot{m}_{3,O_2} = \dot{m}_{1,O_2} + \dot{m}_{2,O_2} \tag{8.242}$$

$$\dot{m}_{3,CO_2} = \dot{m}_{1,CO_2} + \dot{m}_{2,CO_2}. \tag{8.243}$$

Hence, it yields

$$\begin{aligned}&\dot{m}_{1,O_2}\left[h_{O_2}(T_1)-h_{O_2}(T_3)\right]+\dot{m}_{1,N_2}\left[h_{N_2}(T_1)-h_{N_2}(T_3)\right]+\\&+\dot{m}_{2,O_2}\left[h_{O_2}(T_2)-h_{O_2}(T_3)\right]+\dot{m}_{2,N_2}\left[h_{N_2}(T_2)-h_{N_2}(T_3)\right]+\\&+\dot{m}_{2,CO_2}\left[h_{CO_2}(T_2)-h_{CO_2}(T_3)\right]=0.\end{aligned} \tag{8.244}$$

With the caloric equation of state it follows

$$\begin{aligned}&\dot{m}_{1,O_2}c_{p,O_2}\left[T_1-T_3\right]+\dot{m}_{1,N_2}c_{p,N_2}\left[T_1-T_3\right]+\\&+\dot{m}_{2,O_2}c_{p,O_2}\left[T_2-T_3\right]+\dot{m}_{2,N_2}c_{p,N_2}\left[T_2-T_3\right]+\\&+\dot{m}_{2,CO_2}c_{p,CO_2}\left[T_2-T_3\right]=0.\end{aligned} \tag{8.245}$$

Solving for T_3 leads to

$$\begin{aligned}T_3&=\frac{T_1\left(\dot{m}_{1,O_2}c_{p,O_2}+\dot{m}_{1,N_2}c_{p,N_2}\right)+T_2\left(\dot{m}_{2,O_2}c_{p,O_2}+\dot{m}_{2,N_2}c_{p,N_2}+\dot{m}_{2,CO_2}c_{p,CO_2}\right)}{\dot{m}_{1,O_2}c_{p,O_2}+\dot{m}_{1,N_2}c_{p,N_2}+\dot{m}_{2,O_2}c_{p,O_2}+\dot{m}_{2,N_2}c_{p,N_2}+\dot{m}_{2,CO_2}c_{p,CO_2}}\\&=322.2458\,\mathrm{K}.\end{aligned} \tag{8.246}$$

(d) First, the individual gas constants are calculated, i.e.

$$R=\frac{R_M}{M}. \tag{8.247}$$

Hence, it ist

$$R_{O_2}=\frac{R_M}{M_{O_2}}=259.8219\,\frac{\mathrm{J}}{\mathrm{kg\,mol}} \tag{8.248}$$

$$R_{N_2}=\frac{R_M}{M_{N_2}}=296.9393\,\frac{\mathrm{J}}{\mathrm{kg\,mol}} \tag{8.249}$$

$$R_{CO_2}=\frac{R_M}{M_{CO_2}}=188.9614\,\frac{\mathrm{J}}{\mathrm{kg\,mol}}. \tag{8.250}$$

To determine the entropy production, an entropy balance is carried out, i.e.

$$\begin{aligned}\dot{S}_i+\dot{S}_a&+\dot{m}_{1,O_2}s_{O_2}\left(T_1,\,p_{1,O_2}\right)+\dot{m}_{1,N_2}s_{N_2}\left(T_1,\,p_{1,N_2}\right)+\\&+\dot{m}_{2,N_2}s_{N_2}\left(T_2,\,p_{2,N_2}\right)+\dot{m}_{2,O_2}s_{O_2}\left(T_2,\,p_{2,O_2}\right)+\\&+\dot{m}_{2,CO_2}s_{CO_2}\left(T_2,\,p_{2,CO_2}\right)=\dot{m}_{3,N_2}s_{N_2}\left(T_3,\,p_{3,N_2}\right)+\\&+\dot{m}_{3,O_2}s_{O_2}\left(T_3,\,p_{3,O_2}\right)+\dot{m}_{3,CO_2}s_{CO_2}\left(T_3,\,p_{3,CO_2}\right)\end{aligned} \tag{8.251}$$

with $\dot{S}_a=0$. It is $p_{1,O_2}=x_{1,O_2}p_1=0.315$ bar and $p_{1,N_2}=x_{1,N_2}p_1=1.185$ bar. Hence, it yields

$$\begin{aligned}\dot{S}_\mathrm{i} =&\dot{m}_{1,\mathrm{O_2}}\left[s_{\mathrm{O_2}}\left(T_3,\ p_{3,\mathrm{O_2}}\right)-s_{\mathrm{O_2}}\left(T_1,\ p_{1,\mathrm{O_2}}\right)\right]+\\ &+\dot{m}_{1,\mathrm{N_2}}\left[s_{\mathrm{N_2}}\left(T_3,\ p_{3,\mathrm{N_2}}\right)-s_{\mathrm{N_2}}\left(T_1,\ p_{1,\mathrm{N_2}}\right)\right]+\\ &+\dot{m}_{2,\mathrm{O_2}}\left[s_{\mathrm{O_2}}\left(T_3,\ p_{3,\mathrm{O_2}}\right)-s_{\mathrm{O_2}}\left(T_2,\ p_{2,\mathrm{O_2}}\right)\right]+\\ &+\dot{m}_{2,\mathrm{N_2}}\left[s_{\mathrm{N_2}}\left(T_3,\ p_{3,\mathrm{N_2}}\right)-s_{\mathrm{N_2}}\left(T_2,\ p_{2,\mathrm{N_2}}\right)\right]+\\ &+\dot{m}_{2,\mathrm{CO_2}}\left[s_{\mathrm{CO_2}}\left(T_3,\ p_{3,\mathrm{CO_2}}\right)-s_{\mathrm{CO_2}}\left(T_2,\ p_{2,\mathrm{CO_2}}\right)\right].\end{aligned} \tag{8.252}$$

Applying the caloric equation of state yields

$$\begin{aligned}\dot{S}_\mathrm{i} =&\dot{m}_{1,\mathrm{O_2}}\left[c_{p,\mathrm{O_2}}\ln\frac{T_3}{T_1}-R_{\mathrm{O_2}}\ln\frac{p_{3,\mathrm{O_2}}}{p_{1,\mathrm{O_2}}}\right]+\\ &+\dot{m}_{1,\mathrm{N_2}}\left[c_{p,\mathrm{N_2}}\ln\frac{T_3}{T_1}-R_{\mathrm{N_2}}\ln\frac{p_{3,\mathrm{N_2}}}{p_{1,\mathrm{N_2}}}\right]+\\ &+\dot{m}_{2,\mathrm{O_2}}\left[c_{p,\mathrm{O_2}}\ln\frac{T_3}{T_2}-R_{\mathrm{O_2}}\ln\frac{p_{3,\mathrm{O_2}}}{p_{2,\mathrm{O_2}}}\right]+\\ &+\dot{m}_{2,\mathrm{N_2}}\left[c_{p,\mathrm{N_2}}\ln\frac{T_3}{T_2}-R_{\mathrm{N_2}}\ln\frac{p_{3,\mathrm{N_2}}}{p_{2,\mathrm{N_2}}}\right]+\\ &+\dot{m}_{2,\mathrm{CO_2}}\left[c_{p,\mathrm{CO_2}}\ln\frac{T_3}{T_2}-R_{\mathrm{CO_2}}\ln\frac{p_{3,\mathrm{CO_2}}}{p_{2,\mathrm{CO_2}}}\right]=44.6868\,\frac{\mathrm{W}}{\mathrm{K}}.\end{aligned} \tag{8.253}$$

8.20

(a) Determination of water load and specific enthalpies

- State (1)

$$p_{\mathrm{v}_1}=\varphi_1 p_\mathrm{s}\left(\vartheta_1\right)=0.0164\,\mathrm{bar} \tag{8.254}$$

and

$$x_1=0.622\frac{p_{\mathrm{v}_1}}{p-p_{\mathrm{v}_1}}=0.0104 \tag{8.255}$$

and

$$h_{1+x,1}=c_{p,\mathrm{a}}\vartheta_1+x_1\left[c_{p,\mathrm{v}}\vartheta_1+\Delta h_{\mathrm{v},0}\right]=46.3514\,\frac{\mathrm{kJ}}{\mathrm{kg}} \tag{8.256}$$

- State (3)

$$p_{\mathrm{v}_3}=\varphi_3 p_\mathrm{s}\left(\vartheta_3\right)=0.0247\,\mathrm{bar} \tag{8.257}$$

and

$$x_3=0.622\frac{p_{\mathrm{v}_3}}{p-p_{\mathrm{v}_3}}=0.0158 \tag{8.258}$$

and

$$h_{1+x,3}=c_{p,\mathrm{a}}\vartheta_3+x_3\left[c_{p,\mathrm{v}}\vartheta_3+\Delta h_{\mathrm{v},0}\right]=91.0505\,\frac{\mathrm{kJ}}{\mathrm{kg}} \tag{8.259}$$

States (1) and (3) are also marked in a h_{1+x}, x-diagram, see Fig. 8.27. The requested state values can alternatively be taken from the diagram.

(b) If system A is balanced energetically,[11] cf. Fig 8.28a, the water enters the system in liquid form. The thermal energy required to vaporise the liquid water occurs in the system and not at the system boundary.

$$\boxed{\dot{m}_{a,3} h_{1+x,3} + \dot{m}_{liq} h_{liq} = \dot{m}_{a,4} h_{1+x,4} \quad \text{with} \quad \dot{m}_{a,4} = \dot{m}_{a,3}} \tag{8.260}$$

If, on the other hand, system B is balanced as in Fig 8.28b, i.e. only the gas phase, the water enters in vapour form, but the thermal energy required for vaporisation now appears at the system boundary and must be taken into account, i.e.

$$\dot{m}_{a,3} h_{1+x,3} + \dot{m}_v h_v = \dot{Q} + \dot{m}_{a,4} h_{1+x,4} \quad \text{with} \quad \dot{m}_{liq} = \dot{m}_v. \tag{8.261}$$

If one balances according to system boundary C, it becomes clear how large the thermal energy $\dot{Q}$ is for vaporising the water., i.e.

$$\dot{Q} + \dot{m}_{liq} h_{liq} = \dot{m}_v h_v. \tag{8.262}$$

However, a combination of Eqs. 8.261 and 8.262 eventually leads to Eq. 8.260. The balance for system boundary A, i.e. Eq. 8.260 is now followed. It further applies

$$h_{liq} = c_{liq} \vartheta_4 = 123.6050 \, \frac{\text{kJ}}{\text{kg}}. \tag{8.263}$$

The water balance for system A, see Fig. 8.28a obeys

$$\boxed{\dot{m}_{a,3} x_3 + \dot{m}_{liq} = \dot{m}_{a,3} x_4} \tag{8.264}$$

so that

$$\dot{m}_{liq} = \dot{m}_{a,3} \left(x_4 - x_3 \right). \tag{8.265}$$

It further yields for the specific enthalpy

$$h_{1+x,4} = c_{p,a} \vartheta_4 + x_4 \left[c_{p,v} \vartheta_4 + \Delta h_{v,0} \right]. \tag{8.266}$$

Finally, a combination of these equations brings

$$x_4 = \frac{c_{p,a} \vartheta_4 + x_3 c_{liq} \vartheta_4 - h_{1+x,3}}{c_{liq} \vartheta_4 - c_v \vartheta_4 - \Delta h_{v,0}} = 0.0245 \tag{8.267}$$

[11] Note that $\dot{m}_a$ is the mass flow rate of the dry air, which remains constant even though the stream is enriched with water.

respectively

$$h_{1+x,4} = c_{p,\mathrm{a}}\vartheta_4 + x_4\left[c_{p,\mathrm{v}}\vartheta_4 + \Delta h_{\mathrm{v},0}\right] = 92.1274\,\frac{\mathrm{kJ}}{\mathrm{kg}}. \tag{8.268}$$

Alternatively, a graphical solution can be carried out: Liquid water is supplied to the humid air from state (3) so that the direction of the change of state is

$$\frac{\Delta h}{\Delta x} = h_{\mathrm{liq}} = c_{\mathrm{liq}}\vartheta_4 = 123.6050\,\frac{\mathrm{kJ}}{\mathrm{kg}}. \tag{8.269}$$

With the help of the pole construction, this slope is drawn into the h_{1+x}, x-diagram and shifted parallel through state (3), see Fig. 8.27. The intersection of this straight line with ϑ_4 finally results in state (4).

(c) State (2) follows from, cf. Fig. 8.29a:

- Energy balance

$$\boxed{\dot{m}_{\mathrm{a},1}h_{1+x,1} + \dot{m}_{\mathrm{a},4}h_{1+x,4} = \dot{m}_{\mathrm{a},2}h_{1+x,2}} \tag{8.270}$$

- Mass balance of dry air

$$\boxed{\dot{m}_{\mathrm{a},1} + \dot{m}_{\mathrm{a},4} = \dot{m}_{\mathrm{a},2}} \tag{8.271}$$

- Mass balance of water

$$\boxed{\dot{m}_{\mathrm{a},1}x_1 + \dot{m}_{\mathrm{a},4}x_4 = \dot{m}_{\mathrm{a},2}x_2} \tag{8.272}$$

respectively

$$\dot{m}_{\mathrm{a},1}x_1 + \dot{m}_{\mathrm{a},4}x_4 = \dot{m}_{\mathrm{a},1}x_2 + \dot{m}_{\mathrm{a},4}x_2 \tag{8.273}$$

Dividing by $\dot{m}_{\mathrm{a},1}$ brings

$$x_1 + \underbrace{\frac{\dot{m}_{\mathrm{a},4}}{\dot{m}_{\mathrm{a},1}}}_{=\xi} x_4 = x_2 + \frac{\dot{m}_{\mathrm{a},4}}{\dot{m}_{\mathrm{a},1}} x_2. \tag{8.274}$$

The mixing factor ξ finally results in

$$\xi = \frac{\dot{m}_{\mathrm{a},4}}{\dot{m}_{\mathrm{a},1}} = \frac{x_2 - x_1}{x_4 - x_2} = 0.6198. \tag{8.275}$$

It follows from Eq. 8.270 that

$$h_{1+x,2} = \frac{h_{1+x,1} + \xi h_{1+x,4}}{1+\xi} = 63.8663\,\frac{\mathrm{kJ}}{\mathrm{kg}}. \tag{8.276}$$

Since it applies

$$h_{1+x,2} = c_{p,\mathrm{a}}\vartheta_2 + x_2\left[c_{p,\mathrm{v}}\vartheta_2 + \Delta h_{\mathrm{v},0}\right] \quad \text{with} \quad x_2 = x_3 \tag{8.277}$$

the temperature in state (2) is

$$\vartheta_2 = \frac{h_{1+x,2} - x_2\Delta h_{\mathrm{v},0}}{c_{p,\mathrm{a}} + x_2 c_{p,\mathrm{v}}} = 23.6919\,^\circ\mathrm{C}. \tag{8.278}$$

Alternatively, a graphical solution can be carried out, see Fig. 8.27. State (2) must lie on the mixing line $\overline{14}$, as it is an adiabatic mixture of states (1) and (4). Furthermore, $x_2 = x_3$ applies, since the water balance remains unaffected during the change of state in the auxiliary heater from (2) to (3). To determine (2), one therefore connects the states (1) and (4) with a straight line. The intersection of this line with the vertical line $x_2 = x_3$ finally results in state (2).

(d) From Eq. 8.265 it yields

$$\dot{m}_{\mathrm{a},3} = \dot{m}_{\mathrm{a},4} = \dot{m}_{\mathrm{a},2} = \frac{\dot{m}_{\mathrm{liq}}}{x_4 - x_3} = 5.7388\,\frac{\mathrm{kg}}{\mathrm{s}}. \tag{8.279}$$

(e) An energy balance for the heater, see Fig. 8.29b, obeys

$$\dot{Q}_{\mathrm{H}} + \dot{m}_{\mathrm{a},2}h_{1+x,2} = \dot{m}_{\mathrm{a},2}h_{1+x,3}. \tag{8.280}$$

Consequently, the thermal power is

$$\dot{Q}_{\mathrm{H}} = \dot{m}_{\mathrm{a},2}\left(h_{1+x,3} - h_{1+x,2}\right) = 156.0047\,\mathrm{kW}. \tag{8.281}$$

8.21

(a) During the change of state from (1) to (2), the humid air is in a pure gaseous state and the isentropic correlation can be applied, i.e.

$$T_2 = T_1\left(\frac{p_2}{p_1}\right)^{\frac{\kappa-1}{\kappa}} = 283.5502\,\mathrm{K} \tag{8.282}$$

so that

$$\vartheta_2 = 10.4002\,^\circ\mathrm{C}. \tag{8.283}$$

The water content in the saturated state (2) thus obeys

$$x_2 = x_{\mathrm{s}} = 0.622\frac{p_{\mathrm{s}}(\vartheta_2)}{p_2 - p_{\mathrm{s}}(\vartheta_2)} = 0.0089 \quad \text{with} \quad p_{\mathrm{s}}(\vartheta_2) = 0.0126\,\mathrm{bar}. \tag{8.284}$$

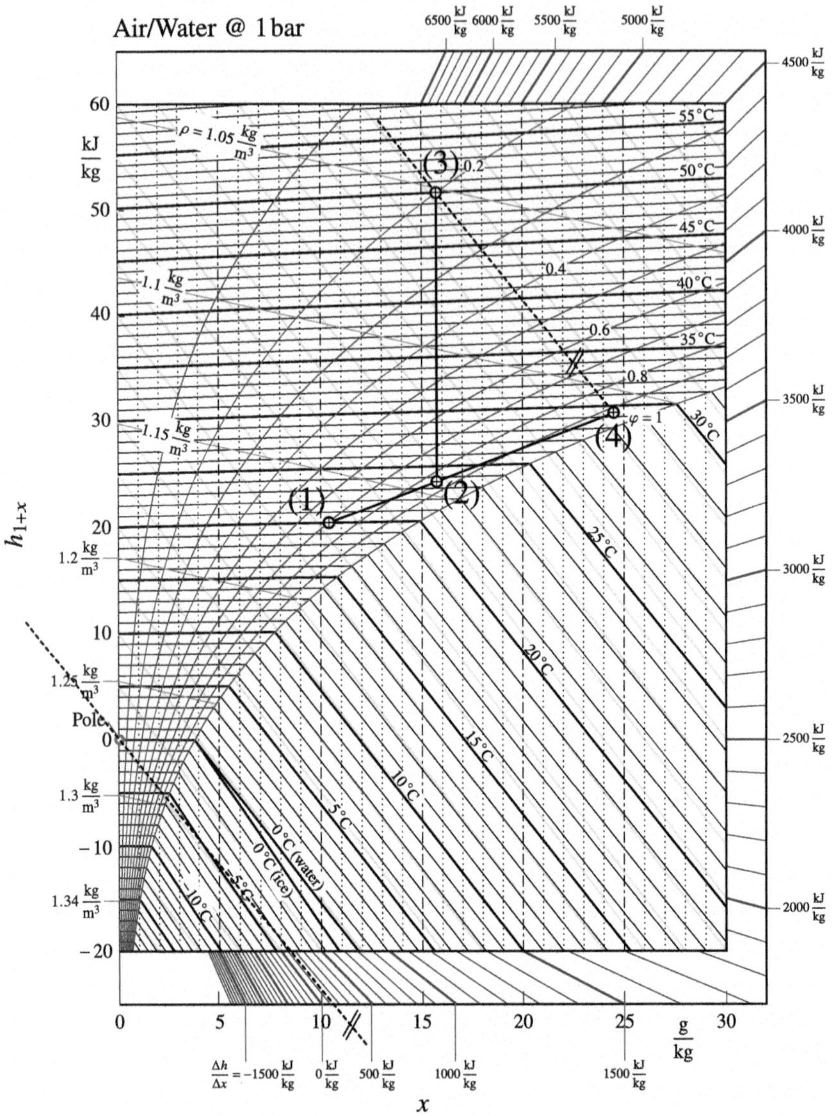

Fig. 8.27 Visualisation in a h_{1+x}, x-diagram

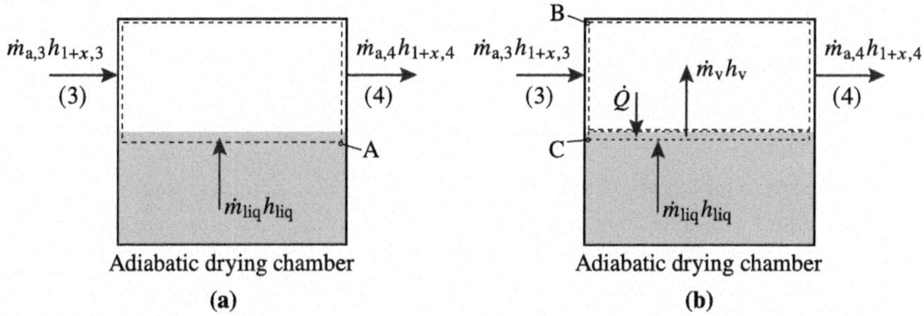

Fig. 8.28 Energy balance

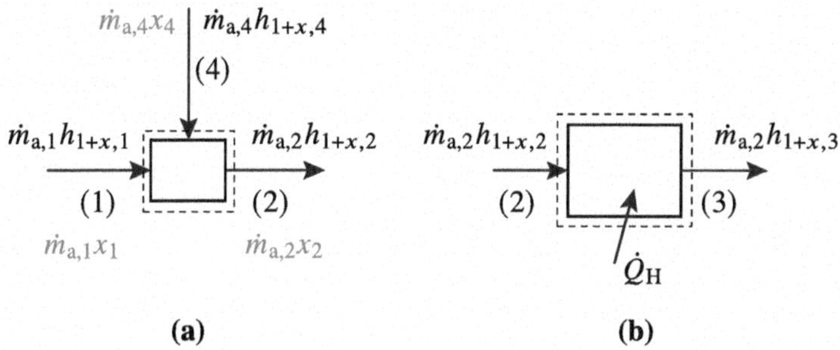

Fig. 8.29 Energy/water balances

It further is $x_1 = x_2$. To determine the height z_2, the first law of thermodynamics is formulated from (1) to (2), see Fig. 8.30, i.e.

$$\dot{m}_a h_{1+x,1} + \overbrace{\dot{m}_{\text{total}} g z_1}^{=0} = \dot{m}_a h_{1+x,2} + \dot{m}_{\text{total}} g z_2 \quad \text{with} \quad \dot{m}_{\text{total}} = \dot{m}_a + \dot{m}_v = \dot{m}_a (1 + x_2) \tag{8.285}$$

so that

$$z_2 = \frac{h_{1+x,1} - h_{1+x,2}}{(1 + x_2)\, g}. \tag{8.286}$$

The specific enthalpies[12] of state (1) and (2) are

$$h_{1+x,1} = c_{p,a}\vartheta_1 + x_1 \left(\Delta h_{v,0} + c_{p,v}\vartheta_1\right) = 42.7705 \times 10^3 \, \frac{\text{J}}{\text{kg}} \tag{8.287}$$

and

$$h_{1+x,2} = c_{p,a}\vartheta_2 + x_2 \left(\Delta h_{v,0} + c_{p,v}\vartheta_2\right) = 32.9726 \times 10^3 \, \frac{\text{J}}{\text{kg}}. \tag{8.288}$$

[12] To obtain the result for the height in m, the specific enthalpy must be given in $\frac{\text{J}}{\text{kg}}$.

The height z_2 at which cloud formation begins is therefore

$$z_2 = \frac{h_{1+x,1} - h_{1+x,2}}{(1 + x_2)\, g} = 989.9159\,\text{m}. \tag{8.289}$$

(b) From state (4) it is known that $p_4 = p_3$ and $\vartheta_4 = \vartheta_3$. After the liquid water has been rained off, state (4) must be saturated humid air. The first law of thermodynamics from (2) to (4), see Fig. 8.30, yields

$$\dot{m}_a h_{1+x,2} + \dot{m}_{\text{total}}\, g z_2 = \dot{m}_a h_{1+x,4} + \dot{m}_{\text{total}}\, g z_4 + \dot{m}_{\text{liq}} c_{\text{liq}} \vartheta_4 \tag{8.290}$$

with

$$\dot{m}_{\text{total}} = \dot{m}_a + \dot{m}_v = \dot{m}_a\,(1 + x_2) \tag{8.291}$$

and

$$\dot{m}_{\text{liq}} = \dot{m}_a\,(x_2 - x_4)\,. \tag{8.292}$$

The first law of thermodynamics simplifies to

$$h_{1+x,2} + (1 + x_2)\, g z_2 = h_{1+x,4} + (1 + x_2)\, g z_4 + (x_2 - x_4)\, c_{\text{liq}} \vartheta_4 \tag{8.293}$$

with

$$h_{1+x,4} = c_{p,a} \vartheta_4 + x_4 \left(\Delta h_{v,0} + c_{p,v} \vartheta_4\right) \tag{8.294}$$

so that

$$\begin{aligned} h_{1+x,2} + (1 + x_2)\, g z_2 = {} & c_{p,a} \vartheta_4 + x_4 \left(\Delta h_{v,0} + c_{p,v} \vartheta_4\right) + \\ & + (1 + x_2)\, g z_4 + (x_2 - x_4)\, c_{\text{liq}} \vartheta_4 . \end{aligned} \tag{8.295}$$

Solving for the unknown water content in state (4) results in

$$x_4 = \frac{h_{1+x,2} + (1 + x_2)\, g z_2 - c_{p,a} \vartheta_4 - (1 + x_2)\, g z_4 - x_2 c_{\text{liq}} \vartheta_4}{\Delta h_{v,0} + c_{p,v} \vartheta_4 - c_{\text{liq}} \vartheta_4} = 0.0071. \tag{8.296}$$

Since state (4) is saturated it is

$$x_4 = 0.622 \frac{p_s(\vartheta_4)}{p_4 - p_s(\vartheta_4)} \quad \text{with} \quad p_s(\vartheta_4) = 0.0087\,\text{bar}. \tag{8.297}$$

Pressure p_3 then yields

$$p_3 = p_4 = p_s(\vartheta_4) + 0.622 \frac{p_s(\vartheta_4)}{x_4} = 0.7688\,\text{bar}. \tag{8.298}$$

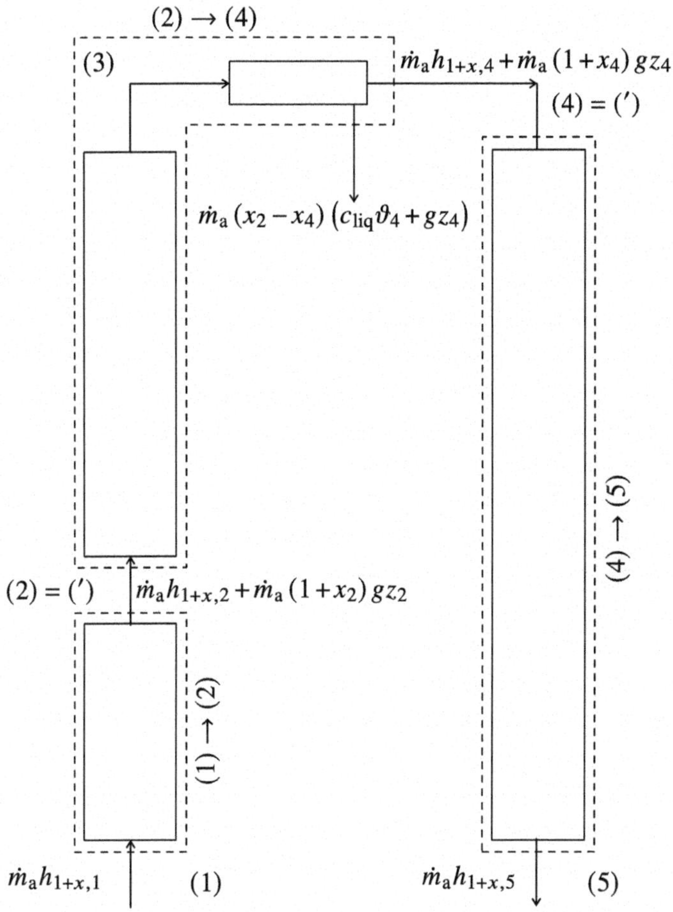

Fig. 8.30 Energy balances

(c) The temperature in state (5) follows from the first law of thermodynamics from state (4) to state (5), see Fig. 8.30, i.e.

$$\dot{m}_a h_{1+x,4} + \dot{m}_{total} g z_4 = \dot{m}_a h_{1+x,5} + \overbrace{\dot{m}_{total} g z_5}^{=0} \tag{8.299}$$

with

$$\dot{m}_{total} = \dot{m}_a + \dot{m}_v = \dot{m}_a (1 + x_4) . \tag{8.300}$$

The specific enthalpies are

$$h_{1+x,4} = c_{p,a}\vartheta_4 + x_4 \left(\Delta h_{v,0} + c_{p,v}\vartheta_4\right) = 22.9373 \, \frac{\text{kJ}}{\text{kg}} . \tag{8.301}$$

and

$$h_{1+x,5} = c_{p,\mathrm{a}}\vartheta_5 + x_5\left(\Delta h_{\mathrm{v},0} + c_{p,\mathrm{v}}\vartheta_5\right) \quad \text{with} \quad x_4 = x_5. \tag{8.302}$$

Hence, the first law of thermodynamics obeys

$$h_{1+x,4} + (1 + x_4)\, g z_4 = c_{p,\mathrm{a}}\vartheta_5 + x_4\left(\Delta h_{\mathrm{v},0} + c_{p,\mathrm{v}}\vartheta_5\right). \tag{8.303}$$

Finally, the temperature in state (5) is

$$\vartheta_5 = \frac{h_{1+x,4} + (1 + x_4)\, g z_4 - x_4 \Delta h_{\mathrm{v},0}}{c_{p,\mathrm{a}} + x_4 c_{p,\mathrm{v}}} = 24.4244\,°\mathrm{C} \tag{8.304}$$

During the change of state from (4) to (5), the humid air is in a pure gaseous state and the isentropic correlation can be applied, i.e.

$$T_5 = T_4\left(\frac{p_5}{p_4}\right)^{\frac{\kappa-1}{\kappa}} \tag{8.305}$$

so that

$$p_5 = p_4\left(\frac{T_5}{T_4}\right)^{\frac{\kappa}{\kappa-1}} = 0.9737\,\mathrm{bar}. \tag{8.306}$$

Chapter 9
Combustion

This concluding chapter deals in particular with material conversion, i.e. technical combustion processes. The first part of the tasks cover the so-called stoichiometry. This clarifies how much combustion air must be supplied to a fuel for complete oxidation. The composition of the resulting exhaust gas and the conditions under which water condenses out of the exhaust gas are investigated as well. For the energy calculation, the heating value method is used on the one hand. This makes it possible, for example, to determine the adiabatic combustion temperature. On the other hand, the method of standard enthalpy and entropy of formation is used for the energetic calculation, which additionally allows an entropic evaluation with regard to reversibility.

9.1 Problems

For the following tasks, the specific heat capacities of the flue gas components can be applied, if necessary, according to Table 9.1.

9.1 Petrol ($c = 0.86$, $h = 0.14$) is burnt with pure oxygen. The lower heating value at 25 °C is determined to be $H_U = 42480\,\frac{kJ}{kg}$. What is the higher heating value of petrol at 25 °C if the enthalpy of vaporisation of water is $\Delta h_v = 2441.7\,\frac{kJ}{kg}$ at 25 °C?

(a) $38410\,\frac{kJ}{kg}$
(b) $42480\,\frac{kJ}{kg}$
(c) $44126\,\frac{kJ}{kg}$
(d) $45557\,\frac{kJ}{kg}$

9.2 How to reduce the adiabatic flame temperature of a superstoichiometric combustion?

(a) By increasing the combustion air temperature.
(b) By increasing the air–fuel ratio λ.

A. Schmidt, *Technical Thermodynamics Workbook for Engineers*,
https://doi.org/10.1007/978-3-031-50172-2_9

Table 9.1 Averaged specific heat capacity $\overline{c_p}|_{0\,°\mathrm{C}}^{\vartheta}$ in $\frac{\mathrm{kJ}}{\mathrm{kg\,K}}$, according to [2]

ϑ [°C]	Air	N_2	O_2	CO_2	H_2O	SO_2
−25	1.0034	1.0393	0.9135	0.8035	1.8567	0.6010
0	1.0037	1.0394	0.9147	0.8173	1.8589	0.6079
25	1.0042	1.0395	0.9163	0.8307	1.8615	0.6149
50	1.0048	1.0397	0.9182	0.8437	1.8646	0.6219
75	1.0055	1.0400	0.9204	0.8563	1.8682	0.6289
100	1.0064	1.0404	0.9229	0.8684	1.8724	0.6359
150	1.0087	1.0416	0.9288	0.8914	1.8820	0.6495
200	1.0116	1.0435	0.9354	0.9128	1.8931	0.6626
400	1.0285	1.0567	0.9649	0.9856	1.9466	0.7078
600	1.0498	1.0763	0.9925	1.0427	2.0083	0.7418
800	1.0712	1.0976	1.0158	1.0885	2.0744	0.7672
1000	1.0910	1.1179	1.0350	1.1257	2.1421	0.7867
2000	1.1614	1.1914	1.0991	1.2378	2.4425	0.8410

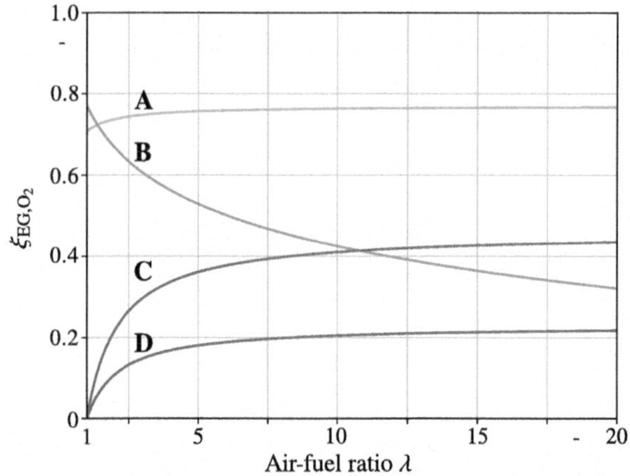

Fig. 9.1 Exhaust gas

(c) By releasing heat from the combustion chamber.
(d) None of the above statements is true.

9.3 A fossil fuel ($c = 0.8, s = 0.1, n = 0.1$) is burnt with air (23 mass—% oxygen, 77 mass—% nitrogen). Which of the following curves, cf. Fig. 9.1, shows the mass concentration of oxygen in the exhaust gas as a function of the air-fuel ratio λ?

(a) A
(b) B
(c) C
(d) D

9.4 The following gas concentrations are known from an exhaust gas flow at a total pressure of 2 bar:

- nitrogen: 65 vol.—%
- oxygen: 5 vol.—%
- vapour: 5 vol.—%
- carbon dioxide: 10 vol.—%
- sulphur dioxide: 15 vol.—%

What is the dew point temperature of the exhaust gas?

(a) 22.88 °C
(b) 45.81 °C
(c) 54.77 °C
(d) 67.32 °C

9.5 For which of these fuels are the lower and higher heating value identical?

(a) Hydrogen
(b) Sulphur
(c) Methane
(d) None of the above statements is true.

9.6 A fossil fuel is burnt isobarically with air. Which of the following curves, see Fig. 9.2, shows the adiabatic flame temperature of the fuel as a function of the air-fuel ratio λ?

(a) A
(b) B
(c) C
(d) D

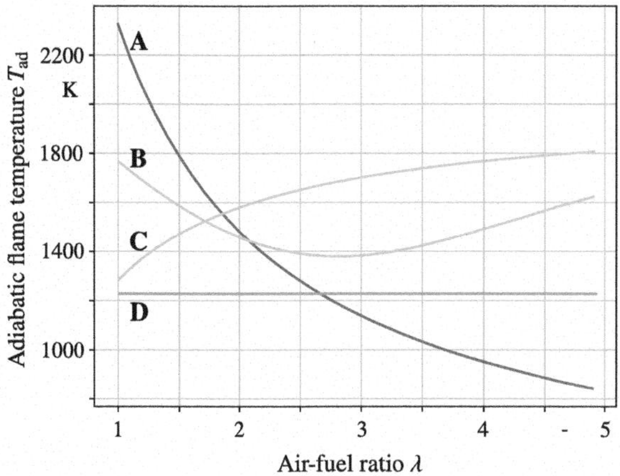

Fig. 9.2 Exhaust gas

9.7 Coal ($c = 0.76$, $h = 0.05$, $o = 0.05$, $w = 0.05$, $a = 0.09$) shall be burnt over-stoichiometrically ($\lambda = 1.2$) with dry air (consisting of nitrogen with $\xi_{N_2} = 0.77$ and oxygen with $\xi_{O_2} = 0.23$). What is the air demand l?

9.8 The exhaust gas of a combustion process has a temperature of 800 °C and a mass flow rate of 2 $\frac{kg}{s}$. Its composition yields

- $\xi_{CO_2} = 0.1$
- $\xi_{N_2} = 0.75$
- $\xi_{O_2} = 0.05$
- $\xi_{H_2O} = 0.1$

The exhaust gas is now cooled isobarically to 25 °C in a heat exchanger. What heat flux is extracted from the exhaust gas?

9.9 An ash-free fuel mass flow of 10 $\frac{g}{s}$ is completely combusted with 191.3 $\frac{g}{s}$ air. The following data are known from the fuel and the exhaust gas

- fuel: $M_F = 16\,\frac{kg}{kmol}$.
- exhaust gas: $M_{EG} = 27.7272\,\frac{kg}{kmol}$, $p = 1$ bar, $\xi_{EG,H_2O} = 0.11176$

What mass flow of water condenses as the exhaust gas is cooled to 50 °C?

9.10 Coal ($c = 0.76$, $h = 0.05$, $o = 0.05$, $w = 0.05$, $a = 0.09$) shall be burnt over-stoichiometrically ($\lambda = 1.2$) with dry air (consisting of nitrogen with $\xi_{N_2} = 0.77$ and oxygen with $\xi_{O_2} = 0.23$). How many kilograms of carbon dioxide per kilogram of fuel are in the exhaust gas?

9.11 Air ($\xi_{O_2} = 0.22$, $\xi_{N_2} = 0.78$) and fuel ($c = 0.88$, $h = 0.12$) are supplied to a combustion chamber at 25 °C. The air-fuel ratio is $\lambda = 1.5$. What is the specific heat released at steady state when the exhaust gas temperature is 800 °C? The lower heating value at 25 °C is $4.3367 \times 10^4\,\frac{kJ}{kg}$.

9.12 A methane mole flow of $\dot{n}_F = 6.25 \times 10^{-4}\,\frac{kmol}{s}$ is combusted over-stoichiometrically with air. The exhaust gas with the following composition

- $\nu_{EG,CO_2} = 1$
- $\nu_{EG,H_2O} = 2$
- $\nu_{EG,N_2} = 8.4174$
- $\nu_{EG,O_2} = 0.2$

is cooled to 50 °C. The total pressure is 2 bar. What molar flux of condensate occurs?

9.13 A mixture of 50 mass − % propane (C_3H_8) and 50 mass − % methane (CH_4) shall be stoichiometrically combusted with air ($\xi_{air,\,N_2} = 0.77$, $\xi_{air,\,O_2} = 0.23$). What is the air demand l?

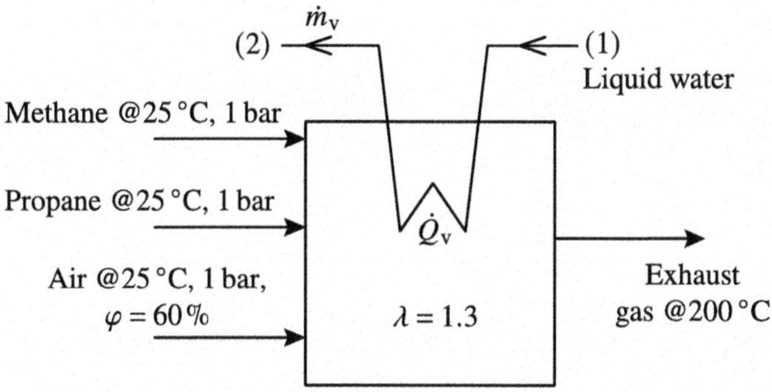

Fig. 9.3 Schematics of the combustion process

9.14 In a combustion chamber, methane (CH_4) and propane (C_3H_8) are burned isobarically with an air-fuel ratio $\lambda = 1.3$. Both combustion gases are fed into the combustion chamber at a temperature of 25 °C and a pressure of 1 bar. Humid air (25 °C, $\varphi = 60\,\%$, mass fractions of dry air: $\xi_{O_2,\text{dry air}} = 0.23$, $\xi_{N_2,\text{dry air}} = 0.77$) serves as combustion air. Cooling of the combustion chamber by $\dot{Q}_v$ results in a flue gas temperature of 200 °C. The pressure at combustion shall be 1 bar. Figure 9.3 shows the operational conditions.

- Methane: Volume flow rate @ Inlet $\dot{V}_M = 0.5\,\frac{m^3}{h}$, Lower heating value $H_{U,M}(25\,°C) = 50.01\,\frac{MJ}{kg}$
- Propane: Volume flow rate @ Inlet $\dot{V}_P = 0.3\,\frac{m^3}{h}$, Lower heating value $H_{U,P}(25\,°C) = 46.35\,\frac{MJ}{kg}$
- Environment: $\vartheta_{env} = 25\,°C$, $p = 1$ bar

Insulation losses of the combustion chamber as well as changes in the outer energies are negligible. Methane, propane and the resulting exhaust gas are to be regarded as ideal gases.

(a) Calculate the mass fluxes of methane and propane.
(b) Conduct an elemental analysis of the fuel (consisting of methane and propane), i.e. determine the mass fractions c and h.
(c) What oxygen demand o_{min} is required for the described combustion? What is the required mass flow of humid air?
(d) Calculate the composition of the resulting flue gas in mass fractions $\xi_{EG,i}$.
(e) How far can the exhaust gas be cooled down without water condensing?
(f) What is the heating value $H_{U,F}(25\,°C)$ of the fuel?
(g) What heat flux $\dot{Q}_v$ is transferred?

9.15 Methanol (CH_3OH) is burnt stoichiometrically with dry air in an adiabatic combustion process, see Fig. 9.4a. The supplied mass flow of methanol is $\dot{m}_F = 2.0\,\frac{kg}{s}$.

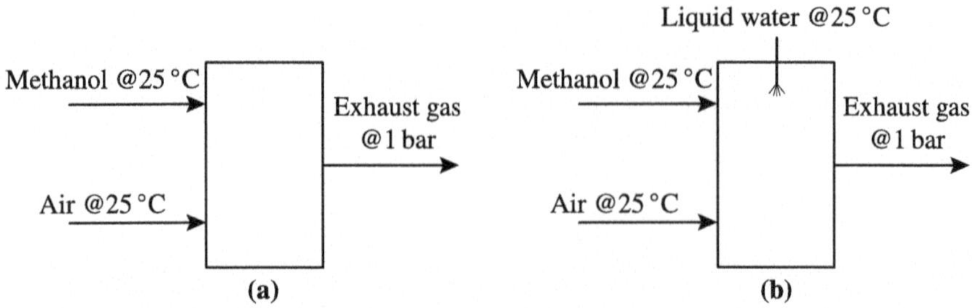

Fig. 9.4 **a** No water injection **b** With water injection

- Lower heating value methanol $H_U(\vartheta_0 = 25\,°C) = 19.9\,\frac{MJ}{kg}$
- Specific heat capacity methanol: $c_F = 2470\,\frac{J}{kgK}$
- Composition dry air: $\xi_{O_2,\text{dry air}} = 0.23$, $\xi_{N_2,\text{dry air}} = 0.77$
- Environment: $\vartheta_{env} = 25\,°C$, $p = 1$ bar
- Water vapour may be treated as an ideal gas and liquid water may be treated as incompressible
- Kinetic and potential energies are to be neglected.

(a) Conduct an elemental analysis of the fuel, i.e. determine the mass fractions c, h, o.
(b) Calculate the minimum oxygen demand o_{min} and the air demand l for the fuel used.
(c) Calculate the exhaust gas concentrations $\xi_{EG,i}$ of the exhaust gas!
(d) What is the adiabatic flame temperature?

To reduce the adiabatic flame temperature, the combustion chamber is modified and a water injection is installed, see Fig. 9.4b. Now, $\dot{m}_W = 0.8\,\frac{kg}{s}$ liquid water with a temperature of 25 °C is injected during combustion. The total pressure in the exhaust gas remains 1 bar, the chamber is still adiabatic.

(e) What is the new adiabatic flame temperature?

9.16 Hydrogen ($T_{H_2} = 298$ K, $p_{H_2} = 1$ bar) is burnt isobarically with air from ambient state ($\dot{m}_a = 28.84\,\frac{kg}{h}$, $T_a = T_{env} = 298$ K, $p_a = p_{env} = 1$ bar, $\lambda = 1$) in an adiabatic combustion chamber, see Fig. 9.5. In the combustion chamber there is a heat exchanger in which boiling water just completely vaporises at a constant pressure of $p_w = 100$ bar, so that the exhaust gas has a temperature of $T_2 = 1000$ K at the outlet. Changes in external energies are negligible. The air can be simplified as a mixture of 79 mol.—% N_2 and 21 mol.—% O_2.

(a) Calculate the molar fluxes of the subsystem *combustion* at inlet and outlet.
(b) Determine the mass flux of the water $\dot{m}_w$.
(c) Calculate the entropy generation and the exergy loss of the entire system.

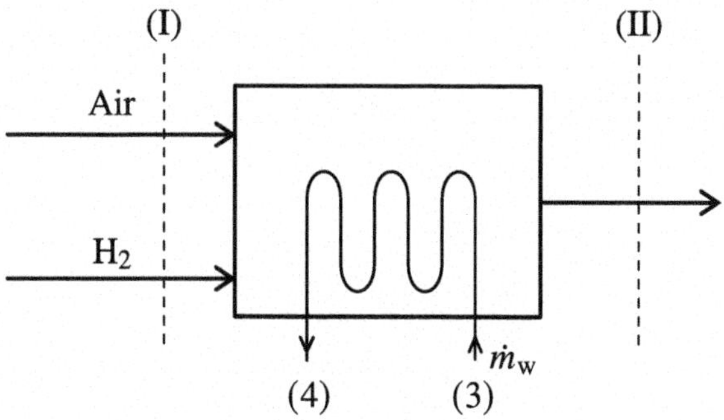

Fig. 9.5 Schematics of the combustion process

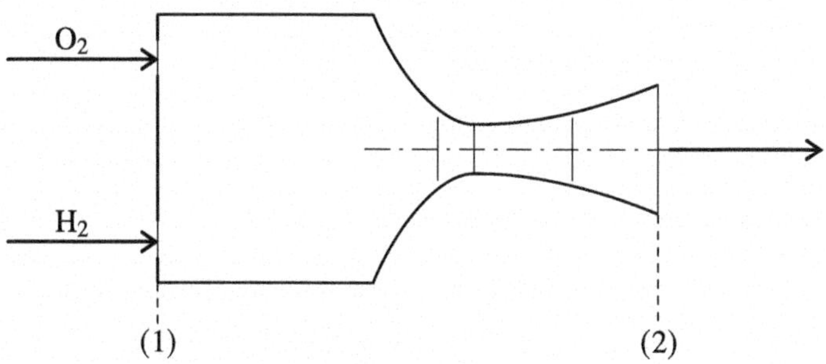

Fig. 9.6 Schematics of the combustion process

(d) Determine the temperature to which the exhaust gas can be cooled in subsequent heat exchangers without condensation occurring.

9.17 A hydrogen mass flow of 50 $\frac{\text{kg}}{\text{h}}$ is burnt stoichiometrically with oxygen. Both reactants are supplied at a temperature of $\vartheta_1 = 120\,°\text{C}$ and a pressure of $p_1 = 7\,\text{bar}$. After the combustion chamber, the exhaust gas passes a laval nozzle and expands to a pressure of $p_2 = 7\,\text{bar}$, see Fig. 9.6. When the gas exits the nozzle, the hydrogen is completely burnt. All components are to be regarded as ideal gases. Kinetic energies at the inlet as well as potential energies are negligible.

(a) What are the molar fluxes of the educts respectively products?
(b) Calculate the velocity at the outlet of the nozzle in case the temperature is $\vartheta_2 = 1700\,°\text{C}$ and a total heat flux of 1800 $\frac{\text{MJ}}{\text{h}}$ is released.
(c) Calculate the the exergy loss of the overall process per kmol H_2, when the heat is released to the environment with $T_{\text{env}} = 300\,\text{K}$.

9.18 A fuel ($c = 0.88$, $h = 0.12$) is to be investigated. Determine the lower heating value of the fuel H_U (25 °C) based on the enthalpies of formation, see Appendix B.

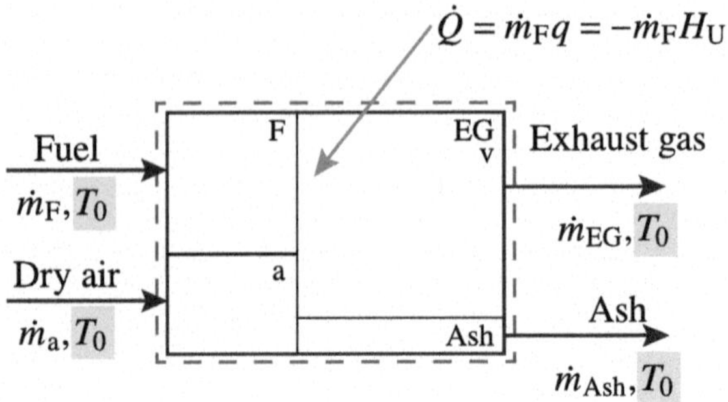

Fig. 9.7 Lower heating value

9.2 Solutions

9.1 **(d)**
In order to describe a chemical transformation energetically, the heating value methodology can be used in addition to the approach for enthalpies of formation.

The heating value is based on isothermal combustion of the fuel with sufficient oxygen, i.e. fuel and oxidant enter the combustion chamber at a uniform temperature T_0 and react with each other. Now it is determined how much heat must be extracted from the system in order to cool the resulting reaction products back to the reference temperature T_0. This amount of heat is released by the chemical transformation and is fuel-specific. If one finally knows the fuel-bound energy at reference temperature, any non-isothermal combustion can be described energetically via caloric approaches, see [1].

However, water in the exhaust gas plays a significant role because it can change its aggregate state:

- When using the *lower* heating value model, the water released during combustion, i.e. e.g. bonded water from the fuel and water resulting from the chemical reaction with oxygen, is theoretically *completely in a gaseous state*,[1] see Fig. 9.7.
- When using the higher heating value model, the water released during combustion, i.e. e.g. the bonded water from the fuel and the water produced during the chemical reaction with oxygen, is theoretically *completely in liquid state*,[2] see Fig. 9.8.

Both models obviously differ energetically from the heat of vaporisation of the water, i.e.

$$\boxed{H_O\left(T_0\right) = H_U\left(T_0\right) + \mu_{EG,H_2O} \cdot \Delta h_v\left(T_0\right)} \tag{9.1}$$

In order to determine the higher heating value, it must be known how large the mass of the exhaust gas water is per mass of the fuel, i.e. μ_{EG,H_2O}. For this purpose, we

[1] Although, for example, if $\vartheta_0 = 25\,°C$ is chosen as the reference temperature, most of the water will condense out of the exhaust gas, cf. [3].

[2] Although the dry exhaust gas absorbs as much water in form of vapour until the gas phase is fully saturated.

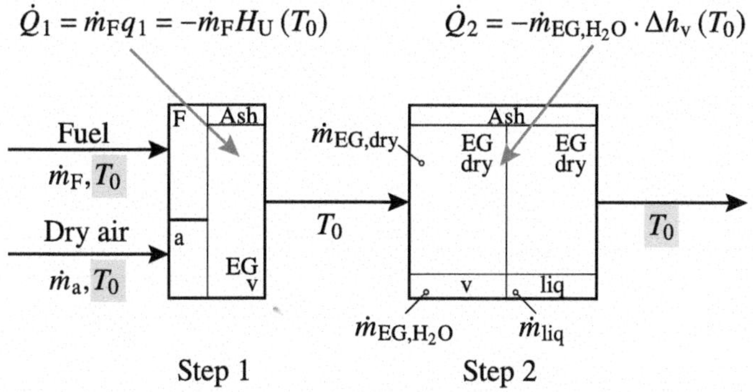

Fig. 9.8 Higher heating value

carry out a stoichiometric evaluation of the two combustible components carbon and hydrogen:

- Carbon:
$$C + O_2 \rightarrow CO_2 \tag{9.2}$$
Water is not produced when the carbon is burnt.
- Hydrogen:
$$2H_2 + O_2 \rightarrow 2H_2O \tag{9.3}$$

Obviously, only the hydrogen contributes to the formation of water. From Eq. 9.3 it can be seen that the following applies to the molar quantity of the product water

$$\frac{n_{EG,H_2O}}{n_{H_2}} = \frac{2}{2} \rightarrow n_{EG,H_2O} = n_{H_2}. \tag{9.4}$$

With the molar masses it yields

$$\frac{m_{EG,H_2O}}{M_{H_2O}} = \frac{m_{H_2}}{M_{H_2}} \tag{9.5}$$

respectively

$$m_{EG,H_2O} = m_{H_2} \frac{M_{H_2O}}{M_{H_2}} = 9 m_{H_2}. \tag{9.6}$$

Dividing by the fuel mass m_F, we finally obtain

$$\mu_{EG,H_2O} = \frac{m_{EG,H_2O}}{m_F} = 9 \frac{m_{H_2}}{m_F} = 9 \cdot h = 1.26. \tag{9.7}$$

Consequently, the higher heating value follows

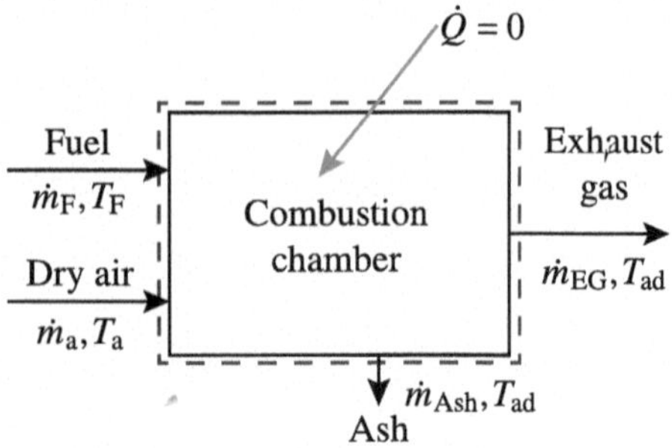

Fig. 9.9 Adiabatic flame temperature with dry air, i.e. $x = 0$

$$H_O(T_0) = H_U(T_0) + \mu_{EG,H_2O} \cdot \Delta h_v(T_0) = 45557 \, \frac{kJ}{kg}. \tag{9.8}$$

9.2 **(b)**

Figure 9.9 shows the boundary conditions for determining the adiabatic flame temperature using the example of the open, steady state system for combustion with dry air. The adiabatic flame temperature T_{ad} refers to the temperature of the reaction products that occurs when fuel and air react with each other and no heat is exchanged across the system boundary. Hence, answer (c) does not make sense.

The energy balance leads to the adiabatic flame temperature, see [1], i.e.

$$\vartheta_{ad} = \vartheta_0 + \frac{H_U(T_0) + \overline{c_{p,F}}|_{\vartheta_0}^{\vartheta_F}(\vartheta_F - \vartheta_0) + \lambda l_{min} \overline{c_{p,a}}|_{\vartheta_0}^{\vartheta_a}(\vartheta_a - \vartheta_0)}{\mu_{EG} \cdot \overline{c_{p,EG}}|_{\vartheta_0}^{\vartheta_{ad}} + a \cdot \overline{c_{Ash}}|_{\vartheta_0}^{\vartheta_{ad}}} \tag{9.9}$$

with

$$\mu_{EG} = (1 - a) + \lambda \cdot l_{min}. \tag{9.10}$$

Obviously, the resulting adiabatic flame temperature depends on the following parameters:

- Combustion air temperature:
  The higher the temperature of the incoming combustion air ϑ_a, the more energy is supplied to the system and the combustion temperature rises. Thus, answer (a) is wrong.
- Fuel temperature:
  The higher the temperature of the incoming fuel ϑ_F, the more energy is supplied to the system and the combustion temperature rises.
- Air-fuel ratio:
  If the air-fuel ration λ is increased, more air is supplied to the system than is required for stoichiometric combustion. The chemically bonded energy released by the chemical reaction must now heat a larger, passive mass of air.[3] As a consequence, the temperature rise during combustion is lower than if there were less passive

[3] This excess air is not involved in the chemical conversion, but only has a caloric effect.

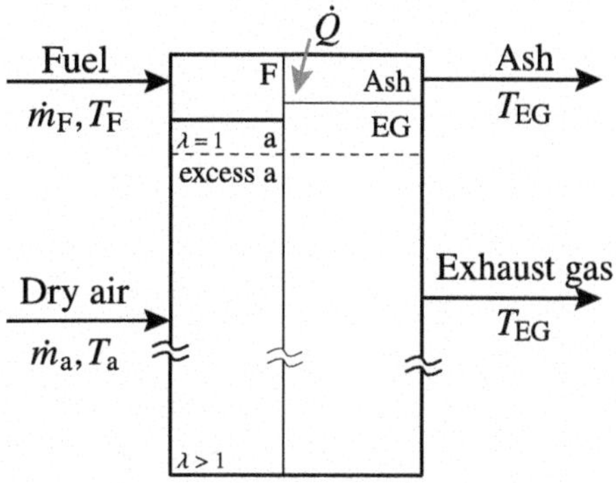

Fig. 9.10 Exhaust gas composition

mass. The combustion temperature can therefore be reduced by increasing the air-fuel ratio. Answer (b) is correct.

- Lower heating value:
  The more chemically bonded energy H_U is supplied to the combustion chamber, the higher the combustion temperature.
- Chemical composition of air and fuel

9.3 **(d)**
The larger the air-fuel ratio, the more excess air not needed for combustion is in the exhaust gas. For small air-fuel ratio, exhaust gas concentrations occur that are determined by the chemical conversion. However, the higher the air-fuel ratio, the more the chemical composition of the combustion air dominates, cf. Fig. 9.10. In the present case, for large λ the concentration of oxygen in the exhaust gas must strive towards the concentration of oxygen in the combustion air, i.e.

$$\lim_{\lambda\to\infty} \xi_{EG,O_2} = \xi_{air,O_2} = 0.23. \tag{9.11}$$

In addition, at $\lambda = 1$ there must be no oxygen in the exhaust gas, as combustion is stoichiometric, i.e. $\xi_{EG,O_2} = 0$. Therefore, graph D is correct.

9.4 **(b)**
The following applies to the concentrations of the exhaust gas

$$\sigma_{EG,i} = x_{EG,i} = \pi_{EG,i}. \tag{9.12}$$

This means that the concentration of the vapour in the exhaust gas yields

$$\pi_{EG,H_2O} = \sigma_{EG,H_2O} = \frac{p_{EG,H_2O}}{p} \tag{9.13}$$

respectively

$$p_{EG,H_2O} = \sigma_{EG,H_2O} \cdot p = 0.1\,\text{bar}. \tag{9.14}$$

In the saturation case to be considered, the partial pressure just reaches the temperature-dependent saturation pressure, i.e.

$$p_{\mathrm{EG,H_2O}} = 0.1\,\mathrm{bar} = p_{\mathrm{s}}\left(\vartheta_{\tau}\right). \tag{9.15}$$

From the steam table, see Appendix A, it can be seen that the dew point temperature is

$$\vartheta_{\tau} = 45.8075\,^{\circ}\mathrm{C}. \tag{9.16}$$

If the exhaust gas falls below this temperature, water condenses out. Answer (b) is correct.

9.5 **(b)**
The topic has already been treated in Problem 9.1. According to Eq. 9.1, higher and lower heating value deviate from each other if water occurs in the exhaust gas due to the combustion process. This can result from a chemical reaction with the bonded hydrogen, or can be water already carried in the fuel. In such a case it yields

$$\mu_{\mathrm{EG,H_2O}} > 0. \tag{9.17}$$

The following applies to the fuels listed

- Hydrogen:
  The combustion of hydrogen obeys

$$2\mathrm{H_2} + \mathrm{O_2} \rightarrow 2\mathrm{H_2O}. \tag{9.18}$$

  This means that water occurs in the exhaust gas, i.e. $\mu_{\mathrm{EG,H_2O}} > 0$. Consequently, higher and lower heating value differ. Answer (a) is wrong.
- Hydrogen:
  The combustion of sulphur obeys

$$\mathrm{S} + \mathrm{O_2} \rightarrow \mathrm{S_2O}. \tag{9.19}$$

  This means that no water occurs in the exhaust gas, i.e. $\mu_{\mathrm{EG,H_2O}} = 0$. Consequently, higher and lower heating value are identical. Answer (b) is correct.
- Methane:
  The combustion of methane obeys

$$\mathrm{CH_4} + 2\mathrm{O_2} \rightarrow 2\mathrm{H_2O} + \mathrm{CO_2}. \tag{9.20}$$

  This means that water occurs in the exhaust gas, i.e. $\mu_{\mathrm{EG,H_2O}} > 0$. Consequently, higher and lower heating value differ. Answer (c) is wrong.

9.6 **(a)**
The most significant parameters influencing the adiabatic combustion temperature have already been discussed in Problem 9.2. In particular, it has been shown that an

increase in the air-fuel ratio is accompanied by a reduction in the adiabatic combustion temperature.
This is due to the fact that with the supplied energy of the fuel, a higher mass absorbs the energy released during combustion and thus the temperature rise is lower than, for example, with stoichiometric combustion. Such behaviour is only shown by Graph A. Therefore, answer (a) is correct.

9.7 First take a look at the chemical reactions of the fuel components

- Carbon:

$$C + O_2 \rightarrow CO_2 \tag{9.21}$$

From this equation it can be seen that the following applies to the molar quantity of the required oxygen

$$\frac{n_{O_2(C)}}{n_C} = \frac{1}{1} \rightarrow n_{O_2(C)} = n_C. \tag{9.22}$$

With the molar masses it yields

$$\frac{m_{O_2(C)}}{M_{O_2}} = \frac{m_C}{M_C} \tag{9.23}$$

respectively

$$m_{O_2(C)} = m_C \frac{M_{O_2}}{M_C} = 2.667 \cdot m_C. \tag{9.24}$$

- Hydrogen:

$$2H_2 + O_2 \rightarrow 2H_2O \tag{9.25}$$

From this equation it can be seen that the following applies to the molar quantity of the required oxygen

$$\frac{n_{O_2(H_2)}}{n_{H_2}} = \frac{1}{2} \rightarrow n_{O_2(H_2)} = \frac{1}{2} n_{H_2}. \tag{9.26}$$

With the molar masses it yields

$$\frac{m_{O_2(H_2)}}{M_{O_2}} = \frac{1}{2} \frac{m_{H_2}}{M_{H_2}} \tag{9.27}$$

respectively

$$m_{O_2(H_2)} = \frac{1}{2} m_{H_2} \frac{M_{O_2}}{M_{H_2}} = 8 \cdot m_{H_2}. \tag{9.28}$$

- Water (fuel):
No further chemical reaction.

- Ash (fuel):
  No further chemical reaction.

Consequently, the total mass of oxygen required for complete stoichiometric combustion of the fuel is the sum of $m_{O_2(C)}$ and $m_{O_2(H_2)}$. However, the mass of oxygen to be added is reduced because the fuel contains oxygen according to gravimetry, i.e. $m_{O_2(\mathrm{Fuel})}$. Thus the result for the minimum oxygen demand is

$$m_{O_2,\min} = m_{O_2(C)} + m_{O_2(H_2)} - m_{O_2(\mathrm{Fuel})} = 2.667 \cdot m_C + 8 \cdot m_{H_2} - m_{O_2(\mathrm{Fuel})}. \tag{9.29}$$

Dividing by the fuel mass m_F, one gets

$$o_{\min} = \frac{m_{O_2,\min}}{m_F} = 2.667c + 8h - o = 2.3769. \tag{9.30}$$

Alternatively, use the minimum oxygen demand equation from the formulary in Appendix E.
If oxygen from the ambient air is used, the minimum air demand is higher than the minimum oxygen demand because the air consists of only 23 mass—% oxygen. It applies

$$l = \lambda \frac{o_{\min}}{\xi_{O_2}} = 12.4013. \tag{9.31}$$

This means that at least 12.4013 kg of air are required per kilogram of fuel for complete combustion. If more air is supplied, combustion is still complete, but oxygen is additionally found in the exhaust gas.

9.8 The first law of thermodynamics for the heat exchanger, cf. Fig. 9.11, obeys

$$\dot{m}_{EG} h_1 + \dot{Q} = \dot{m}_{EG} h_2 \tag{9.32}$$

respectively

$$\dot{Q} = \dot{m}_{EG} \left(h_2 - h_1 \right). \tag{9.33}$$

Specific heat capacities are usually temperature-dependent. This dependence is often neglected in small temperature ranges, but in the present case it must be taken into account due to the large temperature span, i.e. $\overline{c_p}|_{25\,°\mathrm{C}}^{800\,°\mathrm{C}}$ must be determined, so that

Fig. 9.11 Heat exchanger

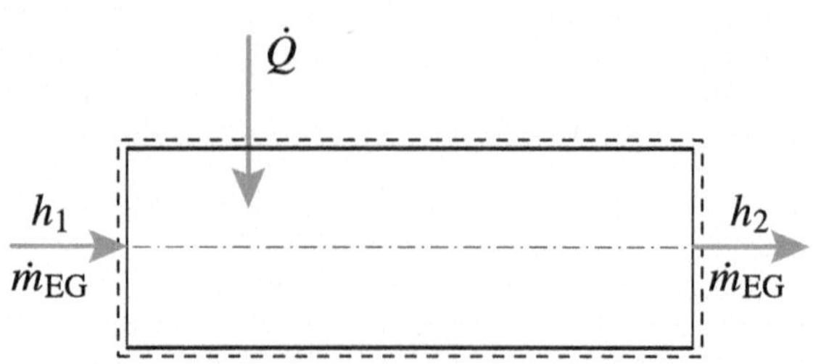

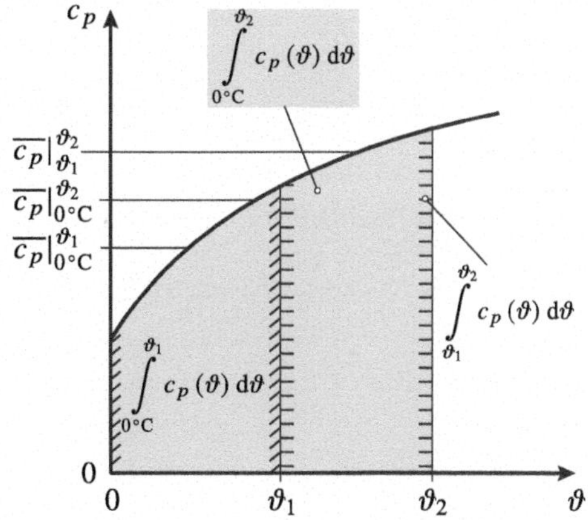

Fig. 9.12 Specific heat capacity

$$\dot{Q} = \dot{m}_{EG} \overline{c_p}|_{25\,°C}^{800\,°C} (\vartheta_2 - \vartheta_1)\,. \tag{9.34}$$

As an example, the specific heat capacity as a function of temperature is shown in Fig. 9.12. Average values for the specific heat capacity of gases, i.e. flue gas components, are usually given with reference to a lower temperature of 0 °C, as it is not very useful to give all possible temperature ranges by tabular works. However, in order to cover the required temperature range of 25 °C to 800 °C, one proceeds as follows,[4] see Fig. 9.12:

$$\overline{c_p}|_{\vartheta_1}^{\vartheta_2} (\vartheta_2 - \vartheta_1) = \overline{c_p}|_{0\,°C}^{\vartheta_2} (\vartheta_2 - 0\,°C) - \overline{c_p}|_{0\,°C}^{\vartheta_1} (\vartheta_1 - 0\,°C) \tag{9.35}$$

respectively

$$\boxed{\overline{c_p}|_{\vartheta_1}^{\vartheta_2} = \frac{\overline{c_p}|_{0\,°C}^{\vartheta_2} \vartheta_2 - \overline{c_p}|_{0\,°C}^{\vartheta_1} \vartheta_1}{\vartheta_2 - \vartheta_1}} \tag{9.36}$$

As already mentioned, functions $\overline{c_p}|_{0\,°C}^{\vartheta}$ are listed, see Table 9.1. According to

$$\overline{c_{p,i}}|_{25\,°C}^{800\,°C} = \frac{\overline{c_{p,i}}|_{0\,°C}^{800\,°C} \cdot 800\,°C - \overline{c_{p,i}}|_{0\,°C}^{25\,°C} \cdot 25\,°C}{800\,°C - 25\,°C} \tag{9.37}$$

the average specific heat capacities can now be determined for the exhaust gas components, i.e.

- $\overline{c_{p,CO_2}}|_{25\,°C}^{800\,°C} = 1.0968\,\frac{kJ}{kg\,K}$
- $\overline{c_{p,N_2}}|_{25\,°C}^{800\,°C} = 1.0995\,\frac{kJ}{kg\,K}$
- $\overline{c_{p,O_2}}|_{25\,°C}^{800\,°C} = 1.0190\,\frac{kJ}{kg\,K}$
- $\overline{c_{p,H_2O}}|_{25\,°C}^{800\,°C} = 2.0813\,\frac{kJ}{kg\,K}$.

[4] The following formula results from the area correlations in Fig. 9.12.

Furthermore, the following applies to an ideal gas mixture

$$\overline{c_p}\Big|_{25\,°\mathrm{C}}^{800\,°\mathrm{C}} = \sum_i \xi_i \cdot \overline{c_{p,i}}\Big|_{25\,°\mathrm{C}}^{800\,°\mathrm{C}} = 1.1934\,\frac{\mathrm{kJ}}{\mathrm{kg\,K}} \tag{9.38}$$

Finally, the exchanged heat flux is

$$\dot{Q} = \dot{m}_{\mathrm{EG}}\overline{c_p}\Big|_{25\,°\mathrm{C}}^{800\,°\mathrm{C}}\left(\vartheta_2 - \vartheta_1\right) = -1.8497 \times 10^3\,\mathrm{kW}. \tag{9.39}$$

9.9 The mass flux of the exhaust gas results from the mass balance for combustion, i.e.

$$\dot{m}_{\mathrm{EG}} = \dot{m}_{\mathrm{F}} + \dot{m}_{\mathrm{a}} = 201.3\,\frac{\mathrm{g}}{\mathrm{s}}. \tag{9.40}$$

Since the molar mass of the exhaust gas is known, the molar flux can also be determined according to

$$\dot{n}_{\mathrm{EG}} = \frac{\dot{m}_{\mathrm{EG}}}{M_{\mathrm{EG}}} = 0.0073\,\frac{\mathrm{kmol}}{\mathrm{s}}. \tag{9.41}$$

To calculate the amount of condensed water during cooling, it is first assumed that the water is in the vapour state. Figure 9.13 shows the mass/molar balance for the cooling process and the associated condensate release. According to Fig. 9.13a, condensation removes as much liquid water from the gas phase until the gas phase is completely saturated with vapour. This gaseous phase therefore has a lower molar flux, i.e.

$$\dot{n}_2 = \dot{n}_{\mathrm{EG}} - \dot{n}_{\mathrm{liq}}. \tag{9.42}$$

The next step is to focus on the water balance, see Fig. 9.13b. The mass of water in the exhaust gas can be easily calculated from the concentration specification

$$\dot{m}_{\mathrm{EG,H_2O}} = \xi_{\mathrm{EG,H_2O}}\dot{m}_{\mathrm{EG}} = 22.49\,\frac{\mathrm{g}}{\mathrm{s}}. \tag{9.43}$$

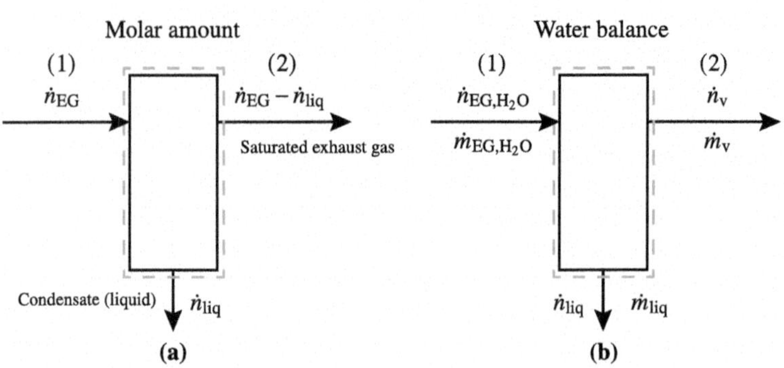

Fig. 9.13 Condensation

The molar quantity of the entire product water follows accordingly

$$\dot{n}_{EG,H_2O} = \frac{\dot{n}_{EG}}{M_{H_2O}} = 0.0012\,\frac{kmol}{s}. \tag{9.44}$$

The molar flux of the vapour in saturated state (2) obeys

$$\dot{n}_v = \dot{n}_{EG,H_2O} - \dot{n}_{liq}. \tag{9.45}$$

For the molar concentration of the vapour in state (2), since the state is saturated, it yields

$$x_v = x_s = \pi_v = \frac{p_s\,(50\,°C)}{p} = 0.1235 \quad \text{with} \quad p_s\,(50\,°C) = 0.1235\,\text{bar}. \tag{9.46}$$

The molar concentration further reads as

$$x_v = x_s = \frac{\dot{n}_v}{\dot{n}_2} = \frac{\dot{n}_{EG,H_2O} - \dot{n}_{liq}}{\dot{n}_{EG} - \dot{n}_{liq}}. \tag{9.47}$$

Finally, the molar flux of the condensed water is

$$\dot{n}_{liq} = \frac{\dot{n}_{EG,H_2O} - x_s\dot{n}_{EG}}{1 - x_s} = 4.0291 \times 10^{-4}\,\frac{kmol}{s} \tag{9.48}$$

and the mass flux follows with the molar mass of water, i.e.

$$\dot{m}_{liq} = \dot{n}_{liq} \cdot M_{H_2O} = 7.2524\,\frac{g}{s}. \tag{9.49}$$

9.10 The only contribution to carbon dioxide in the exhaust gas is due to the carbon bound in the fuel. This follows the chemical equation

$$C + O_2 \rightarrow CO_2 \tag{9.50}$$

From this equation it can be seen that the following applies to the molar quantity of the carbon dioxide

$$\frac{n_{EG,CO_2}}{n_C} = \frac{1}{1} \rightarrow n_{EG,CO_2} = n_C. \tag{9.51}$$

With the molar masses it yields

$$\frac{m_{EG,CO_2}}{M_{CO_2}} = \frac{m_C}{M_C} \tag{9.52}$$

respectively

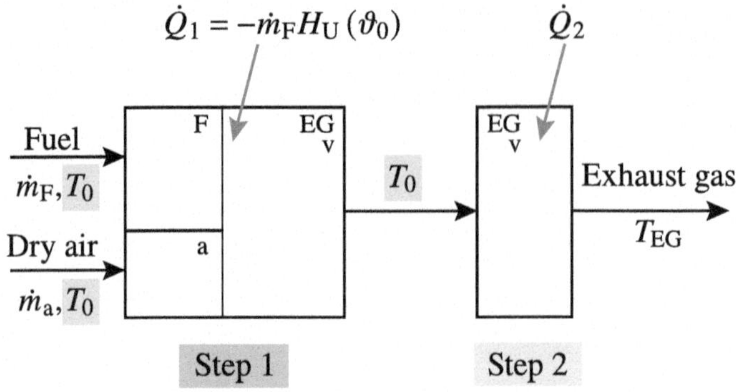

Fig. 9.14 Two-step equivalent model

$$m_{EG,CO_2} = m_C \frac{M_{CO_2}}{M_C} = 3.667 \cdot m_C. \tag{9.53}$$

Finally, by diving with the mass of the fuel

$$\mu_{EG,CO_2} = \frac{m_{EG,CO_2}}{m_F} = 3.667 \frac{m_C}{m_F} = 3.667 \cdot c = 2.7869. \tag{9.54}$$

9.11 The energy balance for the problem is outlined in Fig. 9.14. Since fuel and air are already supplied at the reference temperature ϑ_0 of the lower heating value, the energetic consideration of the chemical reaction based on the heating value already takes place in step 1. In the second and last step, the exhaust gas is lifted to the outlet temperature. The following therefore applies to the total heat

$$\dot{Q} = \dot{Q}_1 + \dot{Q}_2 \quad \text{with} \quad \dot{Q}_1 = -\dot{m}_F H_U(\vartheta_0). \tag{9.55}$$

The heat flux $\dot{Q}_2$ follows from an energy balance for the second step, i.e.

$$\dot{Q}_2 = \dot{m}_{EG}\left[h_{EG}(T_{EG}) - h_{EG}(T_0)\right] = \dot{m}_{EG} \overline{c_{p,EG}}\Big|_{\vartheta_0}^{\vartheta_{EG}} (\vartheta_{EG} - \vartheta_0) \quad \text{with} \quad \vartheta_0 = 25\,^\circ\text{C}. \tag{9.56}$$

It further is, cf. Problem 9.8,

$$\overline{c_{p,EG}}\Big|_{\vartheta_0}^{\vartheta_{EG}} (\vartheta_{EG} - \vartheta_0) = \overline{c_{p,EG}}\Big|_{0\,^\circ\text{C}}^{\vartheta_{EG}} \cdot \vartheta_{EG} - \overline{c_{p,EG}}\Big|_{0\,^\circ\text{C}}^{\vartheta_0} \cdot \vartheta_0 \tag{9.57}$$

so that Eq. 9.55 in specific notation obeys

$$\boxed{q = -H_U(\vartheta_0) + \underbrace{\frac{\dot{m}_{EG}}{\dot{m}_F}}_{=\mu_{EG}} \cdot \left[\overline{c_{p,EG}}\Big|_{0\,^\circ\text{C}}^{\vartheta_{EG}} \cdot \vartheta_{EG} - \overline{c_{p,EG}}\Big|_{0\,^\circ\text{C}}^{\vartheta_0} \cdot \vartheta_0\right]} \tag{9.58}$$

In order to determine the specific heat flux, the stoichiometric consideration now follows. The minimum oxygen demand is

$$o_{\text{min}} = 2.667 \cdot c + 8 \cdot h + s - o = 3.3070 \tag{9.59}$$

and the air demand follows

$$l = \lambda \frac{o_{\text{min}}}{\xi_{O_2}} = 22.5475. \tag{9.60}$$

From the mass balance

$$m_{\text{EG}} = m_{\text{F}} + m_{\text{a}} \tag{9.61}$$

it follows by division with $\dot{m}_{\text{F}}$

$$\mu_{\text{EG}} = 1 + l = 23.5475. \tag{9.62}$$

The composition of the exhaust gas is

- Carbon dioxide:

$$\mu_{\text{EG,CO}_2} = 3.667 \cdot c = 3.2270 \rightarrow \xi_{\text{EG,CO}_2} = \frac{\mu_{\text{EG,CO}_2}}{\mu_{\text{EG}}} = 0.1370 \tag{9.63}$$

- Vapour:

$$\mu_{\text{EG,H}_2\text{O}} = 9 \cdot h = 1.0800 \rightarrow \xi_{\text{EG,H}_2\text{O}} = \frac{\mu_{\text{EG,H}_2\text{O}}}{\mu_{\text{EG}}} = 0.0459 \tag{9.64}$$

- Nitrogen:

$$\mu_{\text{EG,N}_2} = l \cdot \xi_{\text{N}_2} = 17.587 \rightarrow \xi_{\text{EG,N}_2} = \frac{\mu_{\text{EG,N}_2}}{\mu_{\text{EG}}} = 0.7469 \tag{9.65}$$

- Oxygen:

$$\mu_{\text{EG,O}_2} = o_{\text{min}} \cdot (\lambda - 1) = 1.6535 \rightarrow \xi_{\text{EG,O}_2} = \frac{\mu_{\text{EG,O}_2}}{\mu_{\text{EG}}} = 0.0702 \tag{9.66}$$

Now the specific heat capacities can be determined analogous to Problem 9.8 with Table 9.1. This results in

$$\overline{c_{p,\text{EG}}}\Big|_{0\,^\circ\text{C}}^{\vartheta_{\text{EG}}} = \sum_i \xi_{\text{EG},i} \overline{c_{p,\text{EG},i}}\Big|_{0\,^\circ\text{C}}^{\vartheta_{\text{EG}}} = 1.1354 \, \frac{\text{kJ}}{\text{kg K}} \tag{9.67}$$

and

$$\overline{c_{p,\mathrm{EG}}}\Big|_{0\,^\circ\mathrm{C}}^{\vartheta_0} = \sum_i \xi_{\mathrm{EG},i}\overline{c_{p,\mathrm{EG},i}}\Big|_{0\,^\circ\mathrm{C}}^{\vartheta_0} = 1.0399\,\frac{\mathrm{kJ}}{\mathrm{kg\,K}}. \tag{9.68}$$

Finally, the released specific heat obeys

$$q = -H_\mathrm{U}(\vartheta_0) + \mu_\mathrm{EG} \cdot \left[\overline{c_{p,\mathrm{EG}}}\Big|_{0\,^\circ\mathrm{C}}^{\vartheta_\mathrm{EG}} \cdot \vartheta_\mathrm{EG} - \overline{c_{p,\mathrm{EG}}}\Big|_{0\,^\circ\mathrm{C}}^{\vartheta_0} \cdot \vartheta_0\right] = -2.2590 \times 10^4\,\frac{\mathrm{kJ}}{\mathrm{kg}}. \tag{9.69}$$

9.12 The formation of condensate has already been derived in detail in Problem 9.9, see also Fig. 9.13. The following correlation could be derived to determine the condensed water mass

$$\boxed{\dot{n}_\mathrm{liq} = \frac{\dot{n}_{\mathrm{EG,H_2O}} - x_\mathrm{s}\dot{n}_\mathrm{EG}}{1 - x_\mathrm{s}}} \tag{9.70}$$

For the saturated gas state after separation of the condensate, it yields

$$x_\mathrm{s} = \frac{p_\mathrm{s}(50\,^\circ\mathrm{C})}{p} = 0.0618 \quad \text{with} \quad p_\mathrm{s}(50\,^\circ\mathrm{C}) = 0.1235\,\mathrm{bar}. \tag{9.71}$$

From stoichiometric considerations it is

$$\nu_\mathrm{EG} = \sum_i \nu_{\mathrm{EG},i} = 11.6174. \tag{9.72}$$

It further applies

$$\nu_{\mathrm{EG,H_2O}} = \frac{\dot{n}_{\mathrm{EG,H_2O}}}{\dot{n}_\mathrm{F}} \quad \text{and} \quad \nu_\mathrm{EG} = \frac{\dot{n}_\mathrm{EG}}{\dot{n}_\mathrm{F}} \tag{9.73}$$

so that finally

$$\dot{n}_\mathrm{liq} = \dot{n}_\mathrm{F}\frac{\nu_{\mathrm{EG,H_2O}} - x_\mathrm{s}\nu_\mathrm{EG}}{1 - x_\mathrm{s}} = 8.5436 \times 10^{-4}\,\frac{\mathrm{kmol}}{\mathrm{s}}. \tag{9.74}$$

9.13 According to the structural formula C_3H_8, 1 kmol of propane contains 3 kmol of carbon. With the help of the molar masses, the mass fraction of carbon in propane can therefore be determined, i.e.

$$c_{\mathrm{C_3H_8}} = \frac{3M_\mathrm{C}}{3M_\mathrm{C} + 8M_\mathrm{H}} = 0.8182. \tag{9.75}$$

Analogously, 1 kmol of methane contains 1 kmol of carbon. The mass fraction carbon in methane obeys

$$c_{\mathrm{CH_4}} = \frac{M_\mathrm{C}}{M_\mathrm{C} + 4M_\mathrm{H}} = 0.75. \tag{9.76}$$

Via the mass fractions of the gas mixture, the total carbon fraction results according to

$$c = \xi_{C_3H_8} \cdot c_{C_3H_8} + \xi_{CH_4} \cdot c_{CH_4} = 0.7841 \quad \text{with} \quad \xi_{C_3H_8} = \xi_{CH_4} = 0.5. \tag{9.77}$$

The hydrogen fraction then yields

$$h = 1 - c = 0.2159. \tag{9.78}$$

Now that the fuel mixture is known gravimetrically, the air demand can be determined easily, i.e.

$$l = \lambda \cdot l_{\min} = \frac{2.667 \cdot c + 8 \cdot h}{\xi_{\text{air, O}_2}} = 16.6019 \quad \text{with} \quad \lambda = 1. \tag{9.79}$$

9.14

(a) In order to calculate the mass flow, the supplied fuel gases are treated as ideal gases and determined by means of the thermal equation of state, i.e.

$$\dot{m} = \frac{p\dot{V}}{RT} \tag{9.80}$$

with

$$R = \frac{R_M}{M}. \tag{9.81}$$

The molar masses are $M_M = 16\,\frac{\text{kg}}{\text{kmol}}$ for methane and $M_P = 44\,\frac{\text{kg}}{\text{kmol}}$ for propane. Hence, the mass flows are

$$\dot{m}_M = \frac{p\dot{V}_M M_M}{R_M T} = 8.9645 \times 10^{-5}\,\frac{\text{kg}}{\text{s}} \tag{9.82}$$

respectively

$$\dot{m}_P = \frac{p\dot{V}_P M_P}{R_M T} = 1.4791 \times 10^{-4}\,\frac{\text{kg}}{\text{s}}. \tag{9.83}$$

(b) The elemental analysis for methane yields

$$c_M = \frac{1 \cdot M_C}{M_{CH_4}} = 0.75 \quad \text{and} \quad h_M = \frac{4 \cdot M_H}{M_{CH_4}} = 0.25 \tag{9.84}$$

and for propane accordingly

$$c_P = \frac{3 \cdot M_C}{M_{C_3H_8}} = 0.8182 \quad \text{and} \quad h_P = \frac{8 \cdot M_H}{M_{C_3H_8}} = 0.1818. \tag{9.85}$$

The total mass fraction yields

$$c = \frac{\dot{m}_{\mathrm{P}} c_{\mathrm{P}} + \dot{m}_{\mathrm{M}} c_{\mathrm{M}}}{\dot{m}_{\mathrm{P}} + \dot{m}_{\mathrm{M}}} = 0.7925 \quad \text{and} \quad h = \frac{\dot{m}_{\mathrm{P}} h_{\mathrm{P}} + \dot{m}_{\mathrm{M}} h_{\mathrm{M}}}{\dot{m}_{\mathrm{P}} + \dot{m}_{\mathrm{M}}} = 0.2075. \tag{9.86}$$

(c) Since the mass fractions of the fuel are now known, the minimum oxygen demand can be calculated easily, i.e.

$$o_{\mathrm{min}} = 2.667c + 8h = 3.7738. \tag{9.87}$$

The minimum demand of dry air obeys

$$l_{\mathrm{min}} = \frac{o_{\mathrm{min}}}{\xi_{\mathrm{O_2,dry\ air}}} = 16.408. \tag{9.88}$$

The total flow of dry air follows

$$l = \lambda \cdot l_{\mathrm{min}} = 21.33, \tag{9.89}$$

i.e. per $\frac{\mathrm{kg}}{\mathrm{s}}$ fuel 21.33 $\frac{\mathrm{kg}}{\mathrm{s}}$ of dry air are required. This means that the total mass flow of dry air is equal to

$$\dot{m}_{\mathrm{a}} = \dot{m}_{\mathrm{F}} \cdot l \quad \text{with} \quad \dot{m}_{\mathrm{F}} = \dot{m}_{\mathrm{M}} + \dot{m}_{\mathrm{P}}. \tag{9.90}$$

To determine the mass flow of the supplied humid air, the vapour must be taken into account as well, i.e.

$$\dot{m}_{\mathrm{A}} = \dot{m}_{\mathrm{a}} + \dot{m}_{\mathrm{v}} = \dot{m}_{\mathrm{a}}\,(1 + x)\,. \tag{9.91}$$

The water content of the unsaturated humid combustion air is

$$x = 0.622 \frac{p_{\mathrm{v}}}{p - p_{\mathrm{v}}} \quad \text{with} \quad p_{\mathrm{v}} = \varphi \cdot p_{\mathrm{s}}\,(25\,^{\circ}\mathrm{C})\,. \tag{9.92}$$

The saturation pressure at 25 °C can be taken from the steam table and is $p_{\mathrm{s}}\,(25\,^{\circ}\mathrm{C}) = 0.0317$ bar, so that the water content equals

$$x = 0.0121. \tag{9.93}$$

Finally, the humid air mass rate is

$$\dot{m}_{\mathrm{A}} = \dot{m}_{\mathrm{a}}\,(1 + x) = 0.0051\,\frac{\mathrm{kg}}{\mathrm{s}}. \tag{9.94}$$

(d) The following applies to the exhaust gas composition:

$$\mu_{EG,CO_2} = 3.667c = 2.9059 \tag{9.95}$$

$$\mu_{EG,H_2O} = 9h + \lambda l_{min} x = 2.1252 \tag{9.96}$$

$$\mu_{EG,O_2} = o_{min} (\lambda - 1) = 1.1322 \tag{9.97}$$

$$\mu_{EG,N_2} = \lambda l_{min} \xi_{N_2,\text{dry air}} = 16.4244 \tag{9.98}$$

so that

$$\mu_{EG} = \sum_i \mu_{EG,i} = 22.5877. \tag{9.99}$$

This results in the concentrations of the exhaust gas

$$\xi_{EG,CO_2} = \frac{\mu_{EG,CO_2}}{\mu_{EG}} = 0.1287 \tag{9.100}$$

$$\xi_{EG,H_2O} = \frac{\mu_{EG,H_2O}}{\mu_{EG}} = 0.0941 \tag{9.101}$$

$$\xi_{EG,O_2} = \frac{\mu_{EG,O_2}}{\mu_{EG}} = 0.0501 \tag{9.102}$$

$$\xi_{EG,N_2} = \frac{\mu_{EG,N_2}}{\mu_{EG}} = 0.7271. \tag{9.103}$$

(e) It applies to the maximum molar concentration of the vapour in the exhaust gas that

$$x_{EG,H_2O} = \frac{p_s(\vartheta_{crit})}{p}. \tag{9.104}$$

The partial pressure of the vapour in saturated state thus obeys

$$p_s(\vartheta_{crit}) = x_{EG,H_2O} \cdot p. \tag{9.105}$$

The molar concentration can be calculated from the mass concentration, i.e.

$$x_{EG,H_2O} = \xi_{EG,H_2O} \frac{M_{EG}}{M_{H_2O}} = \xi_{EG,H_2O} \frac{R_M}{R_{EG} M_{H_2O}}. \tag{9.106}$$

The individual gas constant of the exhaust gas follows

$$R_{EG} = \sum_i \xi_{EG,i} R_i = \sum_i \xi_{EG,i} \frac{R_M}{M_i} = 296.7 \frac{\text{J}}{\text{kgK}}, \tag{9.107}$$

so that

$$x_{\mathrm{EG,H_2O}} = \xi_{\mathrm{EG,H_2O}} \frac{R_{\mathrm{M}}}{R_{\mathrm{EG}} M_{\mathrm{H_2O}}} = 0.1465. \tag{9.108}$$

For the maximum partial pressure of the vapour it follows

$$p_{\mathrm{s}} (\vartheta_{\mathrm{crit}}) = x_{\mathrm{EG,H_2O}} \cdot p = 0.1465\,\mathrm{bar}. \tag{9.109}$$

Finally, the associated saturation temperature can be determined from the steam table, i.e.

$$\vartheta_{\mathrm{crit}} = 53.48\,^{\circ}\mathrm{C}. \tag{9.110}$$

(f) The chemically bound energy supplied in total is combined into a new lower heating value $H_{\mathrm{U,F}}$, see Fig. 9.15. Hence, it is

$$\dot{m}_{\mathrm{M}} H_{\mathrm{U,M}} + \dot{m}_{\mathrm{P}} H_{\mathrm{U,P}} = (\dot{m}_{\mathrm{M}} + \dot{m}_{\mathrm{P}})\, H_{\mathrm{U,F}} \tag{9.111}$$

Solving for the overall lower heating value results in

$$H_{\mathrm{U,F}} = \frac{\dot{m}_{\mathrm{M}} H_{\mathrm{U,M}} + \dot{m}_{\mathrm{P}} H_{\mathrm{U,P}}}{\dot{m}_{\mathrm{M}} + \dot{m}_{\mathrm{P}}} = 47.731\,\frac{\mathrm{MJ}}{\mathrm{kg}}. \tag{9.112}$$

(g) The first law of thermodynamics obeys

$$\begin{aligned}\dot{Q}_{\mathrm{v}} = & -\dot{m}_{\mathrm{F}} \overbrace{\left[h_{\mathrm{F}}(T_{\mathrm{F}}) - h_{\mathrm{F}}(T_0)\right]}^{=0} - \lambda l_{\mathrm{min}} \dot{m}_{\mathrm{F}} \overbrace{\left[h_{1+x}(T_{\mathrm{A}}) - h_{1+x}(T_0)\right]}^{=0} + \\ & - \dot{m}_{\mathrm{F}} H_{\mathrm{U,F}}(T_0) + \dot{m}_{\mathrm{EG}} \left[h_{\mathrm{EG}}(T_{\mathrm{EG}}) - h_{\mathrm{EG}}(T_0)\right]\end{aligned} \tag{9.113}$$

respectively

$$\dot{Q}_{\mathrm{v}} = -\dot{m}_{\mathrm{F}} H_{\mathrm{U,F}}(T_0) + \dot{m}_{\mathrm{EG}} \left[h_{\mathrm{EG}}(T_{\mathrm{EG}}) - h_{\mathrm{EG}}(T_0)\right]. \tag{9.114}$$

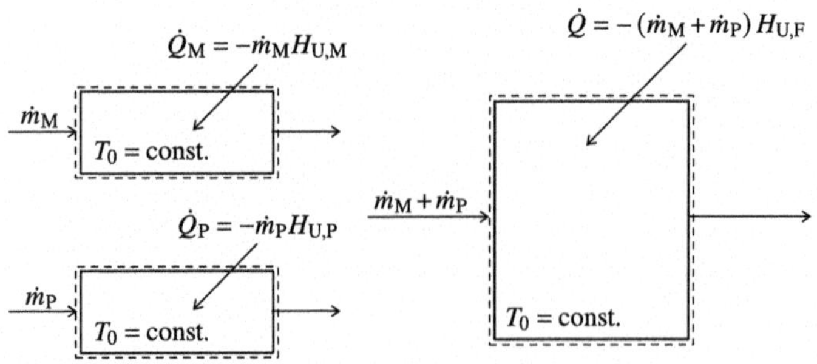

Fig. 9.15 Equivalent model

The specific enthalpy of the exhaust gas is substituted by a caloric equation of state, i.e. with the temperature-averaged specific heat capacities:

$$\dot{Q}_{\mathrm{v}} = -\dot{m}_{\mathrm{F}} H_{\mathrm{U,F}}(T_0) + \dot{m}_{\mathrm{EG}} \left[\overline{c_{p,\mathrm{EG}}}\Big|_{0\,^{\circ}\mathrm{C}}^{200\,^{\circ}\mathrm{C}} \cdot 200\,^{\circ}\mathrm{C} - \overline{c_{p,\mathrm{EG}}}\Big|_{0\,^{\circ}\mathrm{C}}^{25\,^{\circ}\mathrm{C}} \cdot 25\,^{\circ}\mathrm{C} \right]. \tag{9.115}$$

The exhaust gas is considered to be an ideal gas mixture, so that

$$\overline{c_{p,\mathrm{EG}}} = \sum_i \xi_{\mathrm{EG},i} \overline{c_{p,i}}. \tag{9.116}$$

For the temperatures considered, the following applies with Table 9.1:

$$\overline{c_{p,\mathrm{EG}}}\Big|_{0\,^{\circ}\mathrm{C}}^{200\,^{\circ}\mathrm{C}} = \ldots = 1.1012\,\frac{\mathrm{kJ}}{\mathrm{kg\,K}} \tag{9.117}$$

and

$$\overline{c_{p,\mathrm{EG}}}\Big|_{0\,^{\circ}\mathrm{C}}^{25\,^{\circ}\mathrm{C}} = \ldots = 1.0838\,\frac{\mathrm{kJ}}{\mathrm{kg\,K}} \tag{9.118}$$

Finally, the heat flux equals

$$\dot{Q}_{\mathrm{v}} = -\dot{m}_{\mathrm{F}} H_{\mathrm{U,F}}(T_0) + \dot{m}_{\mathrm{EG}} \left[\overline{c_{p,\mathrm{EG}}}\Big|_{0\,^{\circ}\mathrm{C}}^{200\,^{\circ}\mathrm{C}} \cdot 200\,^{\circ}\mathrm{C} - \overline{c_{p,\mathrm{EG}}}\Big|_{0\,^{\circ}\mathrm{C}}^{25\,^{\circ}\mathrm{C}} \cdot 25\,^{\circ}\mathrm{C} \right] = -10.3\,\mathrm{kW}. \tag{9.119}$$

9.15

(a) The molar mass of the fuel used is calculated as follows

$$M_{\mathrm{F}} = 1 \cdot M_{\mathrm{C}} + 4 \cdot M_{\mathrm{H}} + 1 \cdot M_{\mathrm{O}} = 32\,\frac{\mathrm{kg}}{\mathrm{kmol}}. \tag{9.120}$$

This results in the elementary analysis

$$c = \frac{1 \cdot M_{\mathrm{C}}}{M_{\mathrm{F}}} = 0.375 \quad \text{and} \quad h = \frac{4 \cdot M_{\mathrm{H}}}{M_{\mathrm{F}}} = 0.125 \quad \text{and} \quad o = \frac{1 \cdot M_{\mathrm{O}}}{M_{\mathrm{F}}} = 0.5. \tag{9.121}$$

(b) The minimum air demand is

$$o_{\mathrm{min}} = 2.667c + 8h + s - o = 1.5. \tag{9.122}$$

For the air demand under the given boundary conditions, one finally obtains

$$l = \lambda l_{\mathrm{min}} = \lambda \frac{o_{\mathrm{min}}}{\xi_{\mathrm{O_2,dry\ air}}} = 6.5223. \tag{9.123}$$

(c) The following applies to the exhaust gas composition:

$$\mu_{\mathrm{EG,CO_2}} = 3.667c = 1.3751 \tag{9.124}$$

$$\mu_{\mathrm{EG,H_2O}} = 9h = 1.125 \tag{9.125}$$

$$\mu_{\mathrm{EG,O_2}} = o_{\mathrm{min}}\,(\lambda - 1) = 0 \tag{9.126}$$

$$\mu_{\mathrm{EG,N_2}} = \lambda l_{\mathrm{min}} \xi_{\mathrm{N_2,dry\ air}} = 5.0222 \tag{9.127}$$

so that

$$\mu_{\mathrm{EG}} = \sum_i \mu_{\mathrm{EG},i} = 7.5223. \tag{9.128}$$

This results in the concentrations of the exhaust gas

$$\xi_{\mathrm{EG,CO_2}} = \frac{\mu_{\mathrm{EG,CO_2}}}{\mu_{\mathrm{EG}}} = 0.1828 \tag{9.129}$$

$$\xi_{\mathrm{EG,H_2O}} = \frac{\mu_{\mathrm{EG,H_2O}}}{\mu_{\mathrm{EG}}} = 0.1496 \tag{9.130}$$

$$\xi_{\mathrm{EG,O_2}} = \frac{\mu_{\mathrm{EG,O_2}}}{\mu_{\mathrm{EG}}} = 0 \tag{9.131}$$

$$\xi_{\mathrm{EG,N_2}} = \frac{\mu_{\mathrm{EG,N_2}}}{\mu_{\mathrm{EG}}} = 0.6676. \tag{9.132}$$

(d) The first law of thermodynamics for an adiabatic combustion obeys

$$\begin{aligned} 0 = &- \dot{m}_{\mathrm{F}} c_{p,\mathrm{F}} \overbrace{[T_{\mathrm{F}} - T_0]}^{=0} - l \cdot \dot{m}_{\mathrm{F}} c_{p,\mathrm{A}} \overbrace{[T_{\mathrm{A}} - T_0]}^{=0} - \dot{m}_{\mathrm{F}} H_{\mathrm{U}}(T_0) + \\ &+ \dot{m}_{\mathrm{F}} \mu_{\mathrm{EG}} \left[\overline{c_{p,\mathrm{EG}}}\Big|_{0\,^\circ\mathrm{C}}^{\vartheta_{\mathrm{ad}}} \cdot \vartheta_{\mathrm{ad}} - \overline{c_{p,\mathrm{EG}}}\Big|_{0\,^\circ\mathrm{C}}^{25\,^\circ\mathrm{C}} \cdot 25\,^\circ\mathrm{C} \right]. \end{aligned} \tag{9.133}$$

Solving for the adiabatic flame temperature

$$\vartheta_{\mathrm{ad}} = \frac{1}{\overline{c_{p,\mathrm{EG}}}\Big|_{0\,^\circ\mathrm{C}}^{\vartheta_{\mathrm{ad}}}} \left[\frac{H_{\mathrm{U}}(T_0)}{\mu_{\mathrm{EG}}} + \overline{c_{p,\mathrm{EG}}}\Big|_{0\,^\circ\mathrm{C}}^{25\,^\circ\mathrm{C}} \cdot 25\,^\circ\mathrm{C} \right] \tag{9.134}$$

with

$$\overline{c_{p,\mathrm{EG}}}\Big|_{0\,^\circ\mathrm{C}}^{25\,^\circ\mathrm{C}} = \sum_i \xi_{\mathrm{EG},i} \overline{c_{p,i}}\Big|_{0\,^\circ\mathrm{C}}^{25\,^\circ\mathrm{C}} = \ldots = 1.1243\, \frac{\mathrm{kJ}}{\mathrm{kg\,K}}. \tag{9.135}$$

- 0. Iteration (initial guess)

$$\vartheta_{\mathrm{ad,0}} = 2000\,^\circ\mathrm{C} \tag{9.136}$$

$$\left.\overline{c_{p,\mathrm{EG}}}\right|_{0\,^{\circ}\mathrm{C}}^{\vartheta_{\mathrm{ad},0}} = \sum_i \xi_{\mathrm{EG},i} \left.\overline{c_{p,i}}\right|_{0\,^{\circ}\mathrm{C}}^{\vartheta_{\mathrm{ad},0}} = \ldots = 1.3870\,\frac{\mathrm{kJ}}{\mathrm{kg\,K}} \tag{9.137}$$

$$\vartheta_{\mathrm{ad},1} = \frac{1}{\left.\overline{c_{p,\mathrm{EG}}}\right|_{0\,^{\circ}\mathrm{C}}^{\vartheta_{\mathrm{ad},0}}} \left[\frac{H_{\mathrm{U}}(T_0)}{\mu_{\mathrm{EG}}} + \left.\overline{c_{p,\mathrm{EG}}}\right|_{0\,^{\circ}\mathrm{C}}^{25\,^{\circ}\mathrm{C}} \cdot 25\,^{\circ}\mathrm{C}\right] = 1927.6\,^{\circ}\mathrm{C} \tag{9.138}$$

- 1. Iteration

$$\left.\overline{c_{p,\mathrm{EG}}}\right|_{0\,^{\circ}\mathrm{C}}^{\vartheta_{\mathrm{ad},1}} = \sum_i \xi_{\mathrm{EG},i} \left.\overline{c_{p,i}}\right|_{0\,^{\circ}\mathrm{C}}^{\vartheta_{\mathrm{ad},1}} = \ldots = 1.3787\,\frac{\mathrm{kJ}}{\mathrm{kg\,K}} \tag{9.139}$$

$$\vartheta_{\mathrm{ad},2} = \frac{1}{\left.\overline{c_{p,\mathrm{EG}}}\right|_{0\,^{\circ}\mathrm{C}}^{\vartheta_{\mathrm{ad},1}}} \left[\frac{H_{\mathrm{U}}(T_0)}{\mu_{\mathrm{EG}}} + \left.\overline{c_{p,\mathrm{EG}}}\right|_{0\,^{\circ}\mathrm{C}}^{25\,^{\circ}\mathrm{C}} \cdot 25\,^{\circ}\mathrm{C}\right] = 1939.2\,^{\circ}\mathrm{C} \tag{9.140}$$

- 2. Iteration

$$\left.\overline{c_{p,\mathrm{EG}}}\right|_{0\,^{\circ}\mathrm{C}}^{\vartheta_{\mathrm{ad},2}} = \sum_i \xi_{\mathrm{EG},i} \left.\overline{c_{p,i}}\right|_{0\,^{\circ}\mathrm{C}}^{\vartheta_{\mathrm{ad},2}} = \ldots = 1.3800\,\frac{\mathrm{kJ}}{\mathrm{kg\,K}} \tag{9.141}$$

$$\vartheta_{\mathrm{ad},3} = \frac{1}{\left.\overline{c_{p,\mathrm{EG}}}\right|_{0\,^{\circ}\mathrm{C}}^{\vartheta_{\mathrm{ad},2}}} \left[\frac{H_{\mathrm{U}}(T_0)}{\mu_{\mathrm{EG}}} + \left.\overline{c_{p,\mathrm{EG}}}\right|_{0\,^{\circ}\mathrm{C}}^{25\,^{\circ}\mathrm{C}} \cdot 25\,^{\circ}\mathrm{C}\right] = 1937.3\,^{\circ}\mathrm{C} \tag{9.142}$$

- 3. Iteration

$$\left.\overline{c_{p,\mathrm{EG}}}\right|_{0\,^{\circ}\mathrm{C}}^{\vartheta_{\mathrm{ad},3}} = \sum_i \xi_{\mathrm{EG},i} \left.\overline{c_{p,i}}\right|_{0\,^{\circ}\mathrm{C}}^{\vartheta_{\mathrm{ad},3}} = \ldots = 1.3798\,\frac{\mathrm{kJ}}{\mathrm{kg\,K}} \tag{9.143}$$

$$\vartheta_{\mathrm{ad},4} = \frac{1}{\left.\overline{c_{p,\mathrm{EG}}}\right|_{0\,^{\circ}\mathrm{C}}^{\vartheta_{\mathrm{ad},3}}} \left[\frac{H_{\mathrm{U}}(T_0)}{\mu_{\mathrm{EG}}} + \left.\overline{c_{p,\mathrm{EG}}}\right|_{0\,^{\circ}\mathrm{C}}^{25\,^{\circ}\mathrm{C}} \cdot 25\,^{\circ}\mathrm{C}\right] = 1937.6\,^{\circ}\mathrm{C} \tag{9.144}$$

- 4. Iteration

$$\left.\overline{c_{p,\mathrm{EG}}}\right|_{0\,^{\circ}\mathrm{C}}^{\vartheta_{\mathrm{ad},4}} = \sum_i \xi_{\mathrm{EG},i} \left.\overline{c_{p,i}}\right|_{0\,^{\circ}\mathrm{C}}^{\vartheta_{\mathrm{ad},4}} = \ldots = 1.3798\,\frac{\mathrm{kJ}}{\mathrm{kg\,K}} \tag{9.145}$$

$$\vartheta_{\mathrm{ad},5} = \frac{1}{\left.\overline{c_{p,\mathrm{EG}}}\right|_{0\,^{\circ}\mathrm{C}}^{\vartheta_{\mathrm{ad},4}}} \left[\frac{H_{\mathrm{U}}(T_0)}{\mu_{\mathrm{EG}}} + \left.\overline{c_{p,\mathrm{EG}}}\right|_{0\,^{\circ}\mathrm{C}}^{25\,^{\circ}\mathrm{C}} \cdot 25\,^{\circ}\mathrm{C}\right] = 1937.6\,^{\circ}\mathrm{C} \tag{9.146}$$

The solution has now converged and the adiabatic flame temperature is 1937.6 °C.

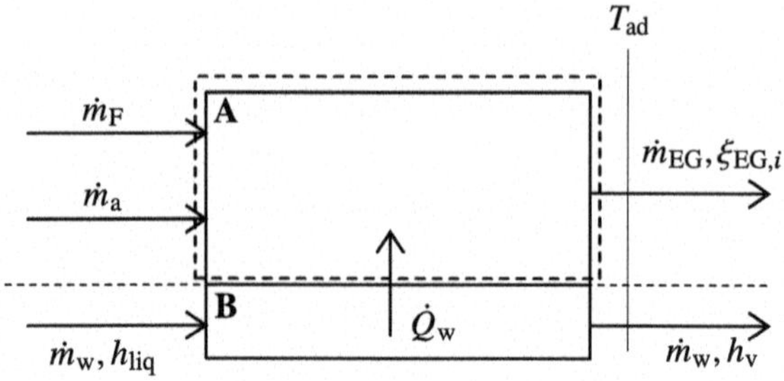

Fig. 9.16 Equivalent thermal model

(e) In such a case the heat flux for vaporising the water is taken from the combustion, cf. system A in Fig. 9.16. The first law of thermodynamics thus obeys

$$\dot{Q}_w = -\dot{m}_F c_{p,F} \overbrace{[T_F - T_0]}^{=0} - l \cdot \dot{m}_F c_{p,A} \overbrace{[T_A - T_0]}^{=0} - \dot{m}_F H_U(T_0) + \\ + \dot{m}_F \mu_{EG} \left[\overline{c_{p,EG}}\Big|_{0\,°C}^{\vartheta_{ad}} \cdot \vartheta_{ad} - \overline{c_{p,EG}}\Big|_{0\,°C}^{25\,°C} \cdot 25\,°C \right]. \tag{9.147}$$

Hence, it is

$$\vartheta_{ad} = \frac{1}{\overline{c_{p,EG}}\Big|_{0\,°C}^{\vartheta_{ad}}} \left[\frac{\left(\frac{\dot{Q}_w}{\dot{m}_F} + H_U(T_0) \right)}{\mu_{EG}} + \overline{c_{p,EG}}\Big|_{0\,°C}^{25\,°C} \cdot 25\,°C \right] \tag{9.148}$$

with

$$\overline{c_{p,EG}}\Big|_{0\,°C}^{25\,°C} = \sum_i \xi_{EG,i} \overline{c_{p,i}}\Big|_{0\,°C}^{25\,°C} = \ldots = 1.1243\,\frac{\text{kJ}}{\text{kg K}}. \tag{9.149}$$

The heat flux $\dot{Q}_w$ can be derived from an energy balance for system B, i.e.

$$\dot{Q}_w = \dot{m}_w \left[h_{liq}(\vartheta_0) - h_v(\vartheta_{ad}) \right]. \tag{9.150}$$

- <u>Alternative 1:</u> Since the vapour is to be considered as an ideal gas, the following caloric correlations could be used, analogous to the procedure for the energetic balancing of humid air:

$$h_{liq}(\vartheta_0) = c_W \vartheta_0 \quad \text{and} \quad h_v(\vartheta_{ad}) = \Delta h_{v,0} + c_{p,v} \vartheta_{ad} \tag{9.151}$$

Thus it applies

$$\dot{Q}_w = \dot{m}_w \left[c_W \vartheta_0 - \left(\Delta h_{v,0} + c_{p,v} \vartheta_{ad} \right) \right] \tag{9.152}$$

– 0. Iteration (initial guess)

$$\vartheta_{\mathrm{ad},0} = 1800\,^{\circ}\mathrm{C} \tag{9.153}$$

$$\left.\overline{c_{p,\mathrm{EG}}}\right|_{0\,^{\circ}\mathrm{C}}^{\vartheta_{\mathrm{ad},0}} = \sum_i \xi_{\mathrm{EG},i} \left.\overline{c_{p,i}}\right|_{0\,^{\circ}\mathrm{C}}^{\vartheta_{\mathrm{ad},0}} = \ldots = 1.3641\,\frac{\mathrm{kJ}}{\mathrm{kg\,K}} \tag{9.154}$$

$$\vartheta_{\mathrm{ad},1} = \frac{1}{\left.\overline{c_{p,\mathrm{EG}}}\right|_{0\,^{\circ}\mathrm{C}}^{\vartheta_{\mathrm{ad},0}}} \left[\frac{\left(\frac{\dot{Q}_{\mathrm{W}}}{\dot{m}_{\mathrm{F}}} + H_{\mathrm{U}}(T_0)\right)}{\mu_{\mathrm{EG}}} + \left.\overline{c_{p,\mathrm{EG}}}\right|_{0\,^{\circ}\mathrm{C}}^{25\,^{\circ}\mathrm{C}} \cdot 25\,^{\circ}\mathrm{C} \right] = 1736.1\,^{\circ}\mathrm{C} \tag{9.155}$$

– 1. Iteration

$$\left.\overline{c_{p,\mathrm{EG}}}\right|_{0\,^{\circ}\mathrm{C}}^{\vartheta_{\mathrm{ad},1}} = \sum_i \xi_{\mathrm{EG},i} \left.\overline{c_{p,i}}\right|_{0\,^{\circ}\mathrm{C}}^{\vartheta_{\mathrm{ad},1}} = \ldots = 1.3568\,\frac{\mathrm{kJ}}{\mathrm{kg\,K}} \tag{9.156}$$

$$\vartheta_{\mathrm{ad},2} = \frac{1}{\left.\overline{c_{p,\mathrm{EG}}}\right|_{0\,^{\circ}\mathrm{C}}^{\vartheta_{\mathrm{ad},1}}} \left[\frac{\left(\frac{\dot{Q}_{\mathrm{W}}}{\dot{m}_{\mathrm{F}}} + H_{\mathrm{U}}(T_0)\right)}{\mu_{\mathrm{EG}}} + \left.\overline{c_{p,\mathrm{EG}}}\right|_{0\,^{\circ}\mathrm{C}}^{25\,^{\circ}\mathrm{C}} \cdot 25\,^{\circ}\mathrm{C} \right] = 1750.1\,^{\circ}\mathrm{C} \tag{9.157}$$

– 2. Iteration

$$\left.\overline{c_{p,\mathrm{EG}}}\right|_{0\,^{\circ}\mathrm{C}}^{\vartheta_{\mathrm{ad},2}} = \sum_i \xi_{\mathrm{EG},i} \left.\overline{c_{p,i}}\right|_{0\,^{\circ}\mathrm{C}}^{\vartheta_{\mathrm{ad},2}} = \ldots = 1.3584\,\frac{\mathrm{kJ}}{\mathrm{kg\,K}} \tag{9.158}$$

$$\vartheta_{\mathrm{ad},3} = \frac{1}{\left.\overline{c_{p,\mathrm{EG}}}\right|_{0\,^{\circ}\mathrm{C}}^{\vartheta_{\mathrm{ad},2}}} \left[\frac{\left(\frac{\dot{Q}_{\mathrm{W}}}{\dot{m}_{\mathrm{F}}} + H_{\mathrm{U}}(T_0)\right)}{\mu_{\mathrm{EG}}} + \left.\overline{c_{p,\mathrm{EG}}}\right|_{0\,^{\circ}\mathrm{C}}^{25\,^{\circ}\mathrm{C}} \cdot 25\,^{\circ}\mathrm{C} \right] = 1747.0\,^{\circ}\mathrm{C} \tag{9.159}$$

– 3. Iteration

$$\left.\overline{c_{p,\mathrm{EG}}}\right|_{0\,^{\circ}\mathrm{C}}^{\vartheta_{\mathrm{ad},3}} = \sum_i \xi_{\mathrm{EG},i} \left.\overline{c_{p,i}}\right|_{0\,^{\circ}\mathrm{C}}^{\vartheta_{\mathrm{ad},3}} = \ldots = 1.3580\,\frac{\mathrm{kJ}}{\mathrm{kg\,K}} \tag{9.160}$$

$$\vartheta_{\mathrm{ad},4} = \frac{1}{\left.\overline{c_{p,\mathrm{EG}}}\right|_{0\,^{\circ}\mathrm{C}}^{\vartheta_{\mathrm{ad},3}}} \left[\frac{\left(\frac{\dot{Q}_{\mathrm{W}}}{\dot{m}_{\mathrm{F}}} + H_{\mathrm{U}}(T_0)\right)}{\mu_{\mathrm{EG}}} + \left.\overline{c_{p,\mathrm{EG}}}\right|_{0\,^{\circ}\mathrm{C}}^{25\,^{\circ}\mathrm{C}} \cdot 25\,^{\circ}\mathrm{C} \right] = 1747.7\,^{\circ}\mathrm{C} \tag{9.161}$$

– 4. Iteration

$$\left.\overline{c_{p,\mathrm{EG}}}\right|_{0\,^{\circ}\mathrm{C}}^{\vartheta_{\mathrm{ad},4}} = \sum_i \xi_{\mathrm{EG},i} \left.\overline{c_{p,i}}\right|_{0\,^{\circ}\mathrm{C}}^{\vartheta_{\mathrm{ad},4}} = \ldots = 1.3581\,\frac{\mathrm{kJ}}{\mathrm{kg\,K}} \tag{9.162}$$

$$\vartheta_{\text{ad},5} = \frac{1}{\left.\overline{c_{p,\text{EG}}}\right|_{0\,°\text{C}}^{\vartheta_{\text{ad},4}}} \left[\frac{\left(\frac{\dot{Q}_{\text{W}}}{\dot{m}_{\text{F}}} + H_{\text{U}}(T_0)\right)}{\mu_{\text{EG}}} + \left.\overline{c_{p,\text{EG}}}\right|_{0\,°\text{C}}^{25\,°\text{C}} \cdot 25\,°\text{C} \right] = 1747.5\,°\text{C} \tag{9.163}$$

After 4 iterations, the iteration is stopped because the solution no longer changes noticeably. The adiabatic flame temperature is therefore

$$\vartheta_{\text{ad}} = 1747.5\,°\text{C}. \tag{9.164}$$

- Alternative 2: Alternative 1 will be inaccurate because a constant value of $c_{p,v} = 1.86\,\frac{\text{kJ}}{\text{kg K}}$ is usually used for the specific heat capacity of vapour. This procedure is justifiable for humid air because the temperature ranges considered are small. When determining the adiabatic flame temperature, the temperatures to be expected are much higher, so that no constant specific heat capacity should be used for the temperature range of 25 °C to approx. 1500 °C. Therefore, the method of standard enthalpies of formation, respectively absolute enthalpies, is now used to solve the task, i.e. Tables B.6 and B.7 are applied, i.e.

$$\dot{Q}_{\text{w}} = \dot{m}_{\text{w}} \left[h_{\text{liq}}(\vartheta_0) - h_{\text{v}}(\vartheta_{\text{ad}})\right] = \frac{\dot{m}_{\text{w}}}{M_{\text{W}}} \left[H_{\text{m,liq}}(\vartheta_0) - H_{\text{m,v}}(\vartheta_{\text{ad}})\right] \tag{9.165}$$

  - 0. Iteration (initial guess)

$$\vartheta_{\text{ad},0} = 1800\,°\text{C} \tag{9.166}$$

$$\left.\overline{c_{p,\text{EG}}}\right|_{0\,°\text{C}}^{\vartheta_{\text{ad},0}} = \sum_i \xi_{\text{EG},i} \left.\overline{c_{p,i}}\right|_{0\,°\text{C}}^{\vartheta_{\text{ad},0}} = \ldots = 1.3641\,\frac{\text{kJ}}{\text{kg K}} \tag{9.167}$$

$$\vartheta_{\text{ad},1} = \frac{1}{\left.\overline{c_{p,\text{EG}}}\right|_{0\,°\text{C}}^{\vartheta_{\text{ad},0}}} \left[\frac{\left(\frac{\dot{Q}_{\text{W}}}{\dot{m}_{\text{F}}} + H_{\text{U}}(T_0)\right)}{\mu_{\text{EG}}} + \left.\overline{c_{p,\text{EG}}}\right|_{0\,°\text{C}}^{25\,°\text{C}} \cdot 25\,°\text{C} \right] = 1698.2\,°\text{C} \tag{9.168}$$

  - 1. Iteration

$$\left.\overline{c_{p,\text{EG}}}\right|_{0\,°\text{C}}^{\vartheta_{\text{ad},1}} = \sum_i \xi_{\text{EG},i} \left.\overline{c_{p,i}}\right|_{0\,°\text{C}}^{\vartheta_{\text{ad},1}} = \ldots = 1.3524\,\frac{\text{kJ}}{\text{kg K}} \tag{9.169}$$

$$\vartheta_{\mathrm{ad},2} = \frac{1}{\overline{c_{p,\mathrm{EG}}}\Big|_{0\,^\circ\mathrm{C}}^{\vartheta_{\mathrm{ad},1}}} \left[\frac{\left(\frac{\dot{Q}_\mathrm{W}}{\dot{m}_\mathrm{F}} + H_\mathrm{U}(T_0)\right)}{\mu_\mathrm{EG}} + \overline{c_{p,\mathrm{EG}}}\Big|_{0\,^\circ\mathrm{C}}^{25\,^\circ\mathrm{C}} \cdot 25\,^\circ\mathrm{C} \right] = 1724.4\,^\circ\mathrm{C} \tag{9.170}$$

- <u>2. Iteration</u>

$$\overline{c_{p,\mathrm{EG}}}\Big|_{0\,^\circ\mathrm{C}}^{\vartheta_{\mathrm{ad},2}} = \sum_i \xi_{\mathrm{EG},i} \overline{c_{p,i}}\Big|_{0\,^\circ\mathrm{C}}^{\vartheta_{\mathrm{ad},2}} = \ldots = 1.3554\,\frac{\mathrm{kJ}}{\mathrm{kg\,K}} \tag{9.171}$$

$$\vartheta_{\mathrm{ad},3} = \frac{1}{\overline{c_{p,\mathrm{EG}}}\Big|_{0\,^\circ\mathrm{C}}^{\vartheta_{\mathrm{ad},2}}} \left[\frac{\left(\frac{\dot{Q}_\mathrm{W}}{\dot{m}_\mathrm{F}} + H_\mathrm{U}(T_0)\right)}{\mu_\mathrm{EG}} + \overline{c_{p,\mathrm{EG}}}\Big|_{0\,^\circ\mathrm{C}}^{25\,^\circ\mathrm{C}} \cdot 25\,^\circ\mathrm{C} \right] = 1717.7\,^\circ\mathrm{C} \tag{9.172}$$

- <u>3. Iteration</u>

$$\overline{c_{p,\mathrm{EG}}}\Big|_{0\,^\circ\mathrm{C}}^{\vartheta_{\mathrm{ad},3}} = \sum_i \xi_{\mathrm{EG},i} \overline{c_{p,i}}\Big|_{0\,^\circ\mathrm{C}}^{\vartheta_{\mathrm{ad},3}} = \ldots = 1.3547\,\frac{\mathrm{kJ}}{\mathrm{kg\,K}} \tag{9.173}$$

$$\vartheta_{\mathrm{ad},4} = \frac{1}{\overline{c_{p,\mathrm{EG}}}\Big|_{0\,^\circ\mathrm{C}}^{\vartheta_{\mathrm{ad},3}}} \left[\frac{\left(\frac{\dot{Q}_\mathrm{W}}{\dot{m}_\mathrm{F}} + H_\mathrm{U}(T_0)\right)}{\mu_\mathrm{EG}} + \overline{c_{p,\mathrm{EG}}}\Big|_{0\,^\circ\mathrm{C}}^{25\,^\circ\mathrm{C}} \cdot 25\,^\circ\mathrm{C} \right] = 1719.4\,^\circ\mathrm{C} \tag{9.174}$$

- <u>4. Iteration</u>

$$\overline{c_{p,\mathrm{EG}}}\Big|_{0\,^\circ\mathrm{C}}^{\vartheta_{\mathrm{ad},4}} = \sum_i \xi_{\mathrm{EG},i} \overline{c_{p,i}}\Big|_{0\,^\circ\mathrm{C}}^{\vartheta_{\mathrm{ad},4}} = \ldots = 1.3549\,\frac{\mathrm{kJ}}{\mathrm{kg\,K}} \tag{9.175}$$

$$\vartheta_{\mathrm{ad},5} = \frac{1}{\overline{c_{p,\mathrm{EG}}}\Big|_{0\,^\circ\mathrm{C}}^{\vartheta_{\mathrm{ad},4}}} \left[\frac{\left(\frac{\dot{Q}_\mathrm{W}}{\dot{m}_\mathrm{F}} + H_\mathrm{U}(T_0)\right)}{\mu_\mathrm{EG}} + \overline{c_{p,\mathrm{EG}}}\Big|_{0\,^\circ\mathrm{C}}^{25\,^\circ\mathrm{C}} \cdot 25\,^\circ\mathrm{C} \right] = 1718.9\,^\circ\mathrm{C} \tag{9.176}$$

- <u>5. Iteration</u>

$$\overline{c_{p,\mathrm{EG}}}\Big|_{0\,^\circ\mathrm{C}}^{\vartheta_{\mathrm{ad},4}} = \sum_i \xi_{\mathrm{EG},i} \overline{c_{p,i}}\Big|_{0\,^\circ\mathrm{C}}^{\vartheta_{\mathrm{ad},4}} = \ldots = 1.3548\,\frac{\mathrm{kJ}}{\mathrm{kg\,K}} \tag{9.177}$$

$$\vartheta_{\mathrm{ad},5} = \frac{1}{\overline{c_{p,\mathrm{EG}}}\Big|_{0\,^\circ\mathrm{C}}^{\vartheta_{\mathrm{ad},4}}} \left[\frac{\left(\frac{\dot{Q}_\mathrm{W}}{\dot{m}_\mathrm{F}} + H_\mathrm{U}(T_0)\right)}{\mu_\mathrm{EG}} + \overline{c_{p,\mathrm{EG}}}\Big|_{0\,^\circ\mathrm{C}}^{25\,^\circ\mathrm{C}} \cdot 25\,^\circ\mathrm{C} \right] = 1719.1\,^\circ\mathrm{C} \tag{9.178}$$

After 5 iterations, the iteration is stopped because the solution no longer changes noticeably. The adiabatic flame temperature is therefore

$$\vartheta_{\mathrm{ad}} = 1719.1\,^{\circ}\mathrm{C}. \tag{9.179}$$

By injecting water, the adiabatic flame temperature can be lowered by approx. 200 K.

9.16

(a) The chemical equation of reaction for a stoichiometric combustion with air obeys

$$\underbrace{\mathrm{H_2} + \frac{1}{2}\mathrm{O_2} + \frac{1}{2}\cdot\frac{79}{21}\mathrm{N_2}}_{\mathrm{I}} \rightarrow \underbrace{\mathrm{H_2O} + \frac{1}{2}\cdot\frac{79}{21}\mathrm{N_2}}_{\mathrm{II}}. \tag{9.180}$$

The molar mass of air is

$$M_{\mathrm{a}} = \sum_i x_{\mathrm{a},i} M_i = x_{\mathrm{a,N_2}} M_{\mathrm{N_2}} + x_{\mathrm{a,O_2}} M_{\mathrm{O_2}} = 28.84\,\frac{\mathrm{kg}}{\mathrm{kmol}}. \tag{9.181}$$

The molar flux of air follows

$$\dot{n}_{\mathrm{I,a}} = \frac{\dot{m}_{\mathrm{I,a}}}{M_{\mathrm{a}}} = 2.7778 \times 10^{-4}\,\frac{\mathrm{kmol}}{\mathrm{s}}. \tag{9.182}$$

Thus, it follows

- Inlet (I)

$$\dot{n}_{\mathrm{I,N_2}} = x_{\mathrm{a,N_2}} \dot{n}_{\mathrm{I,a}} = 2.1944 \times 10^{-4}\,\frac{\mathrm{kmol}}{\mathrm{s}} \tag{9.183}$$

$$\dot{n}_{\mathrm{I,O_2}} = x_{\mathrm{a,O_2}} \dot{n}_{\mathrm{I,a}} = 5.8333 \times 10^{-5}\,\frac{\mathrm{kmol}}{\mathrm{s}} \tag{9.184}$$

$$\dot{n}_{\mathrm{I,H_2}} = 2\dot{n}_{\mathrm{I,O_2}} = 1.1667 \times 10^{-4}\,\frac{\mathrm{kmol}}{\mathrm{s}} \tag{9.185}$$

- Outlet (II)

$$\dot{n}_{\mathrm{II,N_2}} = \dot{n}_{\mathrm{I,N_2}} = 2.1944 \times 10^{-4}\,\frac{\mathrm{kmol}}{\mathrm{s}} \tag{9.186}$$

$$\dot{n}_{\mathrm{II,H_2O}} = \dot{n}_{\mathrm{I,H_2}} = 1.1667 \times 10^{-4}\,\frac{\mathrm{kmol}}{\mathrm{s}}. \tag{9.187}$$

(b) The first law of thermodynamics for an open system in steady state, cf. Fig. 9.17, reads as

$$\dot{H}_{\mathrm{I}} + \underbrace{\dot{Q} + P}_{=0} + \dot{m}_{\mathrm{w}} h'\,(100\,\mathrm{bar}) = \dot{H}_{\mathrm{II}} + \dot{m}_{\mathrm{w}} h''\,(100\,\mathrm{bar}) \tag{9.188}$$

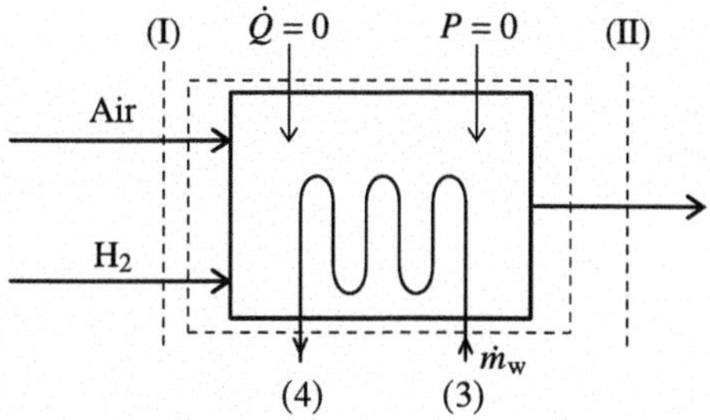

Fig. 9.17 Energy balance

with[5]

$$\dot{H}_{\rm I} = \dot{n}_{\rm I,N_2} H_{\rm m,N_2}(T_{\rm a}) + \dot{n}_{\rm I,O_2} H_{\rm m,O_2}(T_{\rm a}) + \dot{n}_{\rm I,H_2} H_{\rm m,H_2}\left(T_{\rm H_2}\right) = -0.0017\,\text{kW} \tag{9.189}$$

and

$$\dot{H}_{\rm II} = \dot{n}_{\rm II,N_2} H_{\rm m,N_2}(T_{\rm II}) + \dot{n}_{\rm II,H_2O} H_{\rm m,H_2O}(T_{\rm II}) = -20.4695\,\text{kW}. \tag{9.190}$$

With the steam table of water, cf. Appendix A, it further follows

$$h'(100\,\text{bar}) = 1.4079 \times 10^3\,\frac{\text{kJ}}{\text{kg}} \tag{9.191}$$

and

$$h''(100\,\text{bar}) = 2.7255 \times 10^3\,\frac{\text{kJ}}{\text{kg}}. \tag{9.192}$$

The mass flow rate of water then is

$$\dot{m}_{\rm w} = \frac{\dot{H}_{\rm I} - \dot{H}_{\rm II}}{h''(100\,\text{bar}) - h'(100\,\text{bar})} = 0.0155\,\frac{\text{kg}}{\text{s}}. \tag{9.193}$$

(c) The entropy balance for an open system in steady state, cf. Fig. 9.18, reads as

$$\dot{S}_{\rm I} + \underbrace{\dot{S}_{\rm a}}_{=0} + \dot{S}_{\rm i} + \dot{m}_{\rm w} s'(100\,\text{bar}) = \dot{S}_{\rm II} + \dot{m}_{\rm w} s''(100\,\text{bar}) \tag{9.194}$$

In order to calculate $\dot{S}_{\rm II}$ the partial pressures of the components of the air in state (I) are required, i.e.

$$p_{\rm I,N_2} = p_{\rm a} x_{\rm a,N_2} = 0.79\,\text{bar} \tag{9.195}$$

and

$$p_{\rm I,O_2} = p_{\rm a} x_{\rm a,O_2} = 0.21\,\text{bar} \tag{9.196}$$

[5] See Appendix B.

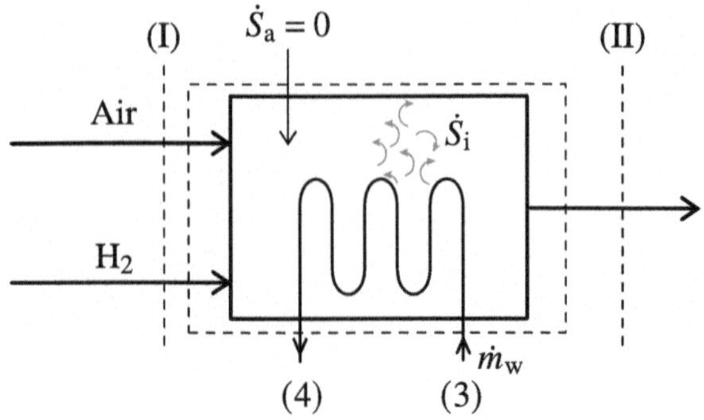

Fig. 9.18 Entropy balance

Molar entropies follow from Appendix B with a pressure correction for the components of the combustion air, so that

$$\begin{aligned}\dot{S}_{\mathrm{I}} &= \dot{n}_{\mathrm{I,N_2}}\left[S_{\mathrm{m,N_2}}\left(T_{\mathrm{a}},\ p_0\right) - R_{\mathrm{M}}\ln\frac{p_{\mathrm{I,N_2}}}{p_0}\right] + \\ &+ \dot{n}_{\mathrm{I,O_2}}\left[S_{\mathrm{m,O_2}}\left(T_{\mathrm{a}},\ p_0\right) - R_{\mathrm{M}}\ln\frac{p_{\mathrm{I,O_2}}}{p_0}\right] + \\ &+ \dot{n}_{\mathrm{I,H_2}} S_{\mathrm{m,H_2}}\left(T_{\mathrm{H_2}},\ p_{\mathrm{I,H_2}} = p_0\right) \\ &= 0.0704\,\frac{\mathrm{kW}}{\mathrm{K}}.\end{aligned} \tag{9.197}$$

In order to calculate $\dot{S}_{\mathrm{II}}$ the partial pressures of the components in state (II) are required, i.e.

$$p_{\mathrm{II,N_2}} = p_{\mathrm{II}}\frac{\dot{n}_{\mathrm{II,N_2}}}{\dot{n}_{\mathrm{II,N_2}} + \dot{n}_{\mathrm{II,H_2O}}} = 0.6529\,\mathrm{bar} \tag{9.198}$$

and

$$p_{\mathrm{II,H_2O}} = p_{\mathrm{II}}\frac{\dot{n}_{\mathrm{II,H_2O}}}{\dot{n}_{\mathrm{II,N_2}} + \dot{n}_{\mathrm{II,H_2O}}} = 0.3471\,\mathrm{bar}. \tag{9.199}$$

The entropy flux for the outlet then obeys[6]

$$\begin{aligned}\dot{S}_{\mathrm{II}} &= \dot{n}_{\mathrm{II,N_2}}\left[S_{\mathrm{m,N_2}}\left(T_{\mathrm{II}},\ p_0\right) - R_{\mathrm{M}}\ln\frac{p_{\mathrm{II,N_2}}}{p_0}\right] + \\ &+ \dot{n}_{\mathrm{II,H_2O}}\left[S_{\mathrm{m,H_2O}}\left(T_{\mathrm{II}},\ p_0\right) - R_{\mathrm{M}}\ln\frac{p_{\mathrm{II,H_2O}}}{p_0}\right] \\ &= 0.0790\,\frac{\mathrm{kW}}{\mathrm{K}}.\end{aligned} \tag{9.200}$$

[6] Since the partial pressures of the components differ from reference pressure a pressure correction is mandatory.

With the steam table of water, cf. Appendix A, it further follows

$$s'\,(100\,\text{bar}) = 3.3603\,\frac{\text{kJ}}{\text{kg K}} \tag{9.201}$$

and

$$s''\,(100\,\text{bar}) = 5.6159\,\frac{\text{kJ}}{\text{kg K}}. \tag{9.202}$$

The generation of entropy follows

$$\dot{S}_{\text{i}} = \dot{S}_{\text{II}} - \dot{S}_{\text{I}} + \dot{m}_{\text{w}}\left[s''\,(100\,\text{bar}) - s'\,(100\,\text{bar})\right] = 0.0436\,\frac{\text{kW}}{\text{K}}. \tag{9.203}$$

Finally, the loss of exergy is

$$\dot{E}_{\text{x,V}} = \dot{S}_{\text{i}} T_{\text{env}} = 12.9997\,\text{kW}. \tag{9.204}$$

(d) The dew point temperature for the pressure $p_{\text{II,H}_2\text{O}}$ can be determined with the steam table, i.e.

$$\vartheta_\tau = \vartheta_{\text{s}}\left(p_{\text{II,H}_2\text{O}}\right) = 72.4855\,^\circ\text{C}. \tag{9.205}$$

9.17

(a) The following applies to the molar flow of the hydrogen

$$\dot{n}_{1,\text{H}_2} = \frac{\dot{m}_{1,\text{H}_2}}{M_{\text{H}_2}} = 0.0069\,\frac{\text{kmol}}{\text{s}}. \tag{9.206}$$

From the chemical reaction

$$\underbrace{\text{H}_2 + \frac{1}{2}\text{O}_2}_{(1)} \rightarrow \underbrace{\text{H}_2\text{O}}_{(2)} \tag{9.207}$$

follow the further molar flow, i.e.

$$\dot{n}_{1,\text{O}_2} = \frac{1}{2}\dot{n}_{1,\text{H}_2} = 0.0035\,\frac{\text{kmol}}{\text{s}} \tag{9.208}$$

and

$$\dot{n}_{2,\text{H}_2\text{O}} = \dot{n}_{1,\text{H}_2} = 0.0069\,\frac{\text{kmol}}{\text{s}}. \tag{9.209}$$

(b) The mass flow at the outlet in state (2) is

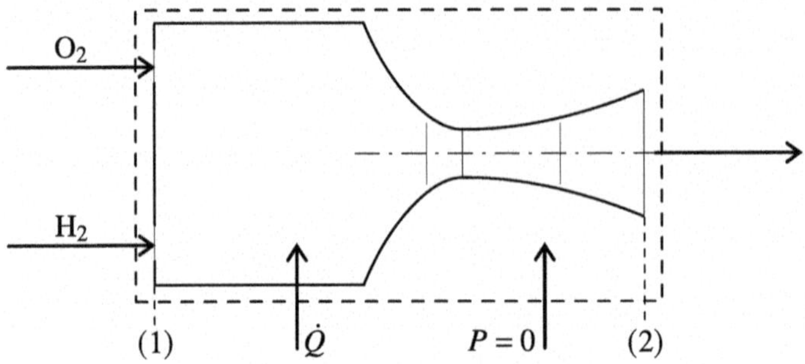

Fig. 9.19 Energy balance

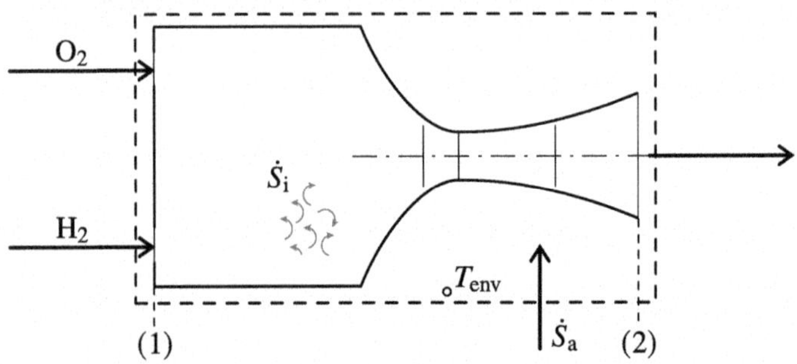

Fig. 9.20 Entropy balance

$$\dot{m}_{2,H_2O} = \dot{n}_{2,H_2O} M_{H_2O} = 0.1250 \, \frac{kg}{s}. \tag{9.210}$$

The energy balance reads as, cf. Fig. 9.19

$$\dot{n}_{1,H_2} H_{m,H_2}(T_1) + \dot{n}_{1,O_2} H_{m,O_2}(T_1) + \dot{Q} = \dot{n}_{2,H_2O} H_{m,H_2O}(T_2) + \frac{1}{2} \dot{m}_{2,H_2O} c_2^2 \tag{9.211}$$

with

$$\dot{Q} = -500 \, kW. \tag{9.212}$$

The velocity at the outlet obeys[7]

$$c_2 = \sqrt{\frac{2 \cdot \left[\dot{n}_{1,H_2} H_{m,H_2}(T_1) + \dot{n}_{1,O_2} H_{m,O_2}(T_1) + \dot{Q} - \dot{n}_{2,H_2O} H_{m,H_2O}(T_2) \right]}{\dot{m}_{2,H_2O}}}$$
$$= 3.3720 \times 10^3 \, \frac{m}{s}. \tag{9.213}$$

(c) The entropy balance follows, cf. Fig. 9.20

[7] Note that $1 \, \frac{J}{kg} = 1 \, \frac{m^2}{s^2}$.

$$\dot{n}_{1,H_2}\overbrace{\left[S_{m,H_2}(T_1,\ p_0)-R_M\ln\frac{p_1}{p_0}\right]}^{S_{m,H_2}(T_1,\,p_1)}+\dot{n}_{1,O_2}\overbrace{\left[S_{m,O_2}(T_1,\ p_0)-R_M\ln\frac{p_1}{p_0}\right]}^{S_{m,O_2}(T_1,\,p_1)}+$$
$$+\dot{S}_a+\dot{S}_i=\dot{n}_{2,H_2O}S_{m,H_2O}(T_2,\ p_2=p_0) \tag{9.214}$$

with

$$\dot{S}_a=\frac{\dot{Q}}{T_{env}}=-1.6667\,\frac{kW}{K}. \tag{9.215}$$

Thus, the generated entropy is

$$\begin{aligned}\dot{S}_i&=\dot{n}_{2,H_2O}S_{m,H_2O}(T_2,\ p_2=p_0)-\dot{S}_a+\\&-\dot{n}_{1,H_2}\left[S_{m,H_2}(T_1,\ p_0)-R_M\ln\frac{p_1}{p_0}\right]+\\&-\dot{n}_{1,O_2}\left[S_{m,O_2}(T_1,\ p_0)-R_M\ln\frac{p_1}{p_0}\right]\\&=1.9659\,\frac{kW}{K}\end{aligned} \tag{9.216}$$

The loss of specific exergy obeys

$$e_{x,Vm}=\frac{\dot{E}_{x,V}}{\dot{n}_{2,H_2}}=\frac{\dot{S}_iT_{env}}{\dot{n}_{2,H_2}}=8.4929\times10^4\,\frac{kJ}{kmol}. \tag{9.217}$$

9.18 Figure 9.21 shows the procedure for determining the lower heating value. The fuel is burnt isothermally, stoichiometrically with pure oxygen. Since the fuel consists only of carbon and hydrogen and is burnt stoichiometrically without excess oxygen, only carbon dioxide and water are produced. The water, because the lower heating value is asked for, is completely present in vapour form. The first law obeys

$$\begin{aligned}\dot{n}_CH_{m,C(gr)}(25\,°C)+\dot{n}_{H_2}H_{m,H_2}(25\,°C)+\dot{n}_{O_2}H_{m,O_2}(25\,°C)+\dot{Q}=\\ \dot{n}_{CO_2}H_{m,CO_2}(25\,°C)+\dot{n}_{H_2O}H_{m,H_2O,v}(25\,°C)\,.\end{aligned} \tag{9.218}$$

The following applies to the flux of carbon

$$\dot{m}_C=c\cdot\dot{m}_F \tag{9.219}$$

respectively

$$\dot{n}_C=\frac{\dot{m}_C}{M_C}=c\cdot\frac{\dot{m}_F}{M_C}. \tag{9.220}$$

The flux of hydrogen is

$$\dot{m}_{H_2}=h\cdot\dot{m}_F \tag{9.221}$$

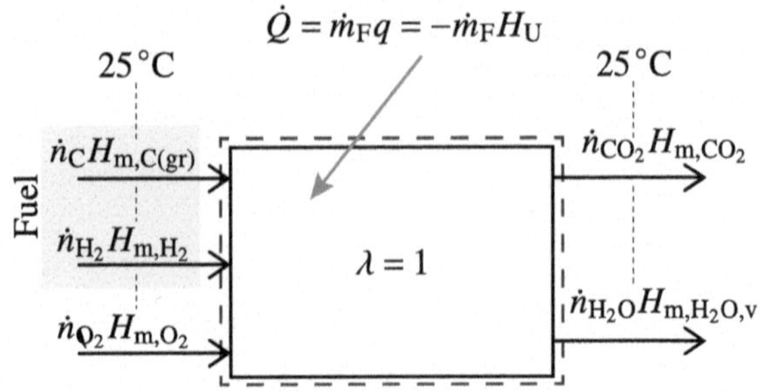

Fig. 9.21 Determination of the lower heating value

respectively

$$\dot{n}_{H_2} = \frac{\dot{m}_{H_2}}{M_{H_2}} = h \cdot \frac{\dot{m}_F}{M_{H_2}}. \tag{9.222}$$

According to the chemical equations

$$C + O_2 \rightarrow CO_2 \tag{9.223}$$

and

$$2H_2 + O_2 \rightarrow 2H_2O \tag{9.224}$$

it applies for

- Oxidant

$$\dot{n}_{O_2} = \dot{n}_C + \frac{1}{2}\dot{n}_{H_2} = c \cdot \frac{\dot{m}_F}{M_C} + \frac{1}{2}h \cdot \frac{\dot{m}_F}{M_{H_2}} \tag{9.225}$$

- Products

$$\dot{n}_{CO_2} = \dot{n}_C = c \cdot \frac{\dot{m}_F}{M_C} \tag{9.226}$$

  and

$$\dot{n}_{H_2O} = \dot{n}_{H_2} = h \cdot \frac{\dot{m}_F}{M_{H_2}}. \tag{9.227}$$

Hence, the first law of thermodynamics, cf. Fig. 9.21, is

$$\begin{aligned} & c \cdot \frac{\dot{m}_F}{M_C} H_{m,C(gr)}\,(25\,°C) + h \cdot \frac{\dot{m}_F}{M_{H_2}} H_{m,H_2}\,(25\,°C) + \\ & + \left(c \cdot \frac{\dot{m}_F}{M_C} + \frac{1}{2}h \cdot \frac{\dot{m}_F}{M_{H_2}}\right) H_{m,O_2}\,(25\,°C) + \dot{Q} = \\ & c \cdot \frac{\dot{m}_F}{M_C} H_{m,CO_2}\,(25\,°C) + h \cdot \frac{\dot{m}_F}{M_{H_2}} H_{m,H_2O,v}\,(25\,°C) \end{aligned} \tag{9.228}$$

respectively after division with $\dot{m}_\mathrm{F}$

$$\frac{\dot{Q}}{\dot{m}_\mathrm{F}} = \frac{c}{M_\mathrm{C}} H_{\mathrm{m,CO_2}}\,(25\,°\mathrm{C}) + \frac{h}{M_{\mathrm{H_2}}} H_{\mathrm{m,H_2O,v}}\,(25\,°\mathrm{C}) - \frac{c}{M_\mathrm{C}} H_{\mathrm{m,C(gr)}}\,(25\,°\mathrm{C}) + \\ - \frac{h}{M_{\mathrm{H_2}}} H_{\mathrm{m,H_2}}\,(25\,°\mathrm{C}) - \left(\frac{c}{M_\mathrm{C}} + \frac{1}{2}\frac{h}{M_{\mathrm{H_2}}}\right) H_{\mathrm{m,O_2}}\,(25\,°\mathrm{C})\,. \tag{9.229}$$

Using the molar enthalpies $H_{\mathrm{m},i}\,(25\,°\mathrm{C})$ from Appendix B, one obtains

$$\frac{\dot{Q}}{\dot{m}_\mathrm{F}} = -4.3367 \times 10^4\,\frac{\mathrm{kJ}}{\mathrm{kg}}. \tag{9.230}$$

By definition, see Fig. 9.21, the lower heating value finally is

$$H_\mathrm{U}\,(25\,°\mathrm{C}) = -\frac{\dot{Q}}{\dot{m}_\mathrm{F}} = 4.3367 \times 10^4\,\frac{\mathrm{kJ}}{\mathrm{kg}}. \tag{9.231}$$

Appendix A
Steam Table (Water) According to IAPWS

See (Tables A.1, A.2, A.3, A.4, A.5, A.6, A.7, A.8, A.9, A.10, A.11 and A.12).[1]

Table A.1 Saturated liquid and saturated steam 1/4

ϑ °C	p_s bar	h'	h''	s'	s''	v'	v''	c'_p	c''_p
		$\frac{kJ}{kg}$		$\frac{kJ}{kgK}$		$\frac{m^3}{kg}$		$\frac{kJ}{kgK}$	
0	0.0061122	−0.041588	2500.9	−0.000155	9.1558	0.0010	206.1431	4.22	1.888
0.01	0.0061166	0.00061178	2500.911	0	9.1555	0.0010002	205.9975	4.2199	1.8882
1	0.0065709	4.1767	2502.7299	0.01526	9.1291	0.0010001	192.4447	4.2165	1.8889
2	0.0070599	8.3916	2504.5666	0.030606	9.1027	0.0010001	179.7636	4.2134	1.8895
3	0.0075808	12.6035	2506.4024	0.045886	9.0765	0.0010001	168.0141	4.2105	1.8902
4	0.0081355	16.8127	2508.2375	0.061101	9.0506	0.0010001	157.1213	4.2078	1.8909
5	0.0087257	21.0194	2510.0717	0.076252	9.0249	0.0010001	147.0169	4.2054	1.8917
6	0.0093535	25.2237	2511.9051	0.09134	8.9994	0.0010001	137.6382	4.2031	1.8924
7	0.010021	29.4258	2513.7377	0.10637	8.9742	0.0010001	128.9281	4.2011	1.8932
8	0.01073	33.626	2515.5693	0.12133	8.9492	0.0010002	120.8344	4.1992	1.894
9	0.011483	37.8244	2517.4001	0.13624	8.9244	0.0010003	113.3092	4.1974	1.8949
10	0.012282	42.0211	2519.2298	0.15109	8.8998	0.0010003	106.3087	4.1958	1.8957
11	0.013129	46.2162	2521.0586	0.16587	8.8755	0.0010004	99.7927	4.1943	1.8966
12	0.014028	50.41	2522.8864	0.18061	8.8514	0.0010005	93.7243	4.193	1.8975
13	0.014981	54.6024	2524.7132	0.19528	8.8275	0.0010007	88.0698	4.1917	1.8985
14	0.015989	58.7936	2526.5389	0.2099	8.8038	0.0010008	82.7981	4.1905	1.8994
15	0.017057	62.9837	2528.3636	0.22447	8.7804	0.0010009	77.8807	4.1894	1.9004
16	0.018188	67.1727	2530.1871	0.23898	8.7571	0.0010011	73.2915	4.1884	1.9014
17	0.019383	71.3608	2532.0094	0.25344	8.7341	0.0010013	69.0063	4.1875	1.9025
18	0.020647	75.5479	2533.8307	0.26785	8.7112	0.0010015	65.0029	4.1866	1.9035
19	0.021982	79.7343	2535.6507	0.2822	8.6886	0.0010016	61.2609	4.1858	1.9046
20	0.023392	83.9199	2537.4695	0.2965	8.6661	0.0010018	57.7615	4.1851	1.9057
21	0.024881	88.1048	2539.287	0.31075	8.6439	0.0010021	54.4873	4.1844	1.9069
22	0.026452	92.289	2541.1033	0.32495	8.6218	0.0010023	51.4225	4.1838	1.908

[1] International Association for the Properties of Water and Steam: The listed data has been generated with XSteam, see [4].

A. Schmidt, *Technical Thermodynamics Workbook for Engineers*,
https://doi.org/10.1007/978-3-031-50172-2

Table A.2 Saturated liquid and saturated steam 2/4

ϑ °C	p_s bar	h'	h''	s'	s''	v'	v''	c'_p	c''_p
		$\frac{kJ}{kg}$		$\frac{kJ}{kgK}$		$\frac{m^3}{kg}$		$\frac{kJ}{kgK}$	
23	0.028109	96.4727	2542.9182	0.3391	8.6	0.0010025	48.5521	4.1832	1.9092
24	0.029856	100.6558	2544.7318	0.3532	8.5783	0.0010028	45.8626	4.1827	1.9104
25	0.031697	104.8384	2546.5441	0.36726	8.5568	0.001003	43.3414	4.1822	1.9116
26	0.033637	109.0205	2548.3549	0.38126	8.5355	0.0010033	40.9768	4.1817	1.9129
27	0.035679	113.2022	2550.1643	0.39521	8.5144	0.0010035	38.7582	4.1813	1.9141
28	0.037828	117.3835	2551.9723	0.40912	8.4934	0.0010038	36.6754	4.1809	1.9154
29	0.040089	121.5645	2553.7788	0.42298	8.4727	0.0010041	34.7194	4.1806	1.9167
30	0.042467	125.7452	2555.5837	0.43679	8.4521	0.0010044	32.8816	4.1803	1.918
31	0.044966	129.9255	2557.3871	0.45056	8.4317	0.0010047	31.154	4.18	1.9194
32	0.047592	134.1057	2559.189	0.46428	8.4115	0.001005	29.5295	4.1798	1.9207
33	0.050351	138.2855	2560.9892	0.47795	8.3914	0.0010054	28.001	4.1795	1.9221
34	0.053247	142.4653	2562.7877	0.49158	8.3715	0.0010057	26.5624	4.1794	1.9235
35	0.056286	146.6448	2564.5846	0.50517	8.3518	0.001006	25.2078	4.1792	1.9249
36	0.059475	150.8242	2566.3797	0.51871	8.3323	0.0010064	23.9318	4.1791	1.9263
37	0.062818	155.0035	2568.1731	0.5322	8.3129	0.0010068	22.7292	4.179	1.9278
38	0.066324	159.1827	2569.9647	0.54566	8.2936	0.0010071	21.5954	4.1789	1.9292
39	0.069997	163.3619	2571.7545	0.55906	8.2746	0.0010075	20.5261	4.1788	1.9307
40	0.073844	167.541	2573.5424	0.57243	8.2557	0.0010079	19.517	4.1788	1.9322
41	0.077873	171.7202	2575.3284	0.58575	8.2369	0.0010083	18.5646	4.1788	1.9337
42	0.08209	175.8993	2577.1125	0.59903	8.2183	0.0010087	17.6652	4.1788	1.9353
43	0.086503	180.0785	2578.8946	0.61227	8.1999	0.0010091	16.8155	4.1788	1.9368
44	0.091118	184.2578	2580.6746	0.62547	8.1816	0.0010095	16.0126	4.1789	1.9384
45	0.095944	188.4372	2582.4526	0.63862	8.1634	0.0010099	15.2534	4.179	1.94
46	0.10099	192.6167	2584.2285	0.65174	8.1454	0.0010103	14.5355	4.1791	1.9416
47	0.10626	196.7963	2586.0023	0.66481	8.1276	0.0010108	13.8562	4.1792	1.9432
48	0.11176	200.9761	2587.7739	0.67785	8.1099	0.0010112	13.2132	4.1794	1.9449
49	0.11751	205.156	2589.5432	0.69084	8.0923	0.0010117	12.6045	4.1796	1.9466
50	0.12351	209.3362	2591.3103	0.70379	8.0749	0.0010121	12.0279	4.1798	1.9482
51	0.12977	213.5166	2593.075	0.71671	8.0576	0.0010126	11.4815	4.18	1.95
52	0.13631	217.6973	2594.8374	0.72958	8.0405	0.0010131	10.9637	4.1802	1.9517
53	0.14312	221.8782	2596.5973	0.74242	8.0235	0.0010136	10.4726	4.1805	1.9534
54	0.15022	226.0594	2598.3548	0.75522	8.0066	0.001014	10.0069	4.1808	1.9552
55	0.15761	230.241	2600.1098	0.76798	7.9899	0.0010145	9.5649	4.1811	1.957
56	0.16532	234.4229	2601.8622	0.7807	7.9733	0.001015	9.1454	4.1814	1.9588
57	0.17335	238.6052	2603.6121	0.79339	7.9568	0.0010155	8.7471	4.1818	1.9607
58	0.18171	242.7878	2605.3592	0.80603	7.9405	0.0010161	8.3688	4.1821	1.9625
59	0.19041	246.9709	2607.1037	0.81864	7.9243	0.0010166	8.0093	4.1825	1.9644
60	0.19946	251.1544	2608.8454	0.83122	7.9082	0.0010171	7.6677	4.1829	1.9664
61	0.20887	255.3383	2610.5843	0.84375	7.8922	0.0010176	7.3428	4.1834	1.9683
62	0.21866	259.5228	2612.3203	0.85625	7.8764	0.0010182	7.0338	4.1838	1.9703
63	0.22884	263.7077	2614.0534	0.86872	7.8607	0.0010187	6.7399	4.1843	1.9723
64	0.23942	267.8931	2615.7836	0.88115	7.8451	0.0010193	6.4601	4.1848	1.9743
65	0.25041	272.0791	2617.5107	0.89354	7.8296	0.0010199	6.1938	4.1853	1.9764
66	0.26183	276.2657	2619.2347	0.9059	7.8142	0.0010204	5.9402	4.1859	1.9785
67	0.27368	280.4528	2620.9556	0.91823	7.799	0.001021	5.6986	4.1864	1.9806
68	0.28599	284.6405	2622.6733	0.93052	7.7839	0.0010216	5.4684	4.187	1.9828
69	0.29876	288.8289	2624.3877	0.94277	7.7689	0.0010222	5.249	4.1876	1.985
70	0.31201	293.0179	2626.0988	0.95499	7.754	0.0010228	5.0397	4.1882	1.9873

Table A.3 Saturated liquid and saturated steam 3/4

ϑ °C	p_S bar	h'	h''	s'	s''	v'	v''	c'_p	c''_p
		$\frac{kJ}{kg}$		$\frac{kJ}{kgK}$		$\frac{m^3}{kg}$		$\frac{kJ}{kgK}$	
71	0.32575	297.2076	2627.8066	0.96718	7.7392	0.0010234	4.8402	4.1889	1.9895
72	0.34	301.398	2629.5109	0.97933	7.7245	0.001024	4.6498	4.1896	1.9919
73	0.35478	305.5891	2631.2117	0.99146	7.71	0.0010246	4.4681	4.1902	1.9942
74	0.37009	309.781	2632.909	1.0035	7.6955	0.0010252	4.2947	4.191	1.9966
75	0.38595	313.9736	2634.6026	1.0156	7.6812	0.0010258	4.1291	4.1917	1.999
76	0.40239	318.1669	2636.2926	1.0276	7.6669	0.0010265	3.9709	4.1924	2.0015
77	0.41941	322.3611	2637.9788	1.0396	7.6528	0.0010271	3.8198	4.1932	2.0041
78	0.43703	326.5561	2639.6612	1.0516	7.6388	0.0010277	3.6754	4.194	2.0066
79	0.45527	330.752	2641.3397	1.0635	7.6248	0.0010284	3.5373	4.1948	2.0092
80	0.47415	334.9487	2643.0143	1.0754	7.611	0.001029	3.4053	4.1956	2.0119
81	0.49368	339.1463	2644.6849	1.0873	7.5973	0.0010297	3.279	4.1965	2.0146
82	0.51387	343.3448	2646.3515	1.0991	7.5837	0.0010304	3.1582	4.1974	2.0174
83	0.53476	347.5443	2648.0139	1.1109	7.5701	0.001031	3.0426	4.1983	2.0202
84	0.55636	351.7447	2649.672	1.1227	7.5567	0.0010317	2.9319	4.1992	2.0231
85	0.57867	355.9461	2651.3259	1.1344	7.5434	0.0010324	2.8259	4.2001	2.026
86	0.60174	360.1485	2652.9755	1.1461	7.5301	0.0010331	2.7244	4.2011	2.029
87	0.62556	364.3519	2654.6206	1.1578	7.517	0.0010338	2.6272	4.202	2.0321
88	0.65017	368.5563	2656.2613	1.1694	7.5039	0.0010345	2.5341	4.203	2.0352
89	0.67559	372.7618	2657.8973	1.1811	7.4909	0.0010352	2.4448	4.2041	2.0383
90	0.70182	376.9684	2659.5288	1.1927	7.4781	0.0010359	2.3591	4.2051	2.0415
91	0.7289	381.1762	2661.1555	1.2042	7.4653	0.0010367	2.2771	4.2062	2.0448
92	0.75685	385.385	2662.7775	1.2158	7.4526	0.0010374	2.1983	4.2072	2.0482
93	0.78568	389.595	2664.3946	1.2273	7.44	0.0010381	2.1228	4.2083	2.0516
94	0.81542	393.8062	2666.0068	1.2387	7.4275	0.0010389	2.0502	4.2095	2.0551
95	0.84609	398.0185	2667.6139	1.2502	7.415	0.0010396	1.9806	4.2106	2.0586
96	0.87771	402.2321	2669.216	1.2616	7.4027	0.0010404	1.9138	4.2118	2.0623
97	0.91031	406.447	2670.813	1.273	7.3904	0.0010411	1.8497	4.213	2.066
98	0.9439	410.6631	2672.4047	1.2844	7.3782	0.0010419	1.788	4.2142	2.0697
99	0.97852	414.8805	2673.991	1.2957	7.3661	0.0010427	1.7288	4.2154	2.0736
100	1.0142	419.0992	2675.572	1.307	7.3541	0.0010435	1.6719	4.2166	2.0775
105	1.209	440.2131	2683.3933	1.3632	7.2951	0.0010474	1.4185	4.2232	2.0983
110	1.4338	461.3634	2691.0676	1.4187	7.238	0.0010516	1.2094	4.2304	2.1212
115	1.6918	482.5528	2698.5848	1.4735	7.1827	0.0010559	1.0359	4.2381	2.1464
120	1.9867	503.7846	2705.9342	1.5278	7.1291	0.0010603	0.8913	4.2464	2.174
125	2.3222	525.0618	2713.1055	1.5815	7.077	0.0010649	0.77011	4.2553	2.2042
130	2.7026	546.3878	2720.0878	1.6346	7.0264	0.0010697	0.66808	4.2648	2.237
135	3.132	567.7661	2726.8708	1.6872	6.9772	0.0010747	0.5818	4.2751	2.2726
140	3.615	589.2003	2733.4439	1.7393	6.9293	0.0010798	0.50852	4.286	2.3109
145	4.1563	610.6941	2739.7968	1.7909	6.8826	0.001085	0.44602	4.2978	2.352
150	4.761	632.2516	2745.9191	1.842	6.837	0.0010905	0.3925	4.3103	2.3959
155	5.4342	653.8769	2751.8005	1.8926	6.7926	0.0010962	0.3465	4.3236	2.4425
160	6.1814	675.5747	2757.4305	1.9428	6.7491	0.001102	0.30682	4.3379	2.4918
165	7.0082	697.3495	2762.7985	1.9926	6.7066	0.001108	0.27246	4.3532	2.5438
170	7.9205	719.2064	2767.8937	2.0419	6.6649	0.0011143	0.24262	4.3695	2.5985
175	8.9245	741.1507	2772.7045	2.0909	6.6241	0.0011207	0.2166	4.3869	2.656
180	10.0263	763.188	2777.2194	2.1395	6.5841	0.0011274	0.19386	4.4056	2.7164
185	11.2327	785.3243	2781.4259	2.1878	6.5447	0.0011343	0.17392	4.4255	2.7798
190	12.5502	807.566	2785.311	2.2358	6.506	0.0011414	0.15638	4.4468	2.8464

Table A.4 Saturated liquid and saturated steam 4/4

ϑ °C	p_S bar	h'	h''	s'	s''	v'	v''	c'_p	c''_p
		$\frac{kJ}{kg}$		$\frac{kJ}{kgK}$		$\frac{m^3}{kg}$		$\frac{kJ}{kgK}$	
195	13.9858	829.9199	2788.861	2.2834	6.4679	0.0011488	0.14091	4.4696	2.9163
200	15.5467	852.3931	2792.0616	2.3308	6.4303	0.0011565	0.12722	4.494	2.99
205	17.2402	874.9933	2794.8974	2.3779	6.3932	0.0011645	0.11509	4.5202	3.0677
210	19.0739	897.7289	2797.3523	2.4248	6.3565	0.0011727	0.1043	4.5482	3.1496
215	21.0555	920.6086	2799.4096	2.4714	6.3202	0.0011813	0.094689	4.5784	3.2363
220	23.1929	943.6417	2801.051	2.5178	6.2842	0.0011902	0.086101	4.6107	3.328
225	25.4942	966.8384	2802.2577	2.5641	6.2485	0.0011994	0.078411	4.6455	3.4252
230	27.9679	990.2095	2803.0093	2.6102	6.2131	0.001209	0.07151	4.6829	3.5283
235	30.6224	1013.7668	2803.2844	2.6561	6.1777	0.001219	0.065304	4.7233	3.6379
240	33.4665	1037.5228	2803.06	2.7019	6.1425	0.0012295	0.05971	4.7668	3.7545
245	36.5091	1061.4911	2802.3114	2.7477	6.1074	0.0012404	0.054658	4.8139	3.8789
250	39.7594	1085.6868	2801.0121	2.7934	6.0722	0.0012517	0.050087	4.865	4.0119
255	43.2267	1110.126	2799.1333	2.8391	6.037	0.0012636	0.045941	4.9204	4.1545
260	46.9207	1134.8266	2796.6436	2.8847	6.0017	0.0012761	0.042175	4.9807	4.308
265	50.8512	1159.8083	2793.5086	2.9304	5.9662	0.0012892	0.038748	5.0466	4.4741
270	55.0284	1185.0928	2789.6899	2.9762	5.9304	0.001303	0.035622	5.1188	4.6547
275	59.4626	1210.7046	2785.1446	3.0221	5.8943	0.0013175	0.032767	5.1982	4.8524
280	64.1646	1236.671	2779.8245	3.0681	5.8578	0.0013328	0.030154	5.2859	5.0701
285	69.1454	1263.023	2773.6747	3.1143	5.8208	0.0013491	0.027758	5.3832	5.3114
290	74.4164	1289.7957	2766.6326	3.1608	5.7832	0.0013663	0.025557	5.4918	5.5806
295	79.9895	1317.0297	2758.6266	3.2076	5.7449	0.0013846	0.023531	5.6137	5.8825
300	85.8771	1344.7713	2749.5737	3.2547	5.7058	0.0014042	0.021663	5.7515	6.2231
305	92.0919	1373.0748	2739.3776	3.3024	5.6656	0.0014252	0.019937	5.9083	6.6096
310	98.6475	1402.0034	2727.9243	3.3506	5.6243	0.0014479	0.018339	6.0883	7.0513
315	105.558	1431.6321	2715.0772	3.3994	5.5816	0.0014724	0.016856	6.2968	7.561
320	112.8386	1462.051	2700.6677	3.4491	5.5373	0.0014991	0.015476	6.5414	8.1575
325	120.5052	1493.3719	2684.483	3.4997	5.4911	0.0015283	0.014189	6.8331	8.8689
330	128.5752	1525.738	2666.2483	3.5516	5.4425	0.0015606	0.012984	7.1888	9.7381
335	137.0673	1559.3407	2645.6023	3.6048	5.391	0.0015967	0.011852	7.6354	10.83
340	146.0018	1594.4466	2622.0667	3.6599	5.3359	0.0016375	0.010784	8.2166	12.2412
345	155.4015	1631.4365	2595.0082	3.7175	5.2763	0.0016846	0.0097698	9.0023	14.112
350	165.2916	1670.8892	2563.6305	3.7783	5.2109	0.0017401	0.0088009	10.102	16.6415
355	175.7012	1713.7096	2526.4499	3.8438	5.1377	0.0018078	0.007866	11.8584	20.7136
360	186.664	1761.4911	2480.9862	3.9164	5.0527	0.0018945	0.006945	14.8744	27.5691
365	198.2216	1817.5893	2422.0051	4.001	4.9482	0.0020156	0.0060043	21.4744	42.0135
370	210.4337	1892.6429	2333.5016	4.1142	4.7996	0.0022221	0.0049462	47.1001	93.4065
373.946	220.6397	2087.5		4.412		0.0031		∞	

Table A.5 Specific volume v of water in $\frac{m^3}{kg}$ 1/2

p in bar	ϑ in °C										
	0	25	50	75	100	125	150	200	250	300	350
0.01	0.0010002	137.5362	149.0961	160.6471	172.1934	183.737	195.2789	218.3599	241.4391	264.5173	287.5949
0.1	0.0010002	0.001003	14.8674	16.0347	17.1967	18.356	19.5136	21.826	24.1365	26.446	28.755
1	0.0010002	0.001003	0.0010121	0.0010258	1.696	1.8173	1.9367	2.1725	2.4062	2.6389	2.871
5	0.00099995	0.0010028	0.0010119	0.0010256	0.0010433	0.0010648	0.0010905	0.42503	0.47443	0.5226	0.57014
10	0.0009997	0.0010026	0.0010117	0.0010254	0.001043	0.0010645	0.0010902	0.206	0.23274	0.25798	0.28249
20	0.00099919	0.0010021	0.0010113	0.0010249	0.0010425	0.0010639	0.0010895	0.0011561	0.11148	0.1255	0.13859
30	0.00099869	0.0010017	0.0010108	0.0010244	0.001042	0.0010633	0.0010888	0.001155	0.070622	0.081175	0.090555
40	0.00099818	0.0010012	0.0010104	0.001024	0.0010415	0.0010628	0.0010881	0.001154	0.0012517	0.058868	0.066474
50	0.00099768	0.0010008	0.0010099	0.0010235	0.001041	0.0010622	0.0010875	0.001153	0.0012499	0.045347	0.051971
60	0.00099718	0.0010003	0.0010095	0.0010231	0.0010405	0.0010616	0.0010868	0.0011521	0.0012481	0.036191	0.042253
70	0.00099668	0.00099987	0.0010091	0.0010226	0.00104	0.0010611	0.0010862	0.0011511	0.0012463	0.029494	0.035265
80	0.00099619	0.00099942	0.0010086	0.0010222	0.0010395	0.0010605	0.0010855	0.0011501	0.0012446	0.02428	0.029978
90	0.00099569	0.00099898	0.0010082	0.0010217	0.001039	0.00106	0.0010849	0.0011491	0.0012429	0.0014024	0.025818
100	0.0009952	0.00099854	0.0010078	0.0010212	0.0010385	0.0010594	0.0010842	0.0011482	0.0012412	0.001398	0.022442
150	0.00099276	0.00099636	0.0010056	0.001019	0.0010361	0.0010567	0.001081	0.0011435	0.001233	0.0013783	0.011481
200	0.00099035	0.0009942	0.0010035	0.0010168	0.0010337	0.001054	0.0010779	0.001139	0.0012254	0.0013611	0.0016649
250	0.00098799	0.00099208	0.0010014	0.0010146	0.0010313	0.0010514	0.0010749	0.0011346	0.0012181	0.0013459	0.0015988
300	0.00098567	0.00098998	0.00099934	0.0010125	0.001029	0.0010488	0.001072	0.0011304	0.0012113	0.0013322	0.0015529
350	0.00098338	0.00098792	0.00099731	0.0010104	0.0010267	0.0010462	0.0010691	0.0011264	0.0012048	0.0013197	0.0015175
400	0.00098113	0.00098588	0.00099531	0.0010083	0.0010245	0.0010438	0.0010663	0.0011224	0.0011986	0.0013083	0.0014884
450	0.00097891	0.00098388	0.00099334	0.0010063	0.0010223	0.0010413	0.0010635	0.0011186	0.0011927	0.0012977	0.0014638
500	0.00097673	0.0009819	0.00099139	0.0010043	0.0010201	0.0010389	0.0010608	0.0011149	0.0011871	0.0012879	0.0014424
600	0.00097247	0.00097802	0.00098758	0.0010003	0.0010159	0.0010343	0.0010555	0.0011077	0.0011764	0.00127	0.0014067
700	0.00096834	0.00097425	0.00098386	0.00099648	0.0010118	0.0010298	0.0010505	0.001101	0.0011666	0.0012541	0.0013773
800	0.00096434	0.00097057	0.00098025	0.00099276	0.0010078	0.0010255	0.0010456	0.0010945	0.0011574	0.0012398	0.0013525
900	0.00096046	0.00096699	0.00097673	0.00098913	0.001004	0.0010212	0.001041	0.0010884	0.0011488	0.0012268	0.0013309
1000	0.00095669	0.00096351	0.00097329	0.0009856	0.0010002	0.0010172	0.0010364	0.0010826	0.0011407	0.0012148	0.0013118

Table A.6 Specific volume v of water in $\frac{m^3}{kg}$ 2/2

p in bar	ϑ in °C								
	400	450	500	550	600	650	700	750	800
0.01	310.6722	333.7492	356.8261	379.9029	402.9796	426.0562	449.1327	472.2092	495.2857
0.1	31.0636	33.372	35.6802	37.9883	40.2963	42.6042	44.9121	47.2199	49.5278
1	3.1027	3.3342	3.5656	3.7968	4.0279	4.259	4.49	4.721	4.952
5	0.61729	0.66421	0.71095	0.75757	0.8041	0.85056	0.89696	0.94333	0.98967
10	0.30659	0.33044	0.35411	0.37766	0.40111	0.4245	0.44783	0.47112	0.49438
20	0.15121	0.16354	0.17568	0.18769	0.19961	0.21146	0.22326	0.23501	0.24674
30	0.099377	0.10788	0.11619	0.12437	0.13244	0.14045	0.1484	0.15631	0.16419
40	0.073432	0.080042	0.086441	0.092699	0.098857	0.10494	0.11097	0.11696	0.12292
50	0.05784	0.063325	0.068583	0.073694	0.078703	0.083637	0.088515	0.09335	0.098151
60	0.047423	0.052168	0.056672	0.061021	0.065264	0.069432	0.073542	0.077609	0.081642
70	0.039962	0.04419	0.048159	0.051966	0.055664	0.059284	0.062847	0.066365	0.069849
80	0.034348	0.038197	0.041769	0.045172	0.048463	0.051673	0.054825	0.057933	0.061005
90	0.029963	0.033528	0.036795	0.039886	0.04286	0.045753	0.048586	0.051374	0.054127
100	0.026439	0.029785	0.032813	0.035655	0.038377	0.041016	0.043594	0.046127	0.048624
150	0.015671	0.018478	0.020828	0.022945	0.024921	0.026803	0.028619	0.030387	0.032118
200	0.0099496	0.01272	0.014793	0.016571	0.018184	0.019694	0.021133	0.02252	0.023869
250	0.0060061	0.0091752	0.011142	0.012735	0.01414	0.01543	0.016643	0.017803	0.018922
300	0.0027982	0.0067381	0.0086903	0.010175	0.011444	0.01259	0.013654	0.014662	0.015629
350	0.0021053	0.0049589	0.0069334	0.0083477	0.0095231	0.010566	0.011524	0.012423	0.01328
400	0.0019109	0.0036921	0.0056249	0.0069853	0.0080891	0.0090538	0.009931	0.010748	0.011523
450	0.0018038	0.002915	0.0046344	0.0059384	0.0069828	0.0078848	0.0086979	0.0094507	0.01016
500	0.001731	0.0024873	0.0038894	0.0051185	0.0061087	0.0069575	0.0077176	0.0084175	0.0090741
600	0.001633	0.0020847	0.0029518	0.0039548	0.0048336	0.0055908	0.0062651	0.0068818	0.0074568
700	0.0015663	0.0018921	0.0024632	0.0032232	0.0039749	0.0046483	0.0052519	0.0058036	0.0063167
800	0.0015162	0.0017739	0.0021881	0.0027601	0.0033837	0.003975	0.0045161	0.0050133	0.0054762
900	0.0014763	0.0016911	0.0020143	0.0024578	0.0029695	0.0034823	0.0039663	0.0044159	0.004836
1000	0.0014431	0.0016282	0.0018934	0.0022498	0.0026723	0.0031145	0.0035462	0.0039532	0.0043355

Table A.7 Specific enthalpy h of water in $\frac{kJ}{kg}$ 1/2

p in bar	ϑ in °C										
	0	25	50	75	100	125	150	200	250	300	350
0.01	−0.041192	2547.5538	2594.4004	2641.3698	2688.536	2735.9492	2783.6463	2879.9959	2977.7344	3076.9529	3177.7157
0.1	−0.032023	104.8447	2591.9936	2639.8034	2687.4309	2735.1334	2783.0201	2879.5902	2977.4457	3076.7343	3177.5435
1	0.059662	104.9281	209.4118	314.0231	2675.7674	2726.6805	2776.5918	2875.4751	2974.5371	3074.5404	3175.8174
5	0.467	105.2985	209.7568	314.3458	419.3985	525.2465	632.2663	2855.8962	2961.1298	3064.5962	3168.0612
10	0.97582	105.7613	210.1879	314.7492	419.7742	525.5915	632.5749	2828.2675	2943.2222	3051.7032	3158.1633
20	1.9923	106.6864	211.0499	315.556	420.5256	526.2821	633.1931	852.5725	2903.2314	3024.2519	3137.6412
30	3.0072	107.6107	211.9116	316.3627	421.2774	526.9734	633.8126	852.9781	2856.5485	2994.3493	3116.0622
40	4.0206	108.5342	212.773	317.1695	422.0295	527.6654	634.4334	853.3874	1085.6861	2961.6515	3093.3182
50	5.0325	109.457	213.634	317.9762	422.7819	528.3581	635.0554	853.8004	1085.662	2925.644	3069.2942
60	6.0429	110.3791	214.4947	318.7829	423.5345	529.0515	635.6786	854.217	1085.6501	2885.4905	3043.8584
70	7.0517	111.3004	215.355	319.5896	424.2875	529.7455	636.3031	854.6371	1085.65	2839.8277	3016.8497
80	8.0591	112.2209	216.215	320.3962	425.0407	530.4402	636.9287	855.0607	1085.6614	2786.3785	2988.061
90	9.0649	113.1408	217.0747	321.2028	425.7942	531.1355	637.5556	855.4876	1085.6839	1344.2693	2957.2186
100	10.0693	114.0599	217.934	322.0094	426.548	531.8315	638.1836	855.9179	1085.7172	1343.0966	2923.9578
150	15.0694	118.6443	222.2256	326.0417	430.3208	535.3207	641.3403	858.1171	1086.0356	1338.0633	2692.9998
200	20.0338	123.2107	226.5087	330.0728	434.0996	538.8246	644.5238	860.3911	1086.5836	1334.1395	1645.9511
250	24.9636	127.7595	230.7833	334.1026	437.8839	542.3422	647.7322	862.734	1087.3336	1331.0633	1623.8646
300	29.8599	132.2909	235.0496	338.1307	441.6732	545.8726	650.9638	865.1402	1088.2629	1328.6604	1608.7975
350	34.7235	136.8054	239.3074	342.1569	445.4671	549.415	654.2174	867.605	1089.3525	1326.8077	1597.5402
400	39.5556	141.3033	243.5569	346.1812	449.265	552.9686	657.4915	870.1243	1090.5863	1325.414	1588.7405
450	44.357	145.7849	247.798	350.2032	453.0667	556.5327	660.7849	872.6941	1091.9506	1324.4102	1581.7011
500	49.1286	150.2505	252.0309	354.2229	456.8716	560.1065	664.0964	875.3112	1093.4335	1323.7424	1575.9832
600	58.5861	159.1351	260.472	362.2545	464.4903	567.2813	670.7697	880.6748	1096.7157	1323.2514	1567.412
700	67.9346	167.9596	268.8806	370.2752	472.1184	574.4884	677.5039	886.1936	1100.3648	1323.6819	1561.5683
800	77.1804	176.7265	277.2572	378.284	479.7538	581.7238	684.2923	891.8495	1104.3275	1324.8526	1557.6673
900	86.3292	185.438	285.6023	386.2804	487.3947	588.9842	691.1293	897.627	1108.5612	1326.6315	1555.2253
1000	95.386	194.0963	293.9166	394.2638	495.0395	596.2666	698.0098	903.5132	1113.031	1328.9193	1553.9225

Table A.8 Specific enthalpy h of water in $\frac{\text{kJ}}{\text{kg}}$ 2/2

p in bar	ϑ in °C								
	400	450	500	550	600	650	700	750	800
0.01	3280.0756	3384.0797	3489.7698	3597.1812	3706.3406	3817.2647	3929.9603	4044.4266	4160.6592
0.1	3279.9362	3383.9646	3489.6734	3597.0995	3706.2707	3817.2044	3929.9078	4044.3805	4160.6185
1	3278.5396	3382.8123	3488.7086	3596.2822	3705.5715	3816.6009	3929.3827	4043.9201	4160.2118
5	3272.292	3377.6695	3484.4082	3592.6421	3702.4586	3813.9149	3927.046	4041.8714	4158.4023
10	3264.3855	3371.19	3479.0037	3588.0739	3698.5556	3810.549	3924.1188	4039.3058	4156.1368
20	3248.2271	3358.0519	3468.0932	3578.8756	3690.7089	3803.7889	3918.2436	4034.1585	4151.5935
30	3231.571	3344.6585	3457.0405	3569.5916	3682.8068	3796.9905	3912.3407	4028.9905	4147.0344
40	3214.3735	3330.9912	3445.8374	3560.2183	3674.8479	3790.1538	3906.4104	4023.8023	4142.4601
50	3196.5917	3317.032	3434.4761	3550.7526	3666.8311	3783.2785	3900.4531	4018.5944	4137.8714
60	3178.183	3302.7635	3422.9493	3541.1913	3658.7552	3776.3644	3894.469	4013.3674	4133.2689
70	3159.1044	3288.1695	3411.2503	3531.5318	3650.6193	3769.4116	3888.4587	4008.1219	4128.6531
80	3139.3106	3273.234	3399.3726	3521.7715	3642.4227	3762.4199	3882.4224	4002.8585	4124.0248
90	3118.7526	3257.9419	3387.3105	3511.9082	3634.1646	3755.3896	3876.3607	3997.5778	4119.3844
100	3097.3753	3242.2779	3375.0584	3501.9399	3625.8446	3748.3207	3870.2739	3992.2803	4114.7328
150	2975.5477	3157.8415	3310.7911	3450.474	3583.3076	3712.4081	3839.4829	3965.5633	4091.3257
200	2816.8362	3061.5343	3241.1865	3396.2412	3539.2259	3675.5941	3808.1522	3938.5205	4067.7254
250	2578.7494	2950.3799	3165.9152	3339.2842	3493.6905	3637.973	3776.3704	3911.233	4044.0049
300	2152.7882	2820.9128	3084.7924	3279.7892	3446.8724	3599.6773	3744.2417	3883.7843	4020.2341
350	1988.2517	2670.967	2998.0171	3218.0815	3399.0162	3560.8734	3711.8836	3856.2609	3996.4806
400	1931.1915	2511.6302	2906.6872	3154.6483	3350.4327	3521.7571	3679.4249	3828.7509	3972.8094
450	1897.6375	2377.2586	2813.3506	3090.1926	3301.4916	3482.5474	3647.0029	3801.3439	3949.2828
500	1874.3676	2284.3776	2722.5198	3025.7028	3252.6142	3443.4808	3614.7605	3774.1295	3925.9604
600	1843.1916	2179.8228	2570.4044	2902.0646	3156.9527	3366.7648	3551.3945	3720.6353	3880.1539
700	1822.9002	2123.3881	2466.3096	2795.013	3067.5056	3293.5699	3490.4519	3668.9601	3835.8142
800	1808.7051	2087.5894	2397.6482	2709.9985	2988.0897	3225.6656	3432.9206	3619.7377	3793.3225
900	1798.473	2062.7176	2350.381	2645.2374	2920.7605	3164.4114	3379.5448	3573.5076	3753.0163
1000	1791.0568	2044.579	2316.2634	2596.0109	2865.0706	3110.6026	3330.7556	3530.6802	3715.1889

Table A.9 Specific entropy s of water in $\frac{\text{kJ}}{\text{kgK}}$ 1/2

p in bar	ϑ in °C										
	0	25	50	75	100	125	150	200	250	300	350
0.01	−0.00015452	9.0921	9.243	9.383	9.5138	9.6368	9.753	9.9682	10.1645	10.3456	10.5142
0.1	−0.00015391	0.36725	8.1741	8.3167	8.4488	8.5725	8.6892	8.9048	9.1014	9.2827	9.4513
1	−0.0001478	0.36723	0.70375	1.0156	7.361	7.4931	7.6147	7.8356	8.0346	8.2171	8.3865
5	−0.00012103	0.36713	0.70357	1.0153	1.3067	1.5813	1.8419	7.0611	7.2726	7.4614	7.6345
10	−8.8423e-05	0.367	0.70334	1.015	1.3063	1.5808	1.8414	6.6955	6.9266	7.1247	7.3028
20	−2.6077e-05	0.36674	0.70287	1.0144	1.3055	1.5798	1.8403	2.3301	6.5474	6.7685	6.9582
30	3.2474e-05	0.36648	0.70241	1.0137	1.3048	1.5789	1.8391	2.3285	6.2893	6.5412	6.7449
40	8.7256e-05	0.36622	0.70195	1.0131	1.304	1.578	1.838	2.3269	2.7933	6.3638	6.5843
50	0.0001383	0.36596	0.70149	1.0125	1.3032	1.577	1.8369	2.3254	2.7909	6.2109	6.4515
60	0.00018563	0.36569	0.70103	1.0119	1.3024	1.5761	1.8358	2.3238	2.7885	6.0702	6.3356
70	0.00022927	0.36543	0.70057	1.0112	1.3017	1.5752	1.8347	2.3223	2.7861	5.9335	6.2303
80	0.00026926	0.36516	0.70011	1.0106	1.3009	1.5743	1.8337	2.3207	2.7837	5.7935	6.1319
90	0.00030561	0.3649	0.69965	1.01	1.3001	1.5734	1.8326	2.3192	2.7814	3.2529	6.0378
100	0.00033836	0.36463	0.69919	1.0094	1.2994	1.5724	1.8315	2.3177	2.7791	3.2484	5.9458
150	0.00044893	0.36328	0.69689	1.0063	1.2956	1.5679	1.8262	2.3102	2.7679	3.2275	5.4435
200	0.00047328	0.3619	0.6946	1.0033	1.2918	1.5635	1.8209	2.303	2.7572	3.2087	3.7288
250	0.00041456	0.36051	0.69232	1.0003	1.2881	1.5591	1.8158	2.2959	2.7469	3.1915	3.6803
300	0.00027584	0.35908	0.69004	0.99729	1.2845	1.5548	1.8107	2.289	2.7371	3.1756	3.6435
350	6.0091e-05	0.35764	0.68777	0.99433	1.2809	1.5505	1.8058	2.2823	2.7276	3.1608	3.6131
400	−0.00022982	0.35618	0.68551	0.99139	1.2773	1.5463	1.8009	2.2758	2.7185	3.1469	3.587
450	−0.00059112	0.35469	0.68325	0.98848	1.2738	1.5422	1.7961	2.2693	2.7097	3.1338	3.5638
500	−0.0010211	0.35319	0.68099	0.98558	1.2703	1.5381	1.7914	2.2631	2.7012	3.1214	3.543
600	−0.0020771	0.35012	0.67649	0.97987	1.2634	1.5301	1.7822	2.2509	2.6848	3.0982	3.5064
700	−0.0033782	0.34698	0.67201	0.97423	1.2567	1.5223	1.7732	2.2392	2.6694	3.0769	3.4747
800	−0.0049067	0.34377	0.66754	0.96866	1.2501	1.5146	1.7645	2.228	2.6548	3.0572	3.4465
900	−0.0066463	0.34049	0.66309	0.96317	1.2436	1.5071	1.756	2.2171	2.6408	3.0388	3.4211
1000	−0.0085823	0.33716	0.65864	0.95774	1.2373	1.4998	1.7477	2.2066	2.6275	3.0215	3.3978

Table A.10 Specific entropy s of water in $\frac{kJ}{kgK}$ 2/2

p in bar	ϑ in °C								
	400	450	500	550	600	650	700	750	800
0.01	10.6722	10.8212	10.9625	11.0971	11.2258	11.3494	11.4682	11.5829	11.6938
0.1	9.6093	9.7584	9.8997	10.0343	10.1631	10.2866	10.4055	10.5202	10.6311
1	8.5451	8.6945	8.8361	8.9709	9.0998	9.2234	9.3424	9.4571	9.5681
5	7.7954	7.9464	8.0891	8.2247	8.3543	8.4784	8.5977	8.7128	8.824
10	7.4668	7.6198	7.764	7.9007	8.0309	8.1557	8.2755	8.3909	8.5024
20	7.129	7.2863	7.4335	7.5723	7.7042	7.8301	7.9509	8.067	8.1791
30	6.9233	7.0853	7.2356	7.3767	7.5102	7.6373	7.759	7.8759	7.9885
40	6.7712	6.9383	7.0919	7.2353	7.3704	7.4989	7.6215	7.7391	7.8523
50	6.6481	6.8208	6.9778	7.1235	7.2604	7.3901	7.5137	7.6321	7.7459
60	6.5431	6.7216	6.8824	7.0306	7.1692	7.3002	7.4248	7.5439	7.6583
70	6.4501	6.6351	6.7997	6.9505	7.0909	7.2232	7.3488	7.4687	7.5837
80	6.3657	6.5577	6.7264	6.8798	7.0221	7.1557	7.2823	7.403	7.5186
90	6.2875	6.4871	6.6601	6.8163	6.9605	7.0955	7.2231	7.3446	7.4608
100	6.2139	6.4217	6.5993	6.7584	6.9045	7.0409	7.1696	7.2918	7.4087
150	5.8817	6.1433	6.3479	6.523	6.6797	6.8235	6.9576	7.0839	7.2039
200	5.5525	5.9041	6.1445	6.339	6.5077	6.6596	6.7994	6.9301	7.0534
250	5.1401	5.6755	5.9642	6.1816	6.3638	6.5246	6.6706	6.8057	6.9324
300	4.4756	5.4419	5.7956	6.0403	6.2374	6.4077	6.5602	6.7	6.8303
350	4.2139	5.1945	5.6331	5.9093	6.1229	6.3032	6.4625	6.6072	6.7411
400	4.1142	4.9446	5.4746	5.7859	6.017	6.2079	6.3743	6.5239	6.6614
450	4.0506	4.7361	5.3209	5.6685	5.9179	6.1197	6.2932	6.4479	6.5891
500	4.0029	4.5892	5.1759	5.5566	5.8245	6.0372	6.218	6.3777	6.5226
600	3.9316	4.4134	4.9357	5.3519	5.6528	5.8867	6.0815	6.2512	6.4034
700	3.8777	4.308	4.7663	5.1786	5.5003	5.7522	5.96	6.139	6.2982
800	3.8338	4.2331	4.6475	5.0391	5.3674	5.6321	5.8509	6.0382	6.2039
900	3.7965	4.1748	4.5593	4.9289	5.254	5.5255	5.7526	5.947	6.1184
1000	3.7638	4.1267	4.49	4.8406	5.158	5.4316	5.664	5.8644	6.0405

Table A.11 Specific heat capacity c_p of water in $\frac{\text{kJ}}{\text{kgK}}$ 1/2

p in bar	ϑ in °C										
	0	25	50	75	100	125	150	200	250	300	350
0.01	4.2199	1.8732	1.8757	1.8823	1.8913	1.902	1.914	1.9405	1.9693	1.9996	2.0311
0.1	4.2199	4.1822	1.9272	1.9058	1.9057	1.9112	1.9201	1.9436	1.9711	2.0007	2.0318
1	4.2194	4.1819	4.1796	4.1915	2.0741	2.0107	1.9857	1.9757	1.9891	2.0121	2.0396
5	4.2174	4.1807	4.1786	4.1907	4.2157	4.2546	4.3102	2.1448	2.0783	2.0657	2.0753
10	4.215	4.1793	4.1775	4.1896	4.2146	4.2533	4.3086	2.4288	2.2116	2.1408	2.1231
20	4.21	4.1764	4.1752	4.1874	4.2123	4.2506	4.3053	4.4914	2.5602	2.3201	2.2301
30	4.2052	4.1736	4.1729	4.1853	4.21	4.248	4.3022	4.4856	3.0772	2.5431	2.3539
40	4.2003	4.1708	4.1706	4.1831	4.2078	4.2455	4.299	4.4799	4.8646	2.8199	2.4967
50	4.1955	4.168	4.1684	4.181	4.2055	4.2429	4.2959	4.4743	4.8511	3.1714	2.661
60	4.1908	4.1652	4.1661	4.1788	4.2033	4.2403	4.2927	4.4687	4.8379	3.6378	2.8504
70	4.1861	4.1625	4.1639	4.1767	4.2011	4.2378	4.2897	4.4632	4.825	4.2919	3.0704
80	4.1814	4.1597	4.1617	4.1746	4.1989	4.2353	4.2866	4.4578	4.8125	5.287	3.3288
90	4.1768	4.157	4.1595	4.1726	4.1967	4.2328	4.2836	4.4525	4.8002	5.7305	3.637
100	4.1723	4.1543	4.1573	4.1705	4.1945	4.2303	4.2806	4.4472	4.7883	5.6816	4.0118
150	4.1501	4.1412	4.1466	4.1603	4.1838	4.2182	4.2659	4.4219	4.7325	5.476	8.7885
200	4.129	4.1285	4.1361	4.1503	4.1734	4.2064	4.2518	4.3982	4.6824	5.3168	8.1062
250	4.109	4.1162	4.126	4.1407	4.1633	4.1951	4.2383	4.3758	4.6371	5.1883	6.98
300	4.0899	4.1044	4.1162	4.1313	4.1534	4.1841	4.2252	4.3547	4.5959	5.0814	6.3935
350	4.0717	4.0931	4.1067	4.1221	4.1439	4.1734	4.2126	4.3347	4.5582	4.9906	6.0154
400	4.0544	4.0821	4.0974	4.1132	4.1346	4.163	4.2005	4.3158	4.5234	4.912	5.7424
450	4.038	4.0716	4.0884	4.1046	4.1255	4.153	4.1888	4.2977	4.4912	4.8431	5.5342
500	4.0225	4.0614	4.0797	4.0961	4.1167	4.1432	4.1774	4.2806	4.4613	4.7819	5.37
600	3.9937	4.0422	4.063	4.0798	4.0997	4.1245	4.1558	4.2485	4.4073	4.6775	5.1244
700	3.9678	4.0242	4.0472	4.0644	4.0836	4.1068	4.1355	4.2191	4.3598	4.5912	4.9456
800	3.9446	4.0076	4.0322	4.0497	4.0683	4.09	4.1164	4.1921	4.3174	4.5181	4.8076
900	3.924	3.9921	4.0181	4.0358	4.0537	4.074	4.0984	4.167	4.2794	4.4553	4.6966
1000	3.9057	3.9777	4.0048	4.0225	4.0397	4.0589	4.0813	4.1437	4.2449	4.4003	4.6048

Table A.12 Specific heat capacity c_p of water in $\frac{\text{kJ}}{\text{kgK}}$ 2/2

p in bar	ϑ in °C								
	400	450	500	550	600	650	700	750	800
0.01	2.0635	2.0968	2.1309	2.1656	2.2008	2.2362	2.2716	2.307	2.3423
0.1	2.0641	2.0972	2.1312	2.1659	2.201	2.2364	2.2718	2.3071	2.3424
1	2.0697	2.1015	2.1345	2.1685	2.2031	2.2381	2.2732	2.3083	2.3434
5	2.0952	2.1206	2.1494	2.1803	2.2126	2.2458	2.2795	2.3135	2.3477
10	2.1284	2.1451	2.1682	2.1951	2.2245	2.2555	2.2875	2.3201	2.3532
20	2.1997	2.1964	2.2069	2.2254	2.2486	2.275	2.3035	2.3333	2.3642
30	2.278	2.2508	2.2472	2.2564	2.2732	2.2948	2.3196	2.3467	2.3753
40	2.3642	2.3088	2.2892	2.2883	2.2982	2.3149	2.336	2.3601	2.3865
50	2.459	2.3705	2.333	2.3212	2.3238	2.3353	2.3525	2.3737	2.3978
60	2.5632	2.4364	2.3787	2.355	2.3499	2.3559	2.3692	2.3874	2.4092
70	2.6778	2.5067	2.4265	2.3899	2.3765	2.3769	2.3861	2.4012	2.4206
80	2.8037	2.5817	2.4765	2.4258	2.4038	2.3983	2.4032	2.4152	2.4322
90	2.9425	2.6617	2.5287	2.4629	2.4316	2.4199	2.4205	2.4293	2.4438
100	3.0958	2.747	2.5833	2.5011	2.46	2.442	2.438	2.4435	2.4555
150	4.1778	3.2687	2.896	2.7112	2.612	2.5575	2.5288	2.5166	2.5155
200	6.3601	4.0074	3.2845	2.9552	2.7812	2.6824	2.6251	2.593	2.5775
250	12.9977	5.086	3.7661	3.2354	2.9679	2.8168	2.7267	2.6725	2.6415
300	25.8263	6.6908	4.3597	3.5529	3.1713	2.9598	2.8331	2.7549	2.7072
350	11.6455	8.9762	5.0715	3.9073	3.3894	3.1103	2.9435	2.8396	2.7744
400	8.7042	10.9505	5.8745	4.2938	3.6194	3.2666	3.057	2.9259	2.8428
450	7.4742	10.8642	6.6881	4.7004	3.8571	3.4264	3.1723	3.0133	2.9118
500	6.7794	9.5666	7.309	5.1031	4.0973	3.5873	3.2881	3.1008	2.9813
600	5.9976	7.5398	7.5221	5.7534	4.5557	3.9007	3.5151	3.273	3.1193
700	5.5542	6.5095	6.9698	6.0369	4.9233	4.1825	3.7267	3.4357	3.2526
800	5.2618	5.9182	6.3751	5.9817	5.1372	4.4081	3.9135	3.5834	3.3765
900	5.0515	5.5326	5.9164	5.7779	5.2063	4.5583	4.0692	3.7127	3.4861
1000	4.8915	5.2584	5.576	5.5489	5.1706	4.6275	4.191	3.8235	3.5762

Appendix B
Selected Absolute Molar Specific Enthalpies/Entropies

The following tables contain absolute molar specific enthalpies H_m and entropies S_m as well as specific heat capacities at constant pressure for selected fluids. However, there are three types of specific heat capacities to be explained in the following:

1. Temperature dependent heat capacities, i.e.

$$\boxed{C_p = C_p\,(\vartheta)} \tag{B.1}$$

2. Arithmetic mean heat capacity
   This is useful, for example, to calculate the enthalpy of an ideal gas, i.e.

$$\mathrm{d}H_\mathrm{m} = C_p\,\mathrm{d}T, \tag{B.2}$$

so that

$$H_\mathrm{m} - H_{\mathrm{m},0} = \int_{T_0}^{T} C_p\,\mathrm{d}T = \overline{C_p}\Big|_{\vartheta_0}^{\vartheta}\,(\vartheta - \vartheta_0)\,. \tag{B.3}$$

This leads to

$$\boxed{\overline{C_p}\Big|_{\vartheta_0}^{\vartheta} = \frac{1}{(\vartheta - \vartheta_0)} \int_{T_0}^{T} C_p\,\mathrm{d}T} \tag{B.4}$$

Note that $H_{\mathrm{m},0} = H_\mathrm{m}\,(\vartheta_0,\,T_0)$ is a reference level. The enthalpy difference between two states (1) and (2) then reads as

$$H_{\mathrm{m},2} - H_{\mathrm{m},0} = \overline{C_p}\Big|_{\vartheta_0}^{\vartheta_2}\,(\vartheta_2 - \vartheta_0) \tag{B.5}$$

and

A. Schmidt, *Technical Thermodynamics Workbook for Engineers*,
https://doi.org/10.1007/978-3-031-50172-2

$$H_{m,1} - H_{m,0} = \overline{C_p}\Big|_{\vartheta_0}^{\vartheta_1} (\vartheta_1 - \vartheta_0)\,. \tag{B.6}$$

Thus, it is

$$H_{m,2} - H_{m,1} = \overline{C_p}\Big|_{\vartheta_0}^{\vartheta_2} (\vartheta_2 - \vartheta_0) - \overline{C_p}\Big|_{\vartheta_0}^{\vartheta_1} (\vartheta_1 - \vartheta_0)\,. \tag{B.7}$$

In the following tables, the reference temperature for the averaged heat capacities is $\vartheta_0 = 0°\text{C}$.

3. Logarithmic mean heat capacity
   This is necessary, for example, to calculate the entropy of an ideal gas. It is

$$\mathrm{d}S_m = C_p \frac{\mathrm{d}T}{T} - R_M \frac{\mathrm{d}p}{p} \tag{B.8}$$

so that

$$S_m - S_{m,0} = \int_{T_0}^{T} C_p \frac{\mathrm{d}T}{T} - \int_{p_0}^{p} R_M \frac{\mathrm{d}p}{p} = \overline{\overline{C_p}}\Big|_{\vartheta_0}^{\vartheta} \ln\frac{T}{T_0} - R_M \ln\frac{p}{p_0}. \tag{B.9}$$

This leads to

$$\boxed{\overline{\overline{C_p}}\Big|_{\vartheta_0}^{\vartheta} = \frac{1}{\ln\frac{T}{T_0}} \int_{T_0}^{T} C_p \frac{\mathrm{d}T}{T}} \tag{B.10}$$

Note that $S_{m,0} = S_m(\vartheta_0, T_0, p_0)$ is a reference level. The entropy difference between two states (1) and (2) then reads as

$$S_{m,2} - S_{m,0} = \overline{\overline{C_p}}\Big|_{\vartheta_0}^{\vartheta_2} \ln\frac{T_2}{T_0} - R_M \ln\frac{p_2}{p_0} \tag{B.11}$$

and

$$S_{m,1} - S_{m,0} = \overline{\overline{C_p}}\Big|_{\vartheta_0}^{\vartheta_1} \ln\frac{T_1}{T_0} - R_M \ln\frac{p_1}{p_0}. \tag{B.12}$$

Thus, it is

$$S_{m,2} - S_{m,1} = \left(\overline{\overline{C_p}}\Big|_{\vartheta_0}^{\vartheta_2} \ln\frac{T_2}{T_0} - R_M \ln\frac{p_2}{p_0}\right) - \left(\overline{\overline{C_p}}\Big|_{\vartheta_0}^{\vartheta_1} \ln\frac{T_1}{T_0} - R_M \ln\frac{p_1}{p_0}\right) \tag{B.13}$$

respectively

$$S_{m,2} - S_{m,1} = \left(\overline{\overline{C_p}}\Big|_{\vartheta_0}^{\vartheta_2} \ln\frac{T_2}{T_0} - \overline{\overline{C_p}}\Big|_{\vartheta_0}^{\vartheta_1} \ln\frac{T_1}{T_0} \right) - R_M \ln\frac{p_2}{p_1}$$
$$\equiv \overline{\overline{C_p}}\Big|_{\vartheta_1}^{\vartheta_2} \ln\frac{T_2}{T_1} - R_M \ln\frac{p_2}{p_1}. \quad \text{(B.14)}$$

Thus, it is

$$\boxed{\overline{\overline{C_p}}\Big|_{\vartheta_1}^{\vartheta_2} = \frac{\overline{\overline{C_p}}\Big|_{\vartheta_0}^{\vartheta_2} \ln\frac{T_2}{T_0} - \overline{\overline{C_p}}\Big|_{\vartheta_0}^{\vartheta_1} \ln\frac{T_1}{T_0}}{\ln\frac{T_2}{T_1}}} \quad \text{(B.15)}$$

In the following tables, the reference temperature for the averaged heat capacities is $\vartheta_0 = 0°C$ (Tables B.1, B.2, B.3, B.4, B.5, B.6, B.7, B.8, B.9, B.10, B.11, B.12, B.13, B.14, B.15, B.16, B.17, B.18, B.19, B.20, B.21, B.22, B.23, B.24, B.25, B.26, B.27, B.28 and B.29).[2]

Table B.1 Absolute molar specific enthalpy and entropy of H_2 at $p_0 = 1$ bar. Reference temperature for averaged heat capacities is $\vartheta_0 = 0\,°C$

ϑ [°C]	C_p $\left[\frac{J}{molK}\right]$	$\overline{C_p}\Big\|_{\vartheta_0}^{\vartheta}$ $\left[\frac{J}{molK}\right]$	$\overline{\overline{C_p}}\Big\|_{\vartheta_0}^{\vartheta}$ $\left[\frac{J}{molK}\right]$	S_m $\left[\frac{J}{molK}\right]$	H_m $\left[\frac{kJ}{mol}\right]$
–20	28.3981	28.5172	28.5157	125.9961	–1.2888
–15	28.4609	28.5463	28.5455	126.5522	–1.1466
–10	28.5196	28.5741	28.5738	127.0987	–1.0042
–5	28.5747	28.6007	28.6007	127.6361	–0.86144
0	28.6262	28.6262	28.6262	128.1644	–0.71844
5	28.6743	28.6505	28.6505	128.6842	–0.57518
10	28.7192	28.6738	28.6735	129.1954	–0.4317
15	28.7611	28.696	28.6954	129.6985	–0.288
20	28.8001	28.7172	28.7162	130.1936	–0.14409
25	28.8363	28.7374	28.7359	130.681	0
50	28.9822	28.8256	28.8207	133.0091	0.72285
100	29.1452	28.9503	28.9371	137.1916	2.1766
150	29.2127	29.0282	29.0078	140.8613	3.6358
200	29.2419	29.0784	29.0527	144.1258	5.0972
250	29.2643	29.1133	29.0836	147.0644	6.5599
300	29.2954	29.1409	29.1076	149.7369	8.0238
350	29.3416	29.166	29.1288	152.1889	9.4897
400	29.4041	29.1917	29.1496	154.4558	10.9583
450	29.4821	29.2195	29.171	156.5652	12.4303
500	29.5742	29.2503	29.1939	158.5393	13.9067

(continued)

[2] The listed data has been taken from NASA Thermo Build, see [5].

Table B.1 (continued)

ϑ [°C]	C_p $\left[\frac{\text{J}}{\text{molK}}\right]$	$\overline{C_p}\Big\vert_{\vartheta_0}^{\vartheta}$ $\left[\frac{\text{J}}{\text{molK}}\right]$	$\overline{\overline{C_p}}\Big\vert_{\vartheta_0}^{\vartheta}$ $\left[\frac{\text{J}}{\text{molK}}\right]$	S_m $\left[\frac{\text{J}}{\text{molK}}\right]$	H_m $\left[\frac{\text{kJ}}{\text{mol}}\right]$
600	29.8007	29.3223	29.2448	162.1494	16.8749
700	30.1045	29.4111	29.3042	165.396	19.8694
800	30.4634	29.5202	29.3741	168.3579	22.8977
900	30.8678	29.6471	29.4526	171.0894	25.9639
1000	31.3025	29.7907	29.5391	173.6319	29.0723
1100	31.7466	29.9483	29.6319	176.0154	32.2247
1200	32.1877	30.1166	29.7293	178.2624	35.4215
1300	32.6185	30.2926	29.8295	180.3904	38.6619
1400	33.0349	30.4737	29.9314	182.4134	41.9447
1500	33.4346	30.6579	30.0339	184.3426	45.2684
1600	33.8167	30.8434	30.1363	186.1875	48.6311
1700	34.1811	31.0291	30.2379	187.9557	52.0311
1800	34.528	31.214	30.3383	189.6542	55.4667
1900	34.8582	31.3972	30.4372	191.2885	58.9362
2000	35.1726	31.5781	30.5344	192.8638	62.4378
2100	35.4721	31.7565	30.6297	194.3845	65.9702
2200	35.7578	31.9319	30.7231	195.8545	69.5318
2300	36.0308	32.1042	30.8145	197.2773	73.1213
2400	36.2922	32.2733	30.9039	198.656	76.7375
2500	36.543	32.4391	30.9913	199.9935	80.3794
2600	36.7843	32.6016	31.0767	201.2923	84.0458
2700	37.017	32.7609	31.1601	202.5548	87.736
2800	37.242	32.9169	31.2417	203.7831	91.449
2900	37.4602	33.0699	31.3215	204.9791	95.1841
3000	37.6723	33.2198	31.3995	206.1447	98.9408
3100	37.879	33.3667	31.4758	207.2815	102.7184
3200	38.0811	33.5109	31.5506	208.3911	106.5165
3300	38.2789	33.6524	31.6238	209.4749	110.3345
3400	38.4731	33.7913	31.6955	210.5341	114.1721
3500	38.664	33.9278	31.7658	211.5701	118.029
3600	38.852	34.062	31.8348	212.5839	121.9048
3700	39.0374	34.194	31.9024	213.5767	125.7993
3800	39.2203	34.3239	31.9689	214.5493	129.7122
3900	39.4009	34.4517	32.0342	215.5028	133.6433
4000	39.5792	34.5777	32.0984	216.4379	137.5923
4500	40.4328	35.1815	32.4042	220.8645	157.5982
5000	41.1931	35.7457	32.6873	224.9308	178.0098

Table B.2 Absolute molar specific enthalpy and entropy of H at $p_0 = 1\,\text{bar}$. Reference temperature for averaged heat capacities is $\vartheta_0 = 0\,°\text{C}$

ϑ [°C]	C_p $\left[\frac{\text{J}}{\text{molK}}\right]$	$\overline{C_p}\Big\vert_{\vartheta_0}^{\vartheta}$ $\left[\frac{\text{J}}{\text{molK}}\right]$	$\overline{\overline{C_p}}\Big\vert_{\vartheta_0}^{\vartheta}$ $\left[\frac{\text{J}}{\text{molK}}\right]$	S_m $\left[\frac{\text{J}}{\text{molK}}\right]$	H_m $\left[\frac{\text{kJ}}{\text{mol}}\right]$
–20	20.7863	20.7863	20.7863	111.3169	217.0634
–15	20.7863	20.7863	20.7863	111.7235	217.1674
–10	20.7863	20.7863	20.7863	112.1222	217.2713
–5	20.7863	20.7863	20.7863	112.5135	217.3752
0	20.7863	20.7863	20.7863	112.8975	217.4792
5	20.7863	20.7863	20.7863	113.2745	217.5831
10	20.7863	20.7863	20.7863	113.6449	217.687
15	20.7863	20.7863	20.7863	114.0087	217.791
20	20.7863	20.7863	20.7863	114.3663	217.8949
25	20.7863	20.7863	20.7863	114.7179	217.9988
50	20.7863	20.7863	20.7863	116.3916	218.5185
100	20.7863	20.7863	20.7863	119.382	219.5578
150	20.7863	20.7863	20.7863	121.9958	220.5971
200	20.7863	20.7863	20.7863	124.3173	221.6364
250	20.7863	20.7863	20.7863	126.4054	222.6757
300	20.7863	20.7863	20.7863	128.3027	223.7151
350	20.7863	20.7863	20.7863	130.0413	224.7544
400	20.7863	20.7863	20.7863	131.6456	225.7937
450	20.7863	20.7863	20.7863	133.1349	226.833
500	20.7863	20.7863	20.7863	134.5246	227.8723
600	20.7863	20.7863	20.7863	137.0529	229.9509
700	20.7863	20.7863	20.7863	139.3068	232.0296
800	20.7863	20.7863	20.7863	141.34	234.1082
900	20.7863	20.7863	20.7863	143.192	236.1868
1000	20.7863	20.7863	20.7863	144.8923	238.2654
1100	20.7863	20.7863	20.7863	146.464	240.3441
1200	20.7863	20.7863	20.7863	147.9252	242.4227
1300	20.7863	20.7863	20.7863	149.2904	244.5013
1400	20.7863	20.7863	20.7863	150.5714	246.58
1500	20.7863	20.7863	20.7863	151.778	248.6586
1600	20.7863	20.7863	20.7863	152.9185	250.7372
1700	20.7863	20.7863	20.7863	153.9996	252.8158
1800	20.7863	20.7863	20.7863	155.0272	254.8945
1900	20.7863	20.7863	20.7863	156.0064	256.9731
2000	20.7863	20.7863	20.7863	156.9415	259.0517
2100	20.7863	20.7863	20.7863	157.8364	261.1303
2200	20.7863	20.7863	20.7863	158.6944	263.209
2300	20.7863	20.7863	20.7863	159.5183	265.2876

(continued)

Table B.2 (continued)

ϑ [°C]	C_p $\left[\frac{\text{J}}{\text{molK}}\right]$	$\overline{C_p}\vert_{\vartheta_0}^{\vartheta}$ $\left[\frac{\text{J}}{\text{molK}}\right]$	$\overline{\overline{C_p}}\vert_{\vartheta_0}^{\vartheta}$ $\left[\frac{\text{J}}{\text{molK}}\right]$	S_m $\left[\frac{\text{J}}{\text{molK}}\right]$	H_m $\left[\frac{\text{kJ}}{\text{mol}}\right]$
2400	20.7863	20.7863	20.7863	160.3108	267.3662
2500	20.7863	20.7863	20.7863	161.0742	269.4449
2600	20.7863	20.7863	20.7863	161.8106	271.5235
2700	20.7863	20.7863	20.7863	162.5217	273.6021
2800	20.7863	20.7863	20.7863	163.2094	275.6807
2900	20.7863	20.7863	20.7863	163.875	277.7594
3000	20.7863	20.7863	20.7863	164.5199	279.838
3100	20.7863	20.7863	20.7863	165.1455	281.9166
3200	20.7863	20.7863	20.7863	165.7528	283.9953
3300	20.7863	20.7863	20.7863	166.3428	286.0739
3400	20.7863	20.7863	20.7863	166.9165	288.1525
3500	20.7863	20.7863	20.7863	167.4749	290.2311
3600	20.7863	20.7863	20.7863	168.0186	292.3098
3700	20.7863	20.7863	20.7863	168.5484	294.3884
3800	20.7863	20.7863	20.7863	169.0651	296.467
3900	20.7863	20.7863	20.7863	169.5693	298.5456
4000	20.7863	20.7863	20.7863	170.0615	300.6243
4500	20.7863	20.7863	20.7863	172.3616	311.0174
5000	20.7863	20.7863	20.7863	174.4324	321.4105

Table B.3 Absolute molar specific enthalpy and entropy of O_2 at $p_0 = 1\,\text{bar}$. Reference temperature for averaged heat capacities is $\vartheta_0 = 0\,°\text{C}$

ϑ [°C]	C_p $\left[\frac{\text{J}}{\text{molK}}\right]$	$\overline{C_p}\vert_{\vartheta_0}^{\vartheta}$ $\left[\frac{\text{J}}{\text{molK}}\right]$	$\overline{\overline{C_p}}\vert_{\vartheta_0}^{\vartheta}$ $\left[\frac{\text{J}}{\text{molK}}\right]$	S_m $\left[\frac{\text{J}}{\text{molK}}\right]$	H_m $\left[\frac{\text{kJ}}{\text{mol}}\right]$
–20	29.2154	29.2427	29.2423	200.358	–1.3179
–15	29.2279	29.2497	29.2495	200.9295	–1.1718
–10	29.2417	29.2572	29.2571	201.4904	–1.0256
–5	29.257	29.2652	29.2652	202.0409	–0.8794
0	29.2736	29.2736	29.2736	202.5816	–0.73308
5	29.2917	29.2825	29.2825	203.1127	–0.58666
10	29.3112	29.2919	29.2918	203.6348	–0.44016
15	29.3322	29.3018	29.3016	204.148	–0.29355
20	29.3546	29.3122	29.3117	204.6528	–0.14683
25	29.3784	29.323	29.3222	205.1495	0
50	29.5176	29.3841	29.3807	207.5203	0.73613
100	29.8826	29.5379	29.5221	211.7912	2.2207
150	30.3289	29.7254	29.6869	215.5757	3.7257

(continued)

Table B.3 (continued)

ϑ [°C]	C_p $\left[\frac{\mathrm{J}}{\mathrm{molK}}\right]$	$\overline{C_p}\Big\|_{\vartheta_0}^{\vartheta}$ $\left[\frac{\mathrm{J}}{\mathrm{molK}}\right]$	$\overline{\overline{C_p}}\Big\|_{\vartheta_0}^{\vartheta}$ $\left[\frac{\mathrm{J}}{\mathrm{molK}}\right]$	S_{m} $\left[\frac{\mathrm{J}}{\mathrm{molK}}\right]$	H_{m} $\left[\frac{\mathrm{kJ}}{\mathrm{mol}}\right]$
200	30.8199	29.9371	29.8659	218.9896	5.2543
250	31.3268	30.1643	30.0519	222.1107	6.808
300	31.8283	30.4	30.2394	224.9928	8.3869
350	32.3098	30.6388	30.4249	227.675	9.9905
400	32.7615	30.8762	30.6055	230.1861	11.6174
450	33.1779	31.1092	30.7795	232.5483	13.2661
500	33.5569	31.3353	30.9459	234.7792	14.9346
600	34.2077	31.7617	31.2537	238.901	18.3239
700	34.7483	32.1507	31.529	242.6397	21.7724
800	35.2233	32.506	31.7764	246.0622	25.2717
900	35.6041	32.8298	31.999	249.2177	28.8137
1000	35.9259	33.1237	32.1992	252.1435	32.3906
1100	36.214	33.3917	32.3805	254.8709	35.9978
1200	36.4828	33.6382	32.546	257.426	39.6328
1300	36.7402	33.867	32.6984	259.8305	43.294
1400	36.9908	34.0812	32.8401	262.1024	46.9806
1500	37.2369	34.2834	32.9727	264.2567	50.692
1600	37.4798	34.4756	33.0977	266.3063	54.4279
1700	37.7198	34.6594	33.2161	268.2618	58.1879
1800	37.957	34.836	33.3288	270.1325	61.9718
1900	38.1914	35.0065	33.4366	271.926	65.7792
2000	38.4224	35.1715	33.54	273.6494	69.6099
2100	38.65	35.3317	33.6394	275.3084	73.4636
2200	38.8738	35.4877	33.7354	276.9083	77.3398
2300	39.0934	35.6397	33.8281	278.4535	81.2382
2400	39.3087	35.7881	33.9179	279.9481	85.1583
2500	39.5195	35.9331	34.005	281.3956	89.0998
2600	39.7255	36.0751	34.0896	282.7993	93.0621
2700	39.9266	36.214	34.1718	284.1618	97.0447
2800	40.1228	36.3501	34.2518	285.4859	101.0472
2900	40.314	36.4835	34.3297	286.7737	105.0691
3000	40.5003	36.6143	34.4056	288.0275	109.1099
3100	40.6815	36.7426	34.4797	289.249	113.169
3200	40.8577	36.8685	34.5519	290.4401	117.246
3300	41.029	36.992	34.6225	291.6023	121.3404

(continued)

Table B.3 (continued)

ϑ [°C]	C_p $\left[\frac{J}{molK}\right]$	$\overline{C_p}\Big\|_{\vartheta_0}^{\vartheta}$ $\left[\frac{J}{molK}\right]$	$\overline{\overline{C_p}}\Big\|_{\vartheta_0}^{\vartheta}$ $\left[\frac{J}{molK}\right]$	S_m $\left[\frac{J}{molK}\right]$	H_m $\left[\frac{kJ}{mol}\right]$
3400	41.1955	37.1132	34.6914	292.7371	125.4516
3500	41.3573	37.2321	34.7588	293.8458	129.5793
3600	41.5144	37.3489	34.8247	294.9297	133.7229
3700	41.6669	37.4635	34.8891	295.9899	137.882
3800	41.8149	37.5761	34.9521	297.0275	142.0562
3900	41.9587	37.6867	35.0138	298.0434	146.2449
4000	42.0982	37.7952	35.0742	299.0386	150.4478
4500	42.7369	38.3097	35.3583	303.7326	171.6605
5000	43.2874	38.7806	35.616	308.0177	193.17

Table B.4 Absolute molar specific enthalpy and entropy of O at p_0 = 1 bar. Reference temperature for averaged heat capacities is $\vartheta_0 = 0\,°C$

ϑ [°C]	C_p $\left[\frac{J}{molK}\right]$	$\overline{C_p}\Big\|_{\vartheta_0}^{\vartheta}$ $\left[\frac{J}{molK}\right]$	$\overline{\overline{C_p}}\Big\|_{\vartheta_0}^{\vartheta}$ $\left[\frac{J}{molK}\right]$	S_m $\left[\frac{J}{molK}\right]$	H_m $\left[\frac{kJ}{mol}\right]$
–20	22.2194	22.1431	22.1441	157.4509	248.1824
–15	22.1801	22.1243	22.1248	157.8851	248.2934
–10	22.1422	22.1059	22.1061	158.3102	248.4042
–5	22.1057	22.088	22.088	158.7266	248.5148
0	22.0704	22.0704	22.0704	159.1347	248.6253
5	22.0364	22.0533	22.0534	159.5347	248.7356
10	22.0035	22.0366	22.0368	159.927	248.8457
15	21.9718	22.0202	22.0207	160.3119	248.9556
20	21.9412	22.0043	22.005	160.6896	249.0654
25	21.9116	21.9887	21.9898	161.0605	249.175
50	21.7781	21.9158	21.9199	162.8193	249.7211
100	21.5698	21.792	21.805	165.937	250.8045
150	21.4184	21.6914	21.715	168.6395	251.879
200	21.3059	21.6084	21.643	171.0251	252.947
250	21.2205	21.539	21.5841	173.161	254.01
300	21.1542	21.4802	21.5351	175.0949	255.0693
350	21.1018	21.4297	21.4938	176.862	256.1257
400	21.0596	21.386	21.4584	178.489	257.1797
450	21.0253	21.3478	21.4277	179.9966	258.2318
500	20.9969	21.3141	21.4009	181.4013	259.2823
600	20.9536	21.2574	21.3563	183.9525	261.3797
700	20.9223	21.2116	21.3206	186.2228	263.4734

(continued)

Table B.4 (continued)

ϑ [°C]	C_p $\left[\frac{\mathrm{J}}{\mathrm{molK}}\right]$	$\overline{C_p}\Big\vert_{\vartheta_0}^{\vartheta}$ $\left[\frac{\mathrm{J}}{\mathrm{molK}}\right]$	$\overline{\overline{C_p}}\Big\vert_{\vartheta_0}^{\vartheta}$ $\left[\frac{\mathrm{J}}{\mathrm{molK}}\right]$	S_m $\left[\frac{\mathrm{J}}{\mathrm{molK}}\right]$	H_m $\left[\frac{\mathrm{kJ}}{\mathrm{mol}}\right]$
800	20.8947	21.1737	21.2911	188.268	265.5642
900	20.8772	21.1416	21.2663	190.1287	267.6528
1000	20.8659	21.1146	21.2453	191.836	269.7399
1100	20.8574	21.0916	21.2273	193.4135	271.826
1200	20.85	21.0718	21.2117	194.8794	273.9114
1300	20.8434	21.0544	21.198	196.2485	275.9961
1400	20.8374	21.0392	21.1859	197.5329	278.0801
1500	20.8324	21.0255	21.175	198.7423	280.1636
1600	20.8285	21.0133	21.1652	199.8852	282.2466
1700	20.8261	21.0024	21.1563	200.9684	284.3293
1800	20.8255	20.9926	21.1482	201.9979	286.4119
1900	20.827	20.9838	21.1409	202.979	288.4945
2000	20.8308	20.976	21.1343	203.9161	290.5774
2100	20.8371	20.9693	21.1283	204.813	292.6607
2200	20.8461	20.9635	21.1229	205.6732	294.7449
2300	20.8577	20.9586	21.1181	206.4998	296.83
2400	20.8721	20.9547	21.1139	207.2953	298.9165
2500	20.8893	20.9517	21.1102	208.0621	301.0046
2600	20.9093	20.9497	21.107	208.8025	303.0945
2700	20.932	20.9486	21.1043	209.5182	305.1865
2800	20.9574	20.9485	21.1021	210.2111	307.281
2900	20.9855	20.9492	21.1004	210.8826	309.3781
3000	21.016	20.951	21.0992	211.5342	311.4781
3100	21.0489	20.9536	21.0984	212.1672	313.5814
3200	21.0842	20.9571	21.098	212.7826	315.688
3300	21.1215	20.9615	21.0981	213.3816	317.7983
3400	21.1609	20.9668	21.0985	213.9652	319.9124
3500	21.2022	20.9729	21.0994	214.5341	322.0305
3600	21.2452	20.9799	21.1006	215.0893	324.1529
3700	21.2897	20.9877	21.1022	215.6314	326.2796
3800	21.3357	20.9962	21.1041	216.1612	328.4109
3900	21.3829	21.0055	21.1064	216.6792	330.5468
4000	21.4312	21.0156	21.109	217.1861	332.6875
4500	21.6829	21.0756	21.1262	219.5712	343.4655
5000	21.9368	21.1491	21.1491	221.7437	354.3707

Table B.5 Absolute molar specific enthalpy and entropy of OH at $p_0 = 1\text{bar}$. Reference temperature for averaged heat capacities is $\vartheta_0 = 0\,°\text{C}$

ϑ [°C]	C_p $\left[\frac{\text{J}}{\text{molK}}\right]$	$\overline{C_p}\|_{\vartheta_0}^{\vartheta}$ $\left[\frac{\text{J}}{\text{molK}}\right]$	$\overline{\overline{C_p}}\|_{\vartheta_0}^{\vartheta}$ $\left[\frac{\text{J}}{\text{molK}}\right]$	S_m $\left[\frac{\text{J}}{\text{molK}}\right]$	H_m $\left[\frac{\text{kJ}}{\text{mol}}\right]$
–20	30.1067	30.0509	30.0516	178.8324	35.9287
–15	30.0778	30.0372	30.0376	179.421	36.0791
–10	30.0501	30.0239	30.024	179.9977	36.2295
–5	30.0237	30.0109	30.0109	180.5631	36.3796
0	29.9983	29.9983	29.9983	181.1175	36.5297
5	29.974	29.9861	29.9861	181.6615	36.6796
10	29.9507	29.9742	29.9743	182.1953	36.8294
15	29.9284	29.9626	29.9629	182.7194	36.9791
20	29.907	29.9514	29.9519	183.234	37.1287
25	29.8864	29.9404	29.9412	183.7397	37.2782
50	29.7948	29.8898	29.8926	186.1423	38.0242
100	29.6575	29.8057	29.8145	190.4185	39.5103
150	29.566	29.74	29.7556	194.1417	40.9907
200	29.511	29.6889	29.711	197.4405	42.4675
250	29.4885	29.6505	29.678	200.4037	43.9423
300	29.497	29.6238	29.6548	203.0955	45.4168
350	29.5362	29.6081	29.6405	205.564	46.8925
400	29.6058	29.6032	29.6343	207.8461	48.371
450	29.7051	29.6087	29.6357	209.9707	49.8536
500	29.8329	29.6245	29.644	211.9607	51.342
600	30.1637	29.6855	29.68	215.6082	54.341
700	30.5673	29.7821	29.7378	218.8999	57.3771
800	31.0007	29.907	29.8121	221.9104	60.4553
900	31.4691	30.0543	29.8988	224.6928	63.5786
1000	31.9442	30.2196	29.9947	227.2863	66.7493
1100	32.4074	30.3976	30.0968	229.7191	69.967
1200	32.8496	30.5836	30.2024	232.0127	73.2301
1300	33.2668	30.7742	30.3095	234.1838	76.5361
1400	33.658	30.9663	30.4167	236.246	79.8826
1500	34.0234	31.1581	30.523	238.2105	83.2668
1600	34.3645	31.348	30.6276	240.0865	86.6864
1700	34.6829	31.5349	30.7301	241.8821	90.139
1800	34.9803	31.7181	30.8301	243.6042	93.6223
1900	35.2587	31.8972	30.9276	245.2586	97.1344
2000	35.5197	32.0719	31.0223	246.8507	100.6735
2100	35.7652	32.242	31.1143	248.3852	104.2378
2200	35.9967	32.4074	31.2036	249.8662	107.826
2300	36.2156	32.5683	31.2903	251.2974	111.4368

(continued)

Table B.5 (continued)

ϑ [°C]	C_p $\left[\frac{J}{molK}\right]$	$\overline{C_p}\vert_{\vartheta_0}^{\vartheta}$ $\left[\frac{J}{molK}\right]$	$\overline{\overline{C_p}}\vert_{\vartheta_0}^{\vartheta}$ $\left[\frac{J}{molK}\right]$	S_m $\left[\frac{J}{molK}\right]$	H_m $\left[\frac{kJ}{mol}\right]$
2400	36.4233	32.7246	31.3743	252.6821	115.0688
2500	36.621	32.8766	31.4559	254.0235	118.7211
2600	36.8096	33.0242	31.5351	255.3241	122.3927
2700	36.9903	33.1678	31.612	256.5866	126.0827
2800	37.1639	33.3074	31.6867	257.8131	129.7905
2900	37.331	33.4433	31.7593	259.0059	133.5153
3000	37.4925	33.5756	31.8299	260.1667	137.2565
3100	37.6488	33.7045	31.8986	261.2973	141.0136
3200	37.8004	33.8301	31.9656	262.3995	144.7861
3300	37.9478	33.9527	32.0308	263.4745	148.5736
3400	38.0913	34.0723	32.0944	264.524	152.3756
3500	38.2313	34.1891	32.1565	265.549	156.1917
3600	38.3678	34.3033	32.2171	266.5508	160.0217
3700	38.5011	34.415	32.2763	267.5306	163.8652
3800	38.6312	34.5242	32.3341	268.4892	167.7218
3900	38.7582	34.6312	32.3907	269.4277	171.5913
4000	38.8821	34.7359	32.4461	270.347	175.4733
4500	39.4505	35.229	32.7061	274.6812	195.0603
5000	39.9103	35.6752	32.941	278.6349	214.9059

Table B.6 Absolute molar specific enthalpy and entropy of H_2O(liq). The dependency of the enthalpy on the pressure is supposed to be insignificant. Liquid water is treated as an incompressible fluid, i.e. the entropy does not need to be corrected. Reference temperature for averaged heat capacities is $\vartheta_0 = 0\,°C$

ϑ [°C]	C_p $\left[\frac{J}{molK}\right]$	$\overline{C_p}\vert_{\vartheta_0}^{\vartheta}$ $\left[\frac{J}{molK}\right]$	$\overline{\overline{C_p}}\vert_{\vartheta_0}^{\vartheta}$ $\left[\frac{J}{molK}\right]$	S_m $\left[\frac{J}{molK}\right]$	H_m $\left[\frac{kJ}{mol}\right]$
0	76.1739	76.1739	76.1739	63.3318	–287.717
5	75.6231	75.8626	75.8634	64.7079	–287.3377
10	75.3967	75.6767	75.6789	66.0529	–286.9602
15	75.3332	75.5699	75.5733	67.372	–286.5835
20	75.3354	75.5103	75.5145	68.6679	–286.2068
25	75.351	75.4769	75.4814	69.9422	–285.8301
50	75.3143	75.408	75.4134	76.0085	–283.9466
75	75.5108	75.4004	75.4044	81.6259	–282.062
100	75.9749	75.4745	75.4688	86.875	–280.1696
125	76.7169	75.6272	75.6004	91.8186	–278.2635

(continued)

Table B.6 (continued)

ϑ [°C]	C_p $\left[\frac{\text{J}}{\text{molK}}\right]$	$\overline{C_p}\Big\vert_{\vartheta_0}^{\vartheta}$ $\left[\frac{\text{J}}{\text{molK}}\right]$	$\overline{\overline{C_p}}\Big\vert_{\vartheta_0}^{\vartheta}$ $\left[\frac{\text{J}}{\text{molK}}\right]$	S_m $\left[\frac{\text{J}}{\text{molK}}\right]$	H_m $\left[\frac{\text{kJ}}{\text{mol}}\right]$
150	77.7611	75.8979	75.8294	96.5227	–276.3322
175	78.9244	76.2418	76.1158	101.0172	–274.3745
200	80.8047	76.6851	76.4778	105.348	–272.3798
225	83.6512	77.2918	76.9624	109.577	–270.3262
250	87.5755	78.1135	77.6052	113.7633	–268.1885
275	93.3648	79.2164	78.4508	117.975	–265.9324
300	103.521	80.7741	79.6195	122.3399	–263.4846
325	123.3365	83.1999	81.3963	127.1319	–260.6769

Table B.7 Absolute molar specific enthalpy and entropy of H_2O(g) at $p_0 = 1\,\text{bar}$. Vapour is treated as an ideal gas. Reference temperature for averaged heat capacities is $\vartheta_0 = 0\,°\text{C}$

ϑ [°C]	C_p $\left[\frac{\text{J}}{\text{molK}}\right]$	$\overline{C_p}\Big\vert_{\vartheta_0}^{\vartheta}$ $\left[\frac{\text{J}}{\text{molK}}\right]$	$\overline{\overline{C_p}}\Big\vert_{\vartheta_0}^{\vartheta}$ $\left[\frac{\text{J}}{\text{molK}}\right]$	S_m $\left[\frac{\text{J}}{\text{molK}}\right]$	H_m $\left[\frac{\text{kJ}}{\text{mol}}\right]$
–20	33.4422	33.4666	33.4663	183.3473	–243.3338
–15	33.4535	33.4729	33.4727	184.0015	–243.1665
–10	33.4658	33.4796	33.4795	184.6434	–242.9993
–5	33.4794	33.4867	33.4867	185.2734	–242.8319
0	33.4942	33.4942	33.4942	185.892	–242.6645
5	33.5103	33.5021	33.5021	186.4998	–242.4969
10	33.5277	33.5105	33.5104	187.0969	–242.3294
15	33.5464	33.5193	33.5191	187.684	–242.1617
20	33.5664	33.5285	33.5281	188.2613	–241.9939
25	33.5877	33.5382	33.5375	188.8291	–241.826
50	33.7132	33.5931	33.59	191.5384	–240.9848
100	34.0481	33.7326	33.7183	196.4108	–239.2912
150	34.4689	33.906	33.8705	200.7174	–237.5786
200	34.9492	34.1057	34.0393	204.5929	–235.8433
250	35.4702	34.326	34.2191	208.1292	–234.083
300	36.0197	34.5621	34.4063	211.3914	–232.2958
350	36.5904	34.8109	34.5983	214.4275	–230.4806
400	37.1778	35.0699	34.7934	217.2739	–228.6365
450	37.7793	35.3374	34.9907	219.9588	–226.7626
500	38.3934	35.6122	35.1893	222.5048	–224.8584

(continued)

Table B.7 (continued)

ϑ [°C]	C_p $\left[\frac{\text{J}}{\text{molK}}\right]$	$\left.\overline{C_p}\right\|_{\vartheta_0}^{\vartheta}$ $\left[\frac{\text{J}}{\text{molK}}\right]$	$\left.\overline{\overline{C_p}}\right\|_{\vartheta_0}^{\vartheta}$ $\left[\frac{\text{J}}{\text{molK}}\right]$	S_m $\left[\frac{\text{J}}{\text{molK}}\right]$	H_m $\left[\frac{\text{kJ}}{\text{mol}}\right]$
600	39.6536	36.1802	35.589	227.2495	–220.9563
700	40.943	36.7683	35.9898	231.6176	–216.9266
800	42.229	37.3706	36.3891	235.6845	–212.768
900	43.5057	37.9816	36.7847	239.503	–208.481
1000	44.7428	38.5963	37.1745	243.1121	–204.0682
1100	45.9212	39.2091	37.5564	246.5397	–199.5344
1200	47.0322	39.8152	37.9285	249.8067	–194.8862
1300	48.0727	40.4108	38.2895	252.9298	–190.1304
1400	49.0429	40.9932	38.6387	255.9224	–185.274
1500	49.945	41.5603	38.9756	258.7956	–180.324
1600	50.782	42.1108	39.3001	261.5588	–175.2872
1700	51.5574	42.644	39.6124	264.2202	–170.1697
1800	52.275	43.1594	39.9125	266.7869	–164.9776
1900	52.9386	43.6568	40.2009	269.2652	–159.7165
2000	53.5521	44.1364	40.4779	271.6608	–154.3916
2100	54.1191	44.5985	40.744	273.9785	–149.0076
2200	54.6431	45.0433	40.9995	276.2232	–143.5692
2300	55.1276	45.4714	41.2449	278.3988	–138.0803
2400	55.5758	45.8832	41.4807	280.5092	–132.5449
2500	55.9907	46.2793	41.7074	282.558	–126.9663
2600	56.3752	46.6603	41.9253	284.5483	–121.3477
2700	56.732	47.0268	42.135	286.4832	–115.6922
2800	57.0638	47.3794	42.3368	288.3655	–110.0022
2900	57.373	47.7187	42.5311	290.1978	–104.2802
3000	57.6617	48.0454	42.7183	291.9824	–98.5283
3100	57.9322	48.36	42.8989	293.7218	–92.7484
3200	58.1863	48.6632	43.073	295.418	–86.9424
3300	58.4261	48.9554	43.2412	297.0731	–81.1116
3400	58.6531	49.2373	43.4037	298.6889	–75.2576
3500	58.869	49.5094	43.5608	300.2673	–69.3814
3600	59.0751	49.7723	43.7129	301.8099	–63.4841
3700	59.273	50.0264	43.8601	303.3183	–57.5666
3800	59.4637	50.2723	44.0027	304.794	–51.6298
3900	59.6483	50.5103	44.1411	306.2386	–45.6741
4000	59.8279	50.7411	44.2754	307.6532	–39.7003
4500	60.6762	51.7984	44.8933	314.3201	–9.5715
5000	61.4803	52.7267	45.4378	320.4044	20.9689

Table B.8 Absolute molar specific enthalpy and entropy of N_2 at $p_0 = 1$ bar. Reference temperature for averaged heat capacities is $\vartheta_0 = 0\,°C$

ϑ [°C]	C_p $\left[\frac{J}{molK}\right]$	$\overline{C_p}\big\|_{\vartheta_0}^{\vartheta}$ $\left[\frac{J}{molK}\right]$	$\overline{\overline{C_p}}\big\|_{\vartheta_0}^{\vartheta}$ $\left[\frac{J}{molK}\right]$	S_m $\left[\frac{J}{molK}\right]$	H_m $\left[\frac{kJ}{mol}\right]$
–20	29.1108	29.1131	29.1131	186.8458	–1.3103
–15	29.1119	29.1137	29.1137	187.4152	–1.1647
–10	29.1131	29.1143	29.1143	187.9737	–1.0191
–5	29.1143	29.115	29.115	188.5217	–0.87357
0	29.1156	29.1156	29.1156	189.0595	–0.72799
5	29.1171	29.1163	29.1163	189.5877	–0.58241
10	29.1186	29.1171	29.1171	190.1065	–0.43682
15	29.1203	29.1179	29.1179	190.6162	–0.29122
20	29.1222	29.1187	29.1187	191.1172	–0.14562
25	29.1244	29.1196	29.1196	191.6097	0
50	29.1389	29.1253	29.125	193.9553	0.72828
100	29.1971	29.1448	29.1428	198.1509	2.1865
150	29.3061	29.1788	29.1724	201.8285	3.6488
200	29.4711	29.2301	29.2151	205.1101	5.118
250	29.6892	29.2993	29.2706	208.081	6.5968
300	29.9526	29.3857	29.3377	210.8025	8.0877
350	30.2512	29.4876	29.4147	213.3198	9.5927
400	30.5741	29.603	29.4998	215.6668	11.1132
450	30.9108	29.7296	29.5911	217.8693	12.6503
500	31.2519	29.8647	29.6867	219.9471	14.2044
600	31.9186	30.152	29.885	223.7885	17.3632
700	32.5384	30.4493	30.0849	227.2829	20.5865
800	33.1019	30.7463	30.2805	230.4933	23.8691
900	33.6008	31.0363	30.4683	233.4649	27.2047
1000	34.0391	31.3152	30.6465	236.2316	30.5872
1100	34.4229	31.5807	30.8145	238.82	34.0107
1200	34.759	31.8318	30.9721	241.2518	37.4702
1300	35.0537	32.0686	31.1197	243.5444	40.9612
1400	35.313	32.2913	31.2579	245.7128	44.4798
1500	35.5418	32.5005	31.3874	247.7694	48.0227
1600	35.7446	32.697	31.5087	249.725	51.5873
1700	35.925	32.8817	31.6225	251.5888	55.1709
1800	36.0863	33.0553	31.7294	253.3689	58.7716
1900	36.2311	33.2187	31.8301	255.0723	62.3876
2000	36.3616	33.3727	31.9249	256.7052	66.0174
2100	36.48	33.5179	32.0144	258.2732	69.6595
2200	36.5876	33.655	32.0991	259.7812	73.313
2300	36.6861	33.7847	32.1793	261.2334	76.9768

(continued)

Table B.8 (continued)

ϑ [°C]	C_p $\left[\frac{\text{J}}{\text{molK}}\right]$	$\overline{C_p}\big\vert_{\vartheta_0}^{\vartheta}$ $\left[\frac{\text{J}}{\text{molK}}\right]$	$\overline{\overline{C_p}}\big\vert_{\vartheta_0}^{\vartheta}$ $\left[\frac{\text{J}}{\text{molK}}\right]$	S_m $\left[\frac{\text{J}}{\text{molK}}\right]$	H_m $\left[\frac{\text{kJ}}{\text{mol}}\right]$
2400	36.7765	33.9075	32.2554	262.6339	80.6499
2500	36.8599	34.0239	32.3277	263.9861	84.3318
2600	36.9371	34.1345	32.3965	265.2932	88.0217
2700	37.0089	34.2396	32.4621	266.5582	91.7191
2800	37.0759	34.3398	32.5247	267.7836	95.4233
2900	37.1387	34.4352	32.5846	268.9718	99.1341
3000	37.1977	34.5263	32.6418	270.1251	102.8509
3100	37.2535	34.6134	32.6967	271.2454	106.5735
3200	37.3063	34.6967	32.7494	272.3345	110.3015
3300	37.3565	34.7766	32.7999	273.3942	114.0347
3400	37.4044	34.8532	32.8486	274.426	117.7728
3500	37.4503	34.9267	32.8954	275.4313	121.5155
3600	37.4943	34.9974	32.9406	276.4115	125.2628
3700	37.5368	35.0655	32.9841	277.3678	129.0143
3800	37.578	35.1311	33.0262	278.3014	132.7701
3900	37.618	35.1943	33.0669	279.2133	136.5299
4000	37.6571	35.2554	33.1062	280.1045	140.2936
4500	37.8455	35.5328	33.2859	284.2818	159.1695
5000	38.0417	35.7737	33.4425	288.0614	178.1404

Table B.9 Absolute molar specific enthalpy and entropy of N at $p_0 = 1\,\text{bar}$. Reference temperature for averaged heat capacities is $\vartheta_0 = 0\,°\text{C}$

ϑ [°C]	C_p $\left[\frac{\text{J}}{\text{molK}}\right]$	$\overline{C_p}\big\vert_{\vartheta_0}^{\vartheta}$ $\left[\frac{\text{J}}{\text{molK}}\right]$	$\overline{\overline{C_p}}\big\vert_{\vartheta_0}^{\vartheta}$ $\left[\frac{\text{J}}{\text{molK}}\right]$	S_m $\left[\frac{\text{J}}{\text{molK}}\right]$	H_m $\left[\frac{\text{kJ}}{\text{mol}}\right]$
–20	20.7863	20.7863	20.7863	149.9011	471.7446
–15	20.7863	20.7863	20.7863	150.3077	471.8485
–10	20.7863	20.7863	20.7863	150.7064	471.9525
–5	20.7863	20.7863	20.7863	151.0977	472.0564
0	20.7863	20.7863	20.7863	151.4817	472.1603
5	20.7863	20.7863	20.7863	151.8588	472.2643
10	20.7863	20.7863	20.7863	152.2291	472.3682
15	20.7863	20.7863	20.7863	152.5929	472.4721
20	20.7863	20.7863	20.7863	152.9505	472.5761
25	20.7863	20.7863	20.7863	153.3021	472.68
50	20.7863	20.7863	20.7863	154.9758	473.1997
100	20.7863	20.7863	20.7863	157.9662	474.239

(continued)

Table B.9 (continued)

ϑ [°C]	C_p $\left[\frac{J}{molK}\right]$	$\overline{C_p}\Big\|_{\vartheta_0}^{\vartheta}$ $\left[\frac{J}{molK}\right]$	$\overline{\overline{C_p}}\Big\|_{\vartheta_0}^{\vartheta}$ $\left[\frac{J}{molK}\right]$	S_m $\left[\frac{J}{molK}\right]$	H_m $\left[\frac{kJ}{mol}\right]$
150	20.7863	20.7863	20.7863	160.58	475.2783
200	20.7863	20.7863	20.7863	162.9015	476.3176
250	20.7863	20.7863	20.7863	164.9896	477.3569
300	20.7863	20.7863	20.7863	166.887	478.3962
350	20.7863	20.7863	20.7863	168.6255	479.4355
400	20.7863	20.7863	20.7863	170.2298	480.4749
450	20.7863	20.7863	20.7863	171.7191	481.5142
500	20.7863	20.7863	20.7863	173.1088	482.5535
600	20.7863	20.7863	20.7863	175.6372	484.6321
700	20.7863	20.7863	20.7863	177.891	486.7107
800	20.7806	20.786	20.7861	179.924	488.7891
900	20.7799	20.7853	20.7857	181.7754	490.8671
1000	20.7828	20.7849	20.7855	183.4753	492.9452
1100	20.7865	20.7849	20.7854	185.0469	495.0237
1200	20.7895	20.7851	20.7855	186.5082	497.1025
1300	20.7913	20.7855	20.7857	187.8737	499.1815
1400	20.792	20.786	20.7859	189.155	501.2607
1500	20.7917	20.7864	20.7861	190.362	503.3399
1600	20.7911	20.7867	20.7863	191.5027	505.419
1700	20.7907	20.7869	20.7864	192.584	507.4981
1800	20.7912	20.7872	20.7865	193.6119	509.5772
1900	20.7935	20.7874	20.7866	194.5913	511.6564
2000	20.7981	20.7878	20.7868	195.5269	513.736
2100	20.8057	20.7885	20.7871	196.4225	515.8161
2200	20.8172	20.7895	20.7876	197.2814	517.8973
2300	20.833	20.791	20.7882	198.1069	519.9797
2400	20.8539	20.7932	20.7891	198.9015	522.064
2500	20.8804	20.7961	20.7904	199.6679	524.1507
2600	20.913	20.8	20.792	200.4081	526.2403
2700	20.9522	20.8049	20.794	201.1243	528.3335
2800	20.9984	20.8109	20.7964	201.8181	530.431
2900	21.052	20.8183	20.7994	202.4914	532.5334
3000	21.1134	20.8271	20.8029	203.1455	534.6416
3100	21.1827	20.8374	20.8071	203.7819	536.7564
3200	21.2604	20.8494	20.8118	204.4019	538.8785
3300	21.3464	20.8631	20.8172	205.0066	541.0087
3400	21.441	20.8787	20.8233	205.5971	543.148

(continued)

Table B.9 (continued)

ϑ [°C]	C_p $\left[\frac{J}{molK}\right]$	$\overline{C_p}\big\vert_{\vartheta_0}^{\vartheta}$ $\left[\frac{J}{molK}\right]$	$\overline{\overline{C_p}}\big\vert_{\vartheta_0}^{\vartheta}$ $\left[\frac{J}{molK}\right]$	S_m $\left[\frac{J}{molK}\right]$	H_m $\left[\frac{kJ}{mol}\right]$
3500	21.5442	20.8962	20.8302	206.1743	545.2972
3600	21.656	20.9158	20.8378	206.7393	547.4572
3700	21.7764	20.9374	20.8461	207.2929	549.6287
3800	21.9053	20.9612	20.8553	207.8357	551.8127
3900	22.0426	20.9871	20.8652	208.3687	554.0101
4000	22.188	21.0153	20.876	208.8924	556.2215
4500	23.0267	21.1906	20.9421	211.3915	567.5179
5000	24.009	21.4224	21.0282	213.7327	579.2723

Table B.10 Absolute molar specific enthalpy and entropy of NO at $p_0 = 1$bar. Reference temperature for averaged heat capacities is $\vartheta_0 = 0\,°C$

ϑ [°C]	C_p $\left[\frac{J}{molK}\right]$	$\overline{C_p}\big\vert_{\vartheta_0}^{\vartheta}$ $\left[\frac{J}{molK}\right]$	$\overline{\overline{C_p}}\big\vert_{\vartheta_0}^{\vartheta}$ $\left[\frac{J}{molK}\right]$	S_m $\left[\frac{J}{molK}\right]$	H_m $\left[\frac{kJ}{mol}\right]$
–20	30.0404	29.9887	29.9893	205.8492	89.9241
–15	30.0129	29.9761	29.9764	206.4365	90.0742
–10	29.9874	29.9641	29.9643	207.012	90.2242
–5	29.9638	29.9528	29.9529	207.5762	90.3741
0	29.9422	29.9422	29.9422	208.1295	90.5239
5	29.9224	29.9321	29.9322	208.6725	90.6735
10	29.9046	29.9228	29.9229	209.2054	90.8231
15	29.8887	29.914	29.9142	209.7288	90.9726
20	29.8746	29.9059	29.9063	210.2428	91.122
25	29.8624	29.8984	29.899	210.748	91.2713
50	29.8275	29.8699	29.8715	213.1508	92.0173
100	29.8758	29.8547	29.8565	217.4436	93.5093
150	30.0516	29.888	29.8841	221.21	95.0071
200	30.3212	29.961	29.9437	224.5804	96.516
250	30.6553	30.0656	30.0268	227.6424	98.0402
300	31.0295	30.1946	30.1266	230.4572	99.5822
350	31.4236	30.3419	30.2378	233.0687	101.1435
400	31.8215	30.502	30.3561	235.5091	102.7247
450	32.2108	30.6704	30.4782	237.803	104.3255
500	32.5828	30.8432	30.6014	239.9689	105.9455
600	33.2577	31.1907	30.8443	243.9732	109.2383
700	33.8438	31.5287	31.0753	247.6113	112.594
800	34.3539	31.851	31.2917	250.947	116.0047
900	34.7822	32.1536	31.4922	254.0271	119.4621

(continued)

Table B.10 (continued)

ϑ [°C]	C_p $\left[\frac{\mathrm{J}}{\mathrm{molK}}\right]$	$\overline{C_p}\Big\vert_{\vartheta_0}^{\vartheta}$ $\left[\frac{\mathrm{J}}{\mathrm{molK}}\right]$	$\overline{\overline{C_p}}\Big\vert_{\vartheta_0}^{\vartheta}$ $\left[\frac{\mathrm{J}}{\mathrm{molK}}\right]$	S_m $\left[\frac{\mathrm{J}}{\mathrm{molK}}\right]$	H_m $\left[\frac{\mathrm{kJ}}{\mathrm{mol}}\right]$
1000	35.1458	32.4351	31.6768	256.8874	122.9589
1100	35.4568	32.696	31.8466	259.5568	126.4895
1200	35.7246	32.9375	32.0029	262.0588	130.0489
1300	35.9571	33.161	32.1469	264.4128	133.6332
1400	36.1602	33.3682	32.28	266.6351	137.2393
1500	36.339	33.5604	32.4032	268.7394	140.8644
1600	36.4974	33.7391	32.5176	270.7375	144.5064
1700	36.6386	33.9056	32.6242	272.6394	148.1633
1800	36.7654	34.061	32.7236	274.4539	151.8337
1900	36.88	34.2064	32.8168	276.1886	155.516
2000	36.9841	34.3427	32.9042	277.8502	159.2093
2100	37.0794	34.4708	32.9864	279.4445	162.9125
2200	37.167	34.5914	33.0639	280.9767	166.6249
2300	37.2481	34.7052	33.1371	282.4516	170.3457
2400	37.3237	34.8127	33.2064	283.8732	174.0744
2500	37.3944	34.9146	33.2722	285.2452	177.8103
2600	37.461	35.0113	33.3348	286.5711	181.5531
2700	37.524	35.1032	33.3944	287.8539	185.3024
2800	37.584	35.1907	33.4512	289.0962	189.0578
2900	37.6414	35.2742	33.5056	290.3006	192.8191
3000	37.6967	35.3541	33.5576	291.4694	196.586
3100	37.7501	35.4305	33.6075	292.6047	200.3584
3200	37.802	35.5038	33.6554	293.7083	204.136
3300	37.8528	35.5742	33.7014	294.782	207.9187
3400	37.9027	35.642	33.7458	295.8275	211.7065
3500	37.952	35.7073	33.7886	296.8463	215.4993
3600	38.001	35.7703	33.8299	297.8397	219.2969
3700	38.0499	35.8312	33.8698	298.809	223.0995
3800	38.0989	35.8903	33.9085	299.7554	226.9069
3900	38.1484	35.9475	33.946	300.6801	230.7192
4000	38.1985	36.0032	33.9824	301.584	234.5366
4500	38.4681	36.2617	34.1505	305.8252	253.7016
5000	38.7953	36.4981	34.301	309.6728	273.0143

Table B.11 Absolute molar specific enthalpy and entropy of NO_2 at $p_0 = 1\,\text{bar}$. Reference temperature for averaged heat capacities is $\vartheta_0 = 0\,°\text{C}$

ϑ [°C]	C_p $\left[\frac{\text{J}}{\text{molK}}\right]$	$\overline{C_p}\Big\vert_{\vartheta_0}^{\vartheta}$ $\left[\frac{\text{J}}{\text{molK}}\right]$	$\overline{\overline{C_p}}\Big\vert_{\vartheta_0}^{\vartheta}$ $\left[\frac{\text{J}}{\text{molK}}\right]$	S_m $\left[\frac{\text{J}}{\text{molK}}\right]$	H_m $\left[\frac{\text{kJ}}{\text{mol}}\right]$
–20	35.8377	36.1226	36.119	234.203	32.5508
–15	35.9783	36.1942	36.1922	234.9053	32.7303
–10	36.1212	36.2666	36.2657	235.5969	32.9106
–5	36.2662	36.3396	36.3394	236.2781	33.0915
0	36.4134	36.4134	36.4134	236.9495	33.2732
5	36.5626	36.4878	36.4876	237.6113	33.4557
10	36.7137	36.5629	36.562	238.2641	33.6389
15	36.8665	36.6386	36.6366	238.908	33.8228
20	37.021	36.7148	36.7113	239.5436	34.0075
25	37.177	36.7917	36.7861	240.171	34.193
50	37.9762	37.183	37.1611	243.1961	35.1324
100	39.6257	37.9905	37.9068	248.7749	37.0723
150	41.2704	38.8108	38.6329	253.8593	39.0948
200	42.8502	39.625	39.3281	258.556	41.1982
250	44.3302	40.4199	39.9865	262.9346	43.3782
300	45.693	41.1868	40.6054	267.0432	45.6293
350	46.9326	41.9206	41.1843	270.9169	47.9454
400	48.0507	42.6182	41.7243	274.5825	50.3205
450	49.0537	43.2786	42.227	278.0615	52.7486
500	49.9506	43.9018	42.6947	281.3712	55.2241
600	51.469	45.0411	43.535	287.5408	60.2979
700	52.6931	46.0498	44.2652	293.1891	65.5081
800	53.6886	46.9446	44.9039	298.3929	70.8289
900	54.508	47.7409	45.4664	303.2133	76.24
1000	55.1909	48.4528	45.9653	307.7006	81.726
1100	55.7674	49.0922	46.411	311.8958	87.2747
1200	56.2615	49.6696	46.8117	315.8336	92.8767
1300	56.6925	50.1936	47.1744	319.543	98.5249
1400	57.0757	50.6717	47.5046	323.0487	104.2136
1500	57.4232	51.1104	47.807	326.3721	109.9388
1600	57.7445	51.5151	48.0856	329.5314	115.6974
1700	58.0474	51.8905	48.3437	332.5425	121.4871
1800	58.3379	52.2407	48.5839	335.4195	127.3065
1900	58.6207	52.5691	48.8086	338.1743	133.1544
2000	58.8994	52.8786	49.0199	340.8178	139.0305

(continued)

Table B.11 (continued)

ϑ [°C]	C_p $\left[\frac{\mathrm{J}}{\mathrm{molK}}\right]$	$\overline{C_p}\Big\vert_{\vartheta_0}^{\vartheta}$ $\left[\frac{\mathrm{J}}{\mathrm{molK}}\right]$	$\overline{\overline{C_p}}\Big\vert_{\vartheta_0}^{\vartheta}$ $\left[\frac{\mathrm{J}}{\mathrm{molK}}\right]$	S_m $\left[\frac{\mathrm{J}}{\mathrm{molK}}\right]$	H_m $\left[\frac{\mathrm{kJ}}{\mathrm{mol}}\right]$
2100	59.1771	53.1719	49.2194	343.3595	144.9343
2200	59.4559	53.4512	49.4085	345.8077	150.8659
2300	59.7374	53.7184	49.5886	348.1699	156.8256
2400	60.0228	53.9751	49.7606	350.4529	162.8135
2500	60.313	54.2228	49.9255	352.6626	168.8303
2600	60.6085	54.4627	50.084	354.8044	174.8763
2700	60.9096	54.6959	50.237	356.8831	180.9522
2800	61.2164	54.9233	50.385	358.9031	187.0584
2900	61.5288	55.1457	50.5284	360.8683	193.1956
3000	61.8466	55.3637	50.6678	362.7823	199.3644
3100	62.1696	55.578	50.8036	364.6483	205.5651
3200	62.4973	55.7891	50.936	366.4694	211.7984
3300	62.8293	55.9974	51.0655	368.2481	218.0647
3400	63.1652	56.2033	51.1922	369.9869	224.3644
3500	63.5044	56.407	51.3164	371.6881	230.6979
3600	63.8464	56.6089	51.4383	373.3537	237.0654
3700	64.1907	56.8092	51.5581	374.9856	243.4672
3800	64.5367	57.008	51.6759	376.5855	249.9036
3900	64.8838	57.2055	51.7918	378.1549	256.3746
4000	65.2317	57.4018	51.906	379.6955	262.8804
4500	66.9647	58.3683	52.4545	387.0081	295.9305
5000	68.6532	59.3129	52.9709	393.7624	329.8375

Table B.12 Absolute molar specific enthalpy and entropy of CO at $p_0 = 1\,\mathrm{bar}$. Reference temperature for averaged heat capacities is $\vartheta_0 = 0\,°\mathrm{C}$

ϑ [°C]	C_p $\left[\frac{\mathrm{J}}{\mathrm{molK}}\right]$	$\overline{C_p}\Big\vert_{\vartheta_0}^{\vartheta}$ $\left[\frac{\mathrm{J}}{\mathrm{molK}}\right]$	$\overline{\overline{C_p}}\Big\vert_{\vartheta_0}^{\vartheta}$ $\left[\frac{\mathrm{J}}{\mathrm{molK}}\right]$	S_m $\left[\frac{\mathrm{J}}{\mathrm{molK}}\right]$	H_m $\left[\frac{\mathrm{kJ}}{\mathrm{mol}}\right]$
–20	29.1201	29.123	29.123	192.894	–111.846
–15	29.1214	29.1238	29.1238	193.4636	–111.7004
–10	29.1229	29.1246	29.1246	194.0222	–111.5548
–5	29.1246	29.1256	29.1256	194.5704	–111.4092
0	29.1266	29.1266	29.1266	195.1085	–111.2635
5	29.1288	29.1277	29.1276	195.6368	–111.1179
10	29.1314	29.1289	29.1288	196.1558	–110.9722
15	29.1343	29.1302	29.1301	196.6658	–110.8266
20	29.1376	29.1316	29.1315	197.167	–110.6809
25	29.1414	29.1332	29.1331	197.6598	–110.5352

(continued)

Table B.12 (continued)

ϑ [°C]	C_p $\left[\frac{\text{J}}{\text{molK}}\right]$	$\overline{C_p}\big\vert_{\vartheta_0}^{\vartheta}$ $\left[\frac{\text{J}}{\text{molK}}\right]$	$\overline{\overline{C_p}}\big\vert_{\vartheta_0}^{\vartheta}$ $\left[\frac{\text{J}}{\text{molK}}\right]$	S_m $\left[\frac{\text{J}}{\text{molK}}\right]$	H_m $\left[\frac{\text{kJ}}{\text{mol}}\right]$
50	29.1673	29.1432	29.1427	200.0072	−109.8064
100	29.263	29.1766	29.1731	204.2093	−108.3459
150	29.4243	29.2304	29.22	207.8982	−106.879
200	29.6494	29.3057	29.283	211.1963	−105.4024
250	29.9295	29.4017	29.3603	214.1882	−103.9131
300	30.2521	29.516	29.4496	216.9343	−102.4087
350	30.6036	29.646	29.5483	219.4789	−100.8874
400	30.971	29.7886	29.6541	221.8549	−99.3481
450	31.3425	29.9406	29.7645	224.0871	−97.7903
500	31.7085	30.0992	29.8776	226.1947	−96.2139
600	32.3986	30.4258	30.1052	230.0933	−93.008
700	33.0206	30.7528	30.3274	233.64	−89.7366
800	33.5746	31.0718	30.5399	236.8973	−86.4061
900	34.0532	31.3771	30.7402	239.9101	−83.0241
1000	34.4673	31.6659	30.9274	242.7128	−79.5976
1100	34.8258	31.9373	31.1016	245.3327	−76.1325
1200	35.1369	32.1913	31.2636	247.7919	−72.634
1300	35.408	32.4285	31.414	250.1086	−69.1064
1400	35.6451	32.65	31.5539	252.2981	−65.5535
1500	35.8535	32.8568	31.6841	254.3734	−61.9784
1600	36.0377	33.0499	31.8056	256.3455	−58.3836
1700	36.2012	33.2306	31.9191	258.2242	−54.7715
1800	36.3472	33.3997	32.0253	260.0175	−51.144
1900	36.4783	33.5584	32.125	261.7329	−47.5026
2000	36.5965	33.7074	32.2187	263.3767	−43.8487
2100	36.7038	33.8476	32.307	264.9546	−40.1836
2200	36.8017	33.9796	32.3902	266.4715	−36.5083
2300	36.8913	34.1043	32.469	267.9321	−32.8236
2400	36.974	34.2222	32.5436	269.3402	−29.1303
2500	37.0504	34.3338	32.6144	270.6995	−25.429
2600	37.1215	34.4397	32.6818	272.0133	−21.7204
2700	37.1879	34.5402	32.7459	273.2845	−18.0049
2800	37.2502	34.6359	32.807	274.5158	−14.2829
2900	37.3088	34.7271	32.8654	275.7095	−10.5549
3000	37.3642	34.8141	32.9213	276.868	−6.8213
3100	37.4167	34.8972	32.9748	277.9932	−3.0822
3200	37.4667	34.9767	33.0261	279.0871	0.66199
3300	37.5144	35.0529	33.0754	280.1513	4.4111

(continued)

Table B.12 (continued)

ϑ [°C]	C_p $\left[\frac{\text{J}}{\text{molK}}\right]$	$\overline{C_p}\Big\vert_{\vartheta_0}^{\vartheta}$ $\left[\frac{\text{J}}{\text{molK}}\right]$	$\overline{\overline{C_p}}\Big\vert_{\vartheta_0}^{\vartheta}$ $\left[\frac{\text{J}}{\text{molK}}\right]$	S_{m} $\left[\frac{\text{J}}{\text{molK}}\right]$	H_{m} $\left[\frac{\text{kJ}}{\text{mol}}\right]$
3400	37.5601	35.126	33.1228	281.1874	8.1648
3500	37.604	35.1962	33.1684	282.1969	11.923
3600	37.6462	35.2636	33.2123	283.1811	15.6855
3700	37.6871	35.3286	33.2548	284.1412	19.4522
3800	37.7267	35.3912	33.2957	285.0785	23.2229
3900	37.7652	35.4515	33.3353	285.994	26.9975
4000	37.8027	35.5099	33.3736	286.8888	30.7759
4500	37.9822	35.7747	33.5484	291.0817	49.7225
5000	38.1628	36.0044	33.7005	294.8742	68.7582

Table B.13 Absolute molar specific enthalpy and entropy of CO_2 at $p_0 = 1\,\text{bar}$. Reference temperature for averaged heat capacities is $\vartheta_0 = 0\,°\text{C}$

ϑ [°C]	C_p $\left[\frac{\text{J}}{\text{molK}}\right]$	$\overline{C_p}\Big\vert_{\vartheta_0}^{\vartheta}$ $\left[\frac{\text{J}}{\text{molK}}\right]$	$\overline{\overline{C_p}}\Big\vert_{\vartheta_0}^{\vartheta}$ $\left[\frac{\text{J}}{\text{molK}}\right]$	S_{m} $\left[\frac{\text{J}}{\text{molK}}\right]$	H_{m} $\left[\frac{\text{kJ}}{\text{mol}}\right]$
–20	34.984	35.4743	35.4682	207.8901	–395.1333
–15	35.2305	35.5967	35.5932	208.5767	–394.9578
–10	35.4755	35.7184	35.7169	209.2549	–394.781
–5	35.7188	35.8396	35.8392	209.9249	–394.603
0	35.9601	35.9601	35.9601	210.5871	–394.4238
5	36.1995	36.08	36.0797	211.2415	–394.2434
10	36.4368	36.1992	36.1978	211.8886	–394.0618
15	36.6719	36.3177	36.3145	212.5284	–393.879
20	36.9048	36.4354	36.4298	213.1613	–393.6951
25	37.1354	36.5524	36.5438	213.7874	–393.51
50	38.2528	37.1257	37.0936	216.8223	–392.5675
100	40.3113	38.2135	38.1003	222.4728	–390.6025
150	42.1524	39.2252	38.9993	227.6573	–388.54
200	43.8051	40.1672	39.8081	232.4573	–386.3904
250	45.2963	41.0464	40.5412	236.9326	–384.1622
300	46.648	41.8692	41.2102	241.129	–381.8631
350	47.8777	42.6411	41.8241	245.0821	–379.4994
400	48.9995	43.3668	42.3902	248.8208	–377.0771
450	50.0246	44.0505	42.9144	252.3683	–374.6011
500	50.9622	44.6955	43.4015	255.7443	–372.0761
600	52.6051	45.8809	44.2797	262.0439	–366.8953
700	53.9797	46.9426	45.0495	267.8233	–361.564

(continued)

Table B.13 (continued)

ϑ [°C]	C_p $\left[\frac{\mathrm{J}}{\mathrm{molK}}\right]$	$\overline{C_p}\Big\|_{\vartheta_0}^{\vartheta}$ $\left[\frac{\mathrm{J}}{\mathrm{molK}}\right]$	$\overline{\overline{C_p}}\Big\|_{\vartheta_0}^{\vartheta}$ $\left[\frac{\mathrm{J}}{\mathrm{molK}}\right]$	S_{m} $\left[\frac{\mathrm{J}}{\mathrm{molK}}\right]$	H_{m} $\left[\frac{\mathrm{kJ}}{\mathrm{mol}}\right]$
800	55.1417	47.8968	45.7299	273.1608	–356.1063
900	56.1138	48.7574	46.3354	278.1175	–350.5421
1000	56.9323	49.5351	46.8771	282.7417	–344.8887
1100	57.6289	50.2401	47.3645	287.0732	–339.1597
1200	58.2275	50.8814	47.8053	291.1456	–333.3662
1300	58.7459	51.4669	48.2061	294.987	–327.5169
1400	59.1981	52.0033	48.5723	298.6215	–321.6192
1500	59.5951	52.4965	48.9083	302.0696	–315.6791
1600	59.9457	52.9513	49.2179	305.3489	–309.7017
1700	60.257	53.3721	49.5042	308.4749	–303.6913
1800	60.5347	53.7624	49.7699	311.4608	–297.6514
1900	60.7837	54.1255	50.0172	314.3184	–291.5853
2000	61.0081	54.4641	50.2482	317.0581	–285.4955
2100	61.2112	54.7807	50.4645	319.689	–279.3844
2200	61.396	55.0772	50.6676	322.2193	–273.2539
2300	61.5651	55.3557	50.8587	324.6563	–267.1057
2400	61.7205	55.6177	51.039	327.0066	–260.9413
2500	61.8643	55.8647	51.2094	329.276	–254.762
2600	61.9982	56.0981	51.3708	331.4699	–248.5688
2700	62.1237	56.319	51.524	333.5932	–242.3626
2800	62.2423	56.5284	51.6697	335.6503	–236.1443
2900	62.3551	56.7274	51.8085	337.6452	–229.9144
3000	62.4633	56.9168	51.9409	339.5817	–223.6734
3100	62.5681	57.0974	52.0675	341.463	–217.4218
3200	62.6703	57.27	52.1888	343.2924	–211.1599
3300	62.771	57.4352	52.305	345.0728	–204.8878
3400	62.8711	57.5936	52.4167	346.8068	–198.6057
3500	62.9713	57.7458	52.5242	348.4969	–192.3136
3600	63.0725	57.8923	52.6277	350.1454	–186.0114
3700	63.1755	58.0337	52.7277	351.7545	–179.699
3800	63.281	58.1704	52.8243	353.3262	–173.3762
3900	63.3898	58.3028	52.9178	354.8623	–167.0427
4000	63.5025	58.4314	53.0084	356.3647	–160.6981
4500	64.15	59.0294	53.4261	363.4253	–128.7917
5000	64.9992	59.5819	53.8004	369.8556	–96.5145

Table B.14 Absolute molar specific enthalpy and entropy of $C_{Graphite}$ The dependency of the enthalpy on the pressure is supposed to be insignificant. Graphite is treated as an incompressible solid, i.e. the entropy does not need to be corrected. Reference temperature for averaged heat capacities is $\vartheta_0 = 0\,°C$

ϑ [°C]	C $\left[\frac{J}{molK}\right]$	$\overline{C}\Big\|_{\vartheta_0}^{\vartheta}$ $\left[\frac{J}{molK}\right]$	$\overline{\overline{C}}\Big\|_{\vartheta_0}^{\vartheta}$ $\left[\frac{J}{molK}\right]$	S_m $\left[\frac{J}{molK}\right]$	H_m $\left[\frac{kJ}{mol}\right]$
–20	6.9366	7.2972	7.2927	4.4712	–0.34829
–15	7.1178	7.3872	7.3847	4.6086	–0.31315
–10	7.2979	7.4768	7.4757	4.7469	–0.27711
–5	7.477	7.5661	7.5658	4.886	–0.24018
0	7.655	7.655	7.655	5.0257	–0.20235
5	7.8318	7.7435	7.7432	5.1662	–0.16363
10	8.0075	7.8316	7.8306	5.3073	–0.12403
15	8.1821	7.9194	7.917	5.449	–0.083554
20	8.3556	8.0068	8.0027	5.5912	–0.042209
25	8.528	8.0938	8.0874	5.734	9.9811e-09
50	9.3742	8.5235	8.4995	6.4545	0.22383
100	10.9941	9.3578	9.271	7.9179	0.73343
150	12.5159	10.1598	9.9826	9.3952	1.3216
200	13.9199	10.927	10.6398	10.8712	1.9831
250	15.1829	11.6544	11.2448	12.3331	2.7113
300	16.2953	12.3372	11.7988	13.7701	3.4988
350	17.271	12.9739	12.3047	15.1742	4.3385
400	18.1227	13.5655	12.7664	16.5404	5.2238
450	18.865	14.114	13.1882	17.8657	6.149
500	19.5133	14.6222	13.5741	19.1489	7.1088
600	20.5818	15.5302	14.2529	21.5888	9.1158
700	21.4178	16.3139	14.8294	23.8668	11.2174
800	22.086	16.9952	15.3247	25.9951	13.3938
900	22.6321	17.5921	15.755	27.9876	15.6306
1000	23.0882	18.1196	16.1328	29.8578	17.9172
1100	23.4768	18.5894	16.4677	31.6185	20.246
1200	23.8139	19.011	16.7672	33.2808	22.6109
1300	24.1109	19.3921	17.0372	34.8546	25.0074
1400	24.3761	19.7388	17.2822	36.3488	27.432
1500	24.6157	20.0561	17.5061	37.7708	29.8818
1600	24.8347	20.348	17.7118	39.1273	32.3544
1700	25.0371	20.6179	17.9019	40.4243	34.8482
1800	25.2259	20.8688	18.0782	41.6667	37.3614
1900	25.4034	21.1028	18.2426	42.8593	39.893
2000	25.5714	21.3221	18.3964	44.0059	42.4418

(continued)

Table B.14 (continued)

ϑ [°C]	C $\left[\frac{J}{molK}\right]$	$\overline{C}\Big\vert_{\vartheta_0}^{\vartheta}$ $\left[\frac{J}{molK}\right]$	$\overline{\overline{C}}\Big\vert_{\vartheta_0}^{\vartheta}$ $\left[\frac{J}{molK}\right]$	S_m $\left[\frac{J}{molK}\right]$	H_m $\left[\frac{kJ}{mol}\right]$
2100	25.7314	21.5283	18.5409	45.1103	45.007
2200	25.8843	21.7228	18.677	46.1755	47.5878
2300	26.0313	21.907	18.8057	47.2044	50.1837
2400	26.1732	22.0818	18.9277	48.1996	52.7939
2500	26.3106	22.2482	19.0436	49.1634	55.4181
2600	26.4442	22.407	19.154	50.0978	58.0559
2700	26.5745	22.559	19.2594	51.0047	60.7069
2800	26.7018	22.7047	19.3602	51.886	63.3707
2900	26.8266	22.8447	19.4569	52.743	66.0472
3000	26.9492	22.9794	19.5497	53.5773	68.736
3100	27.0698	23.1094	19.639	54.3901	71.4369
3200	27.1887	23.2351	19.7251	55.1827	74.1499
3300	27.3061	23.3567	19.8081	55.9561	76.8746
3400	27.4222	23.4745	19.8884	56.7114	79.611
3500	27.5371	23.589	19.9661	57.4495	82.359
3600	27.6509	23.7002	20.0413	58.1713	85.1184
3700	27.7638	23.8085	20.1143	58.8776	87.8892
3800	27.8759	23.9141	20.1852	59.5691	90.6712
3900	27.9873	24.0171	20.2541	60.2466	93.4643
4000	28.098	24.1177	20.3211	60.9106	96.2686
4500	28.6444	24.5904	20.6324	64.0496	110.4546
5000	29.1847	25.0229	20.9109	66.9297	124.912

Table B.15 Absolute molar specific enthalpy and entropy of S at $p_0 = 1\,\text{bar}$. Reference temperature for averaged heat capacities is $\vartheta_0 = 0\,°\text{C}$

ϑ [°C]	C_p $\left[\frac{J}{molK}\right]$	$\overline{C_p}\Big\vert_{\vartheta_0}^{\vartheta}$ $\left[\frac{J}{molK}\right]$	$\overline{\overline{C_p}}\Big\vert_{\vartheta_0}^{\vartheta}$ $\left[\frac{J}{molK}\right]$	S_m $\left[\frac{J}{molK}\right]$	H_m $\left[\frac{kJ}{mol}\right]$
–20	23.7146	23.7203	23.7203	163.9527	276.1031
–15	23.7199	23.7213	23.7213	164.4166	276.2217
–10	23.7222	23.7213	23.7213	164.8717	276.3403
–5	23.7217	23.7205	23.7205	165.3182	276.4589
0	23.7188	23.7188	23.7188	165.7564	276.5775
5	23.7136	23.7164	23.7164	166.1866	276.6961
10	23.7063	23.7133	23.7133	166.609	276.8146
15	23.6971	23.7095	23.7096	167.0239	276.9331
20	23.6861	23.705	23.7052	167.4315	277.0516
25	23.6736	23.7	23.7004	167.832	277.17

(continued)

Table B.15 (continued)

ϑ [°C]	C_p $\left[\frac{J}{molK}\right]$	$\overline{C_p}\|_{\vartheta_0}^{\vartheta}$ $\left[\frac{J}{molK}\right]$	$\overline{\overline{C_p}}\|_{\vartheta_0}^{\vartheta}$ $\left[\frac{J}{molK}\right]$	S_m $\left[\frac{J}{molK}\right]$	H_m $\left[\frac{kJ}{mol}\right]$
50	23.5911	23.6674	23.6692	169.7351	277.7609
100	23.367	23.5754	23.5848	173.1139	278.935
150	23.1147	23.4641	23.4869	176.0367	280.0971
200	22.8671	23.3455	23.3863	178.6046	281.2466
250	22.639	23.2266	23.2884	180.8903	282.3842
300	22.436	23.1114	23.1958	182.9474	283.5109
350	22.2586	23.0019	23.1097	184.8165	284.6282
400	22.105	22.8992	23.0302	186.5284	285.7372
450	21.9721	22.8034	22.9572	188.1074	286.839
500	21.8566	22.7144	22.8901	189.5725	287.9347
600	21.6668	22.5549	22.7718	192.2192	290.1104
700	21.5213	22.4171	22.6711	194.5604	292.2695
800	21.4222	22.2986	22.5852	196.6605	294.4164
900	21.3365	22.1964	22.5115	198.5652	296.5542
1000	21.2665	22.1067	22.4471	200.3077	298.6842
1100	21.2155	22.0279	22.3906	201.9137	300.8082
1200	21.184	21.9587	22.3409	203.4038	302.928
1300	21.1708	21.8985	22.2972	204.7946	305.0456
1400	21.1743	21.8466	22.2589	206.0993	307.1627
1500	21.1925	21.8023	22.2255	207.3289	309.2809
1600	21.2235	21.7651	22.1964	208.4924	311.4016
1700	21.2654	21.7344	22.1714	209.5973	313.526
1800	21.3165	21.7097	22.1499	210.6498	315.655
1900	21.3752	21.6905	22.1316	211.6554	317.7895
2000	21.44	21.6764	22.1162	212.6184	319.9303
2100	21.5098	21.6668	22.1034	213.5429	322.0777
2200	21.5833	21.6613	22.093	214.4322	324.2323
2300	21.6596	21.6595	22.0846	215.2893	326.3945
2400	21.7377	21.6612	22.0782	216.1165	328.5643
2500	21.8169	21.6658	22.0734	216.9163	330.742
2600	21.8965	21.6732	22.0701	217.6906	332.9277
2700	21.9758	21.6829	22.0682	218.4411	335.1213
2800	22.0543	21.6948	22.0675	219.1694	337.3229
2900	22.1316	21.7085	22.0678	219.8768	339.5322
3000	22.2072	21.7239	22.0691	220.5647	341.7491
3100	22.2809	21.7407	22.0712	221.2341	343.9735
3200	22.3522	21.7587	22.074	221.8861	346.2052
3300	22.4209	21.7777	22.0775	222.5215	348.4439

(continued)

Table B.15 (continued)

ϑ [°C]	C_p $\left[\frac{J}{molK}\right]$	$\overline{C_p}\Big\|_{\vartheta_0}^{\vartheta}$ $\left[\frac{J}{molK}\right]$	$\overline{\overline{C_p}}\Big\|_{\vartheta_0}^{\vartheta}$ $\left[\frac{J}{molK}\right]$	S_m $\left[\frac{J}{molK}\right]$	H_m $\left[\frac{kJ}{mol}\right]$
3400	22.487	21.7976	22.0815	223.1413	350.6893
3500	22.5501	21.8182	22.0859	223.7462	352.9412
3600	22.6102	21.8394	22.0908	224.3368	355.1992
3700	22.6673	21.861	22.096	224.9139	357.4631
3800	22.7212	21.8829	22.1015	225.4781	359.7326
3900	22.7719	21.9051	22.1073	226.0298	362.0073
4000	22.8195	21.9273	22.1132	226.5696	364.2869
4500	23.0131	22.0379	22.1444	229.1058	375.7479
5000	23.1466	22.1425	22.176	231.4054	387.2898

Table B.16 Absolute molar specific enthalpy and entropy of S(a) The dependency of the enthalpy on the pressure is supposed to be insignificant. Sulphur(a) is treated as an incompressible solid, i.e. the entropy does not need to be corrected. Reference temperature for averaged heat capacities is $\vartheta_0 = 0\,°C$

ϑ [°C]	C $\left[\frac{J}{molK}\right]$	$\overline{C}\Big\|_{\vartheta_0}^{\vartheta}$ $\left[\frac{J}{molK}\right]$	$\overline{\overline{C}}\Big\|_{\vartheta_0}^{\vartheta}$ $\left[\frac{J}{molK}\right]$	S_m $\left[\frac{J}{molK}\right]$	H_m $\left[\frac{kJ}{mol}\right]$
–40	20.7119	21.384	21.3669	26.7294	–1.4143
–35	20.8907	21.4672	21.4544	27.1708	–1.3102
–30	21.0641	21.5487	21.5396	27.6067	–1.2054
–25	21.2325	21.6287	21.6225	28.0371	–1.0996
–20	21.3961	21.7073	21.7034	28.4623	–0.99304
–15	21.5551	21.7843	21.7822	28.8824	–0.88566
–10	21.71	21.8601	21.8591	29.2973	–0.77749
–5	21.8607	21.9345	21.9343	29.7074	–0.66856
0	22.0076	22.0076	22.0076	30.1126	–0.55889
5	22.1508	22.0795	22.0793	30.5131	–0.44849
10	22.2906	22.1503	22.1494	30.909	–0.33739
15	22.4269	22.2198	22.218	31.3004	–0.22559
20	22.56	22.2883	22.2851	31.6874	–0.11313
25	22.69	22.3557	22.3507	32.07	0
50	23.2971	22.6775	22.6594	33.9216	0.57498
75	23.8403	22.9762	22.9392	35.678	1.1643
95.15	24.2371	23.2018	23.1464	37.0305	1.6488

Table B.17 Absolute molar specific enthalpy and entropy of S(b) The dependency of the enthalpy on the pressure is supposed to be insignificant. Sulphur(b) is treated as an incompressible solid, i.e. the entropy does not need to be corrected

ϑ [°C]	C $\left[\frac{\mathrm{J}}{\mathrm{molK}}\right]$	S_m $\left[\frac{\mathrm{J}}{\mathrm{molK}}\right]$	H_m $\left[\frac{\mathrm{kJ}}{\mathrm{mol}}\right]$
95.16	24.7732	38.12	2.05
100	24.8714	38.4441	2.1701
105	24.9729	38.7758	2.2948
110	25.0744	39.1045	2.4199
115.21	25.1801	39.4439	2.5508

Table B.18 Absolute molar specific enthalpy and entropy of S(liq) The dependency of the enthalpy on the pressure is supposed to be insignificant. Sulphur(liq) is treated as an incompressible liquid, i.e. the entropy does not need to be corrected

ϑ [°C]	C $\left[\frac{\mathrm{J}}{\mathrm{molK}}\right]$	S_m $\left[\frac{\mathrm{J}}{\mathrm{molK}}\right]$	H_m $\left[\frac{\mathrm{kJ}}{\mathrm{mol}}\right]$
115.22	31.7109	43.8761	4.2721
150	34.8325	46.6847	5.4116
200	40.222	51.4219	7.5322
250	36.6963	55.2677	9.4449
300	34.9499	58.5299	11.2311
350	33.9564	61.4095	12.9522
400	33.0559	63.997	14.6282

Table B.19 Absolute molar specific enthalpy and entropy of S_2 at $p_0 = 1\,\mathrm{bar}$. Reference temperature for averaged heat capacities is $\vartheta_0 = 0\,°\mathrm{C}$

ϑ [°C]	C_p $\left[\frac{\mathrm{J}}{\mathrm{molK}}\right]$	$\overline{C_p}\big\vert_{\vartheta_0}^{\vartheta}$ $\left[\frac{\mathrm{J}}{\mathrm{molK}}\right]$	$\overline{\overline{C_p}}\big\vert_{\vartheta_0}^{\vartheta}$ $\left[\frac{\mathrm{J}}{\mathrm{molK}}\right]$	S_m $\left[\frac{\mathrm{J}}{\mathrm{molK}}\right]$	H_m $\left[\frac{\mathrm{kJ}}{\mathrm{mol}}\right]$
−20	31.5876	31.7999	31.7972	222.9236	127.1575
−15	31.6946	31.8528	31.8513	223.5425	127.3157
−10	31.8006	31.9053	31.9047	224.1515	127.4744
−5	31.9055	31.9574	31.9573	224.751	127.6337
0	32.0091	32.0091	32.0091	225.3414	127.7935
5	32.1114	32.0604	32.0602	225.923	127.9538
10	32.2122	32.1112	32.1106	226.496	128.1146
15	32.3115	32.1614	32.1601	227.0607	128.2759
20	32.4093	32.2112	32.2089	227.6174	128.4377
25	32.5054	32.2605	32.2568	228.1664	128.6
50	32.9611	32.4986	32.4853	230.8021	129.4184
100	33.7483	32.9334	32.8881	235.6012	131.0868

(continued)

Table B.19 (continued)

ϑ [°C]	C_p $\left[\frac{\text{J}}{\text{molK}}\right]$	$\overline{C_p}\Big\vert_{\vartheta_0}^{\vartheta}$ $\left[\frac{\text{J}}{\text{molK}}\right]$	$\overline{\overline{C_p}}\Big\vert_{\vartheta_0}^{\vartheta}$ $\left[\frac{\text{J}}{\text{molK}}\right]$	S_{m} $\left[\frac{\text{J}}{\text{molK}}\right]$	H_{m} $\left[\frac{\text{kJ}}{\text{mol}}\right]$
150	34.3854	33.315	33.2281	239.8856	132.7907
200	34.8992	33.6491	33.5165	243.7551	134.5233
250	35.3156	33.9422	33.763	247.2822	136.279
300	35.6571	34.2005	33.9756	250.5216	138.0536
350	35.9424	34.4295	34.1608	253.5161	139.8438
400	36.1867	34.6342	34.3238	256.2997	141.6472
450	36.4019	34.8189	34.4688	258.9002	143.462
500	36.5973	34.9871	34.5994	261.3404	145.287
600	36.9547	35.2857	34.8271	265.8135	148.9649
700	37.2901	35.5483	35.0228	269.8385	152.6773
800	37.6057	35.7858	35.196	273.5013	156.4221
900	37.9326	36.0061	35.3531	276.866	160.1989
1000	38.2636	36.2153	35.4989	279.9823	164.0088
1100	38.5892	36.4164	35.6359	282.8878	167.8515
1200	38.9032	36.6106	35.7656	285.6114	171.7262
1300	39.2019	36.7986	35.8889	288.1763	175.6316
1400	39.483	36.9804	36.0063	290.6009	179.566
1500	39.7454	37.1561	36.1183	292.9005	183.5276
1600	39.9889	37.3256	36.2252	295.0878	187.5145
1700	40.2136	37.489	36.3271	297.1735	191.5247
1800	40.4203	37.6462	36.4245	299.1667	195.5566
1900	40.6098	37.7972	36.5174	301.0753	199.6082
2000	40.7833	37.9423	36.6061	302.9062	203.678
2100	40.9421	38.0814	36.6909	304.6655	207.7644
2200	41.0876	38.2148	36.7719	306.3583	211.866
2300	41.2211	38.3426	36.8494	307.9896	215.9815
2400	41.3442	38.4652	36.9235	309.5636	220.1099
2500	41.4582	38.5826	36.9945	311.0842	224.25
2600	41.5646	38.6953	37.0625	312.5547	228.4012
2700	41.6647	38.8034	37.1277	313.9785	232.5628
2800	41.7598	38.9073	37.1904	315.3584	236.734
2900	41.8513	39.0073	37.2506	316.6971	240.9146
3000	41.9402	39.1036	37.3087	317.997	245.1042
3100	42.0278	39.1965	37.3646	319.2605	249.3026
3200	42.115	39.2863	37.4187	320.4896	253.5097
3300	42.2029	39.3734	37.471	321.6863	257.7256
3400	42.2922	39.4579	37.5218	322.8524	261.9504
3500	42.3838	39.5402	37.571	323.9896	266.1841

(continued)

Table B.19 (continued)

ϑ [°C]	C_p $\left[\frac{\text{J}}{\text{molK}}\right]$	$\overline{C_p}\Big\|_{\vartheta_0}^{\vartheta}$ $\left[\frac{\text{J}}{\text{molK}}\right]$	$\overline{\overline{C_p}}\Big\|_{\vartheta_0}^{\vartheta}$ $\left[\frac{\text{J}}{\text{molK}}\right]$	S_m $\left[\frac{\text{J}}{\text{molK}}\right]$	H_m $\left[\frac{\text{kJ}}{\text{mol}}\right]$
3600	42.4783	39.6205	37.619	325.0995	270.4272
3700	42.5764	39.699	37.6657	326.1836	274.6799
3800	42.6785	39.7761	37.7113	327.2431	278.9426
3900	42.785	39.8519	37.756	328.2796	283.2158
4000	42.8962	39.9266	37.7998	329.294	287.4998
4500	43.5259	40.2904	38.0085	334.0738	309.1003
5000	44.2574	40.6499	38.206	338.4451	331.043

Table B.20 Absolute molar specific enthalpy and entropy of SO at $p_0 = 1\text{bar}$. Reference temperature for averaged heat capacities is $\vartheta_0 = 0\,°\text{C}$

ϑ [°C]	C_p $\left[\frac{\text{J}}{\text{molK}}\right]$	$\overline{C_p}\Big\|_{\vartheta_0}^{\vartheta}$ $\left[\frac{\text{J}}{\text{molK}}\right]$	$\overline{\overline{C_p}}\Big\|_{\vartheta_0}^{\vartheta}$ $\left[\frac{\text{J}}{\text{molK}}\right]$	S_m $\left[\frac{\text{J}}{\text{molK}}\right]$	H_m $\left[\frac{\text{kJ}}{\text{mol}}\right]$
–20	29.6637	29.7673	29.766	217.0489	3.4144
–15	29.7139	29.7935	29.7928	217.6296	3.5628
–10	29.7661	29.8204	29.82	218.2001	3.7115
–5	29.8201	29.8478	29.8477	218.7609	3.8605
0	29.8758	29.8758	29.8758	219.3123	4.0097
5	29.9332	29.9044	29.9043	219.8547	4.1592
10	29.992	29.9334	29.9331	220.3886	4.3091
15	30.0522	29.963	29.9622	220.9141	4.4592
20	30.1137	29.9929	29.9915	221.4316	4.6096
25	30.1764	30.0233	30.0211	221.9414	4.7603
50	30.5032	30.1807	30.1719	224.3841	5.5188
100	31.1907	30.513	30.4786	228.8204	7.061
150	31.8714	30.8532	30.7796	232.7847	8.6377
200	32.5084	31.1885	31.0661	236.3797	10.2474
250	33.0845	31.5112	31.3336	239.6744	11.8875
300	33.5947	31.8168	31.5809	242.7177	13.5548
350	34.0412	32.1034	31.808	245.5464	15.2459
400	34.4302	32.3705	32.0159	248.1889	16.9579
450	34.7695	32.6186	32.2062	250.6681	18.6881
500	35.0674	32.8489	32.3805	253.0027	20.4342
600	35.5675	33.2619	32.6884	257.299	23.9669
700	35.972	33.6212	32.9516	261.1779	27.5445
800	36.322	33.9373	33.1801	264.7137	31.1595
900	36.6203	34.2192	33.3813	267.9631	34.807
1000	36.8912	34.473	33.5606	270.9698	38.4827

(continued)

Table B.20 (continued)

ϑ [°C]	C_p $\left[\frac{\text{J}}{\text{molK}}\right]$	$\overline{C_p}\big\vert_{\vartheta_0}^{\vartheta}$ $\left[\frac{\text{J}}{\text{molK}}\right]$	$\overline{\overline{C_p}}\big\vert_{\vartheta_0}^{\vartheta}$ $\left[\frac{\text{J}}{\text{molK}}\right]$	S_m $\left[\frac{\text{J}}{\text{molK}}\right]$	H_m $\left[\frac{\text{kJ}}{\text{mol}}\right]$
1100	37.1524	34.7047	33.7226	273.769	42.1849
1200	37.4116	34.9195	33.8711	276.3897	45.9131
1300	37.6713	35.1212	34.0087	278.8552	49.6673
1400	37.9315	35.3126	34.1376	281.1847	53.4474
1500	38.191	35.4959	34.2593	283.3941	57.2535
1600	38.4482	35.6724	34.375	285.4964	61.0855
1700	38.7015	35.8431	34.4855	287.5027	64.9431
1800	38.9494	36.0089	34.5913	289.4221	68.8257
1900	39.1906	36.17	34.693	291.2626	72.7327
2000	39.424	36.3269	34.791	293.031	76.6635
2100	39.649	36.4798	34.8855	294.7331	80.6172
2200	39.865	36.6288	34.9768	296.374	84.593
2300	40.0716	36.774	35.065	297.9583	88.5899
2400	40.2688	36.9155	35.1503	299.4899	92.607
2500	40.4565	37.0535	35.2329	300.9722	96.6434
2600	40.6349	37.1878	35.3129	302.4086	100.698
2700	40.8044	37.3186	35.3904	303.8017	104.7701
2800	40.9652	37.446	35.4655	305.1543	108.8586
2900	41.1178	37.57	35.5383	306.4685	112.9628
3000	41.2628	37.6907	35.6089	307.7466	117.0819
3100	41.4006	37.8082	35.6774	308.9904	121.2151
3200	41.5318	37.9225	35.7439	310.2018	125.3618
3300	41.6571	38.0338	35.8085	311.3825	129.5213
3400	41.777	38.1422	35.8713	312.534	133.693
3500	41.8921	38.2477	35.9323	313.6577	137.8765
3600	42.0031	38.3504	35.9916	314.755	142.0713
3700	42.1105	38.4506	36.0494	315.8271	146.277
3800	42.2149	38.5483	36.1056	316.8751	150.4933
3900	42.3169	38.6436	36.1604	317.9002	154.7199
4000	42.4168	38.7367	36.2139	318.9035	158.9566
4500	42.9028	39.1727	36.4631	323.6236	180.287
5000	43.3937	39.5701	36.6878	327.9216	201.8604

Table B.21 Absolute molar specific enthalpy and entropy of SO_2 at $p_0 = 1\,\text{bar}$. Reference temperature for averaged heat capacities is $\vartheta_0 = 0\,°C$

ϑ [°C]	C_p $\left[\frac{J}{molK}\right]$	$\overline{C_p}\Big\vert_{\vartheta_0}^{\vartheta}$ $\left[\frac{J}{molK}\right]$	$\overline{\overline{C_p}}\Big\vert_{\vartheta_0}^{\vartheta}$ $\left[\frac{J}{molK}\right]$	S_m $\left[\frac{J}{molK}\right]$	H_m $\left[\frac{kJ}{mol}\right]$
–20	38.2268	38.5826	38.5781	241.8393	–298.5664
–15	38.4041	38.6717	38.6692	242.5887	–298.3748
–10	38.5822	38.761	38.7599	243.3271	–298.1824
–5	38.7609	38.8505	38.8502	244.055	–297.989
0	38.9402	38.9402	38.9402	244.7727	–297.7948
5	39.12	39.03	39.0297	245.4807	–297.5996
10	39.3002	39.12	39.1189	246.1793	–297.4036
15	39.4807	39.2102	39.2077	246.8688	–297.2066
20	39.6615	39.3004	39.2961	247.5495	–297.0088
25	39.8425	39.3907	39.3841	248.2218	–296.81
50	40.7466	39.8428	39.8175	251.4658	–295.8026
100	42.5188	40.7407	40.6475	257.4531	–293.7207
150	44.1915	41.6156	41.4233	262.9039	–291.5524
200	45.7273	42.4547	42.1418	267.925	–289.3038
250	47.1103	43.2501	42.8031	272.5882	–286.9822
300	48.3394	43.998	43.4097	276.9448	–284.5954
350	49.4223	44.6972	43.965	281.0335	–282.1507
400	50.3715	45.3485	44.4729	284.8849	–279.6554
450	51.202	45.9538	44.9379	288.524	–277.1156
500	51.9293	46.5157	45.364	291.9718	–274.5369
600	53.1318	47.5224	46.1155	298.3629	–269.2813
700	54.0741	48.3937	46.7554	304.1762	–263.9192
800	54.8218	49.1522	47.3058	309.5028	–258.473
900	55.4233	49.8167	47.784	314.4144	–252.9597
1000	55.9157	50.4028	48.2033	318.9687	–247.392
1100	56.326	50.9232	48.5742	323.2124	–241.7793
1200	56.6732	51.3882	48.9049	327.1842	–236.1289
1300	56.9713	51.8065	49.202	330.9163	–230.4463
1400	57.2307	52.1849	49.4707	334.4354	–224.7359
1500	57.459	52.5291	49.7151	337.7643	–219.0012
1600	57.6625	52.8437	49.9387	340.9224	–213.2449
1700	57.8457	53.1326	50.1442	343.9262	–207.4694
1800	58.0125	53.3991	50.3341	346.7901	–201.6763
1900	58.1658	53.646	50.5103	349.5266	–195.8673
2000	58.3081	53.8756	50.6744	352.1467	–190.0435

(continued)

Table B.21 (continued)

ϑ [°C]	C_p $\left[\frac{J}{molK}\right]$	$\overline{C_p}\big\|_{\vartheta_0}^{\vartheta}$ $\left[\frac{J}{molK}\right]$	$\overline{\overline{C_p}}\big\|_{\vartheta_0}^{\vartheta}$ $\left[\frac{J}{molK}\right]$	S_m $\left[\frac{J}{molK}\right]$	H_m $\left[\frac{kJ}{mol}\right]$
2100	58.4412	54.0899	50.8277	354.6598	–184.206
2200	58.5669	54.2906	50.9715	357.0745	–178.3555
2300	58.6864	54.4791	51.1068	359.3984	–172.4928
2400	58.801	54.6568	51.2345	361.6381	–166.6184
2500	58.9114	54.8248	51.3552	363.7997	–160.7327
2600	59.0185	54.9841	51.4698	365.8885	–154.8362
2700	59.123	55.1354	51.5787	367.9095	–148.9291
2800	59.2255	55.2797	51.6825	369.867	–143.0117
2900	59.3263	55.4175	51.7817	371.7651	–137.0841
3000	59.426	55.5494	51.8766	373.6075	–131.1465
3100	59.525	55.6761	51.9675	375.3973	–125.1989
3200	59.6234	55.7979	52.0549	377.1378	–119.2415
3300	59.7217	55.9153	52.139	378.8316	–113.2742
3400	59.82	56.0287	52.2201	380.4814	–107.2971
3500	59.9185	56.1384	52.2983	382.0895	–101.3102
3600	60.0176	56.2448	52.374	383.6581	–95.3134
3700	60.1172	56.3481	52.4472	385.1893	–89.3067
3800	60.2176	56.4486	52.5182	386.6849	–83.29
3900	60.3189	56.5466	52.5872	388.1467	–77.2631
4000	60.4213	56.6422	52.6542	389.5763	–71.2261
4500	60.9528	57.0913	52.9646	396.2907	–40.8841
5000	61.5262	57.5057	53.2428	402.3906	–10.2664

Table B.22 Absolute molar specific enthalpy and entropy of CH_4 at $p_0 = 1\,\text{bar}$. Reference temperature for averaged heat capacities is $\vartheta_0 = 0\,°C$

ϑ [°C]	C_p $\left[\frac{J}{molK}\right]$	$\overline{C_p}\big\|_{\vartheta_0}^{\vartheta}$ $\left[\frac{J}{molK}\right]$	$\overline{\overline{C_p}}\big\|_{\vartheta_0}^{\vartheta}$ $\left[\frac{J}{molK}\right]$	S_m $\left[\frac{J}{molK}\right]$	H_m $\left[\frac{kJ}{mol}\right]$
–20	34.3372	34.5866	34.5833	180.6541	–76.1732
–15	34.4549	34.6503	34.6484	181.3269	–76.0012
–10	34.581	34.7169	34.716	181.989	–75.8286
–5	34.7155	34.7862	34.7859	182.6412	–75.6554
0	34.8582	34.8582	34.8582	183.2838	–75.4815
5	35.0092	34.933	34.9328	183.9175	–75.3068
10	35.1681	35.0105	35.0096	184.5426	–75.1314
15	35.3349	35.0906	35.0885	185.1596	–74.9551
20	35.5094	35.1734	35.1695	185.769	–74.778

(continued)

Table B.22 (continued)

ϑ [°C]	C_p $\left[\frac{\text{J}}{\text{molK}}\right]$	$\overline{C_p}\big\|_{\vartheta_0}^{\vartheta}$ $\left[\frac{\text{J}}{\text{molK}}\right]$	$\overline{\overline{C_p}}\big\|_{\vartheta_0}^{\vartheta}$ $\left[\frac{\text{J}}{\text{molK}}\right]$	S_m $\left[\frac{\text{J}}{\text{molK}}\right]$	H_m $\left[\frac{\text{kJ}}{\text{mol}}\right]$
25	35.6913	35.2586	35.2526	186.3711	–74.6
50	36.704	35.7213	35.6955	189.2841	–73.6954
100	39.1444	36.8031	36.6917	194.7301	–71.8012
150	41.9488	38.0436	37.7843	199.8222	–69.7749
200	44.9423	39.3916	38.9275	204.6702	–67.6032
250	48.0068	40.8077	40.0898	209.3361	–65.2795
300	51.0666	42.263	41.2508	213.8559	–62.8026
350	54.0755	43.7364	42.3972	218.2516	–60.1737
400	57.0066	45.2129	43.521	222.5374	–57.3963
450	59.8452	46.6819	44.6172	226.7229	–54.4746
500	62.5842	48.136	45.6833	230.8151	–51.4135
600	67.7508	50.9804	47.7209	238.7396	–44.8932
700	72.4843	53.7197	49.6318	246.3419	–37.8777
800	76.7676	56.3373	51.4185	253.6414	–30.4116
900	80.6511	58.8267	53.0871	260.6543	–22.5374
1000	84.1528	61.1873	54.6454	267.3955	–14.2942
1100	87.3058	63.4209	56.1012	273.8783	–5.7185
1200	90.1498	65.5318	57.4625	280.116	3.1567
1300	92.7236	67.5261	58.7372	286.1218	12.3025
1400	95.0634	69.4108	59.9328	291.9086	21.6936
1500	97.2011	71.1932	61.0564	297.4894	31.3084
1600	99.1648	72.8809	62.1145	302.8763	41.128
1700	100.9787	74.4811	63.113	308.0812	51.1364
1800	102.6639	76.0005	64.0572	313.1152	61.3195
1900	104.2385	77.4458	64.9521	317.9887	71.6655
2000	105.718	78.8227	65.802	322.7117	82.164
2100	107.1158	80.137	66.6108	327.2931	92.8063
2200	108.4437	81.3938	67.382	331.7417	103.5849
2300	109.7116	82.5976	68.1189	336.0654	114.4931
2400	110.9282	83.7529	68.8243	340.2715	125.5255
2500	112.1013	84.8635	69.5008	344.367	136.6773
2600	113.2374	85.9331	70.1507	348.3583	147.9445
2700	114.3422	86.9649	70.776	352.2514	159.3237
2800	115.4209	87.962	71.3788	356.0518	170.8121
2900	116.4779	88.9271	71.9608	359.7646	182.4072
3000	117.5171	89.8628	72.5234	363.3948	194.107
3100	118.542	90.7715	73.0682	366.9468	205.9101
3200	119.5555	91.6552	73.5965	370.4247	217.8151

(continued)

Table B.22 (continued)

ϑ [°C]	C_p $\left[\frac{\text{J}}{\text{molK}}\right]$	$\overline{C_p}\vert_{\vartheta_0}^{\vartheta}$ $\left[\frac{\text{J}}{\text{molK}}\right]$	$\overline{\overline{C_p}}\vert_{\vartheta_0}^{\vartheta}$ $\left[\frac{\text{J}}{\text{molK}}\right]$	S_m $\left[\frac{\text{J}}{\text{molK}}\right]$	H_m $\left[\frac{\text{kJ}}{\text{mol}}\right]$
3300	120.5605	92.5159	74.1094	373.8326	229.8209
3400	121.5592	93.3554	74.6081	377.174	241.9269
3500	122.5538	94.1755	75.0934	380.4525	254.1326
3600	123.5463	94.9775	75.5665	383.6712	266.4376
3700	124.5383	95.7631	76.028	386.8331	278.8419
3800	125.5314	96.5334	76.4788	389.9411	291.3453
3900	126.5269	97.2897	76.9196	392.9978	303.9482
4000	127.526	98.0331	77.351	396.0057	316.6508
4500	132.6129	101.591	79.3878	410.3918	381.6781
5000	137.9262	104.9566	81.2661	423.8609	449.3016

Table B.23 Absolute molar specific enthalpy and entropy of C_2H_6 at $p_0 = 1\,\text{bar}$. Reference temperature for averaged heat capacities is $\vartheta_0 = 0\,°\text{C}$

ϑ [°C]	C_p $\left[\frac{\text{J}}{\text{molK}}\right]$	$\overline{C_p}\vert_{\vartheta_0}^{\vartheta}$ $\left[\frac{\text{J}}{\text{molK}}\right]$	$\overline{\overline{C_p}}\vert_{\vartheta_0}^{\vartheta}$ $\left[\frac{\text{J}}{\text{molK}}\right]$	S_m $\left[\frac{\text{J}}{\text{molK}}\right]$	H_m $\left[\frac{\text{kJ}}{\text{mol}}\right]$
–20	47.3491	48.44	48.426	221.0732	–86.0957
–15	47.8864	48.7144	48.7065	222.0045	–85.8576
–10	48.4337	48.9919	48.9884	222.9283	–85.6168
–5	48.9904	49.2726	49.2717	223.8452	–85.3733
0	49.5562	49.5562	49.5562	224.7554	–85.1269
5	50.1304	49.8426	49.8417	225.6595	–84.8777
10	50.7125	50.1317	50.1282	226.5578	–84.6256
15	51.302	50.4233	50.4156	227.4507	–84.3705
20	51.8983	50.7174	50.7036	228.3383	–84.1125
25	52.501	51.0138	50.9923	229.2211	–83.8515
50	55.5921	52.5253	52.4407	233.5705	–82.5006
100	61.9938	55.6522	55.3283	242.0156	–79.5617
150	68.4212	58.84	58.1494	250.2078	–76.3009
200	74.6722	62.0217	60.8652	258.1943	–72.7225
250	80.6408	65.1539	63.4572	265.9929	–68.8384
300	86.2822	68.2098	65.9191	273.6098	–64.6639
350	91.5882	71.1745	68.2523	281.0476	–60.2158
400	96.571	74.0409	70.4618	288.3083	–55.5105
450	101.2523	76.807	72.5552	295.3949	–50.5637
500	105.656	79.474	74.5406	302.3113	–45.3899
600	113.7078	84.5221	78.2187	315.6523	–34.4136
700	120.8042	89.2098	81.5516	328.3682	–22.68

(continued)

Table B.23 (continued)

ϑ [°C]	C_p $\left[\frac{J}{molK}\right]$	$\overline{C_p}\Big\vert_{\vartheta_0}^{\vartheta}$ $\left[\frac{J}{molK}\right]$	$\overline{\overline{C_p}}\Big\vert_{\vartheta_0}^{\vartheta}$ $\left[\frac{J}{molK}\right]$	S_m $\left[\frac{J}{molK}\right]$	H_m $\left[\frac{kJ}{mol}\right]$
800	127.0036	93.5544	84.5801	340.4891	–10.2833
900	132.4539	97.5805	87.3413	352.049	2.6956
1000	137.2166	101.3114	89.8665	363.0805	16.1845
1100	141.374	104.7687	92.1819	373.6146	30.1187
1200	145.0061	107.9739	94.3105	383.6816	44.4418
1300	148.1856	110.9475	96.2725	393.3107	59.1049
1400	150.9758	113.7092	98.0857	402.5301	74.0659
1500	153.4314	116.2772	99.7656	411.3662	89.2889
1600	155.5988	118.6684	101.3261	419.8442	104.7426
1700	157.5177	120.8984	102.7791	427.9873	120.4004
1800	159.2218	122.9811	104.1353	435.8173	136.239
1900	160.7396	124.9292	105.404	443.3541	152.2385
2000	162.0956	126.7542	106.5935	450.6164	168.3815
2100	163.3106	128.4666	107.7109	457.6214	184.6529
2200	164.4024	130.0757	108.7629	464.3847	201.0396
2300	165.3864	131.5899	109.755	470.921	217.5298
2400	166.2756	133.0169	110.6924	477.2438	234.1137
2500	167.0814	134.3636	111.5796	483.3654	250.7822
2600	167.8137	135.6363	112.4207	489.2974	267.5275
2700	168.4808	136.8406	113.2193	495.0503	284.3427
2800	169.0901	137.9816	113.9788	500.634	301.2217
2900	169.648	139.0641	114.7021	506.0576	318.159
3000	170.1601	140.0922	115.3918	511.3294	335.1498
3100	170.6311	141.0699	116.0504	516.4574	352.1897
3200	171.0655	142.0005	116.68	521.4488	369.2748
3300	171.4667	142.8874	117.2826	526.3103	386.4017
3400	171.8382	143.7335	117.8601	531.0483	403.5672
3500	172.1827	144.5415	118.4141	535.6687	420.7684
3600	172.5027	145.3138	118.946	540.1768	438.0029
3700	172.8006	146.0527	119.4574	544.578	455.2682
3800	173.0781	146.7603	119.9494	548.8768	472.5623
3900	173.3371	147.4385	120.4232	553.0779	489.8832
4000	173.579	148.089	120.8799	557.1854	507.2292
4500	174.5775	150.9797	122.9383	576.45	594.2816
5000	175.3104	153.3779	124.6888	593.8794	681.7626

Table B.24 Absolute molar specific enthalpy and entropy of C_3H_8 at $p_0 = 1\,\text{bar}$. Reference temperature for averaged heat capacities is $\vartheta_0 = 0\,°\text{C}$

ϑ [°C]	C_p $\left[\frac{\text{J}}{\text{molK}}\right]$	$\overline{C_p}\Big\|_{\vartheta_0}^{\vartheta}$ $\left[\frac{\text{J}}{\text{molK}}\right]$	$\overline{\overline{C_p}}\Big\|_{\vartheta_0}^{\vartheta}$ $\left[\frac{\text{J}}{\text{molK}}\right]$	S_m $\left[\frac{\text{J}}{\text{molK}}\right]$	H_m $\left[\frac{\text{kJ}}{\text{mol}}\right]$
–20	64.9696	66.8275	66.8038	259.0084	–107.795
–15	65.8893	67.2939	67.2806	260.2881	–107.4678
–10	66.8204	67.7638	67.7579	261.561	–107.1361
–5	67.7622	68.2372	68.2357	262.8275	–106.7996
0	68.7137	68.7137	68.7137	264.0881	–106.4584
5	69.6741	69.1932	69.1917	265.3432	–106.1125
10	70.6426	69.6755	69.6697	266.5931	–105.7617
15	71.6185	70.1603	70.1474	267.8382	–105.406
20	72.6008	70.6475	70.6246	269.0787	–105.0455
25	73.5888	71.1369	71.1013	270.3149	–104.68
50	78.5904	73.6098	73.4714	276.4383	–102.7779
100	88.6646	78.6217	78.1021	288.4528	–98.5962
150	98.4837	83.6172	82.5272	300.2107	–93.9158
200	107.8063	88.511	86.7106	311.7262	–88.7562
250	116.5293	93.2527	90.6426	322.9919	–83.1452
300	124.633	97.8159	94.3286	333.9975	–77.1136
350	132.145	102.1903	97.7821	344.7355	–70.6918
400	139.1159	106.3757	101.0207	355.2034	–63.9082
450	145.6039	110.3781	104.0629	365.4033	–56.7883
500	151.664	114.207	106.9268	375.3404	–49.3549
600	162.6661	121.3863	112.1859	394.4578	–33.6266
700	172.2816	127.987	116.9069	412.6204	–16.8675
800	180.6221	134.0556	121.1647	429.8817	0.78609
900	187.934	139.6452	125.0246	446.3024	19.2222
1000	194.3052	144.7999	128.5388	461.9387	38.3415
1100	199.8548	149.5585	131.7495	476.8427	58.0559
1200	204.6954	153.9561	134.6926	491.0636	78.2889
1300	208.9273	158.0255	137.3988	504.6479	98.9747
1400	212.6373	161.7967	139.8948	517.6394	120.0569
1500	215.8996	165.2971	142.2035	530.0786	141.4872
1600	218.7772	168.5515	144.345	542.0036	163.224
1700	221.3233	171.5824	146.3366	553.4491	185.2316
1800	223.5832	174.4098	148.1936	564.4474	207.4792
1900	225.5954	177.0518	149.9291	575.028	229.94
2000	227.3923	179.5247	151.555	585.2181	252.5911
2100	229.002	181.8432	153.0814	595.0427	275.4122
2200	230.448	184.0202	154.5173	604.5248	298.386
2300	231.7509	186.0676	155.8709	613.6855	321.4971

(continued)

Table B.24 (continued)

ϑ [°C]	C_p $\left[\frac{J}{molK}\right]$	$\overline{C_p}\Big\|_{\vartheta_0}^{\vartheta}$ $\left[\frac{J}{molK}\right]$	$\overline{\overline{C_p}}\Big\|_{\vartheta_0}^{\vartheta}$ $\left[\frac{J}{molK}\right]$	S_m $\left[\frac{J}{molK}\right]$	H_m $\left[\frac{kJ}{mol}\right]$
2400	232.9281	187.996	157.1492	622.544	344.732
2500	233.9947	189.815	158.3585	631.1184	368.079
2600	234.9638	191.5331	159.5045	639.425	391.5277
2700	235.8465	193.1583	160.5923	647.4791	415.0689
2800	236.6526	194.6975	161.6264	655.2946	438.6945
2900	237.3906	196.1571	162.6109	662.8846	462.3972
3000	238.0679	197.543	163.5494	670.2609	486.1706
3100	238.691	198.8604	164.4453	677.4349	510.0089
3200	239.2653	200.1142	165.3017	684.4167	533.9071
3300	239.7959	201.3088	166.1212	691.2159	557.8605
3400	240.2871	202.4481	166.9064	697.8416	581.865
3500	240.7425	203.5358	167.6594	704.3021	605.9168
3600	241.1656	204.5752	168.3824	710.605	630.0124
3700	241.5593	205.5696	169.0773	716.7576	654.1489
3800	241.9262	206.5215	169.7457	722.7667	678.3234
3900	242.2685	207.4338	170.3894	728.6387	702.5333
4000	242.5883	208.3087	171.0097	734.3795	726.7764
4500	243.9078	212.1948	173.8048	761.2987	848.418
5000	244.8764	215.4169	176.1808	785.647	970.626

Table B.25 Absolute molar specific enthalpy and entropy of C_2H_5OH at $p_0 = 1\,\text{bar}$. Reference temperature for averaged heat capacities is $\vartheta_0 = 0\,°C$

ϑ [°C]	C_p $\left[\frac{J}{molK}\right]$	$\overline{C_p}\Big\|_{\vartheta_0}^{\vartheta}$ $\left[\frac{J}{molK}\right]$	$\overline{\overline{C_p}}\Big\|_{\vartheta_0}^{\vartheta}$ $\left[\frac{J}{molK}\right]$	S_m $\left[\frac{J}{molK}\right]$	H_m $\left[\frac{kJ}{mol}\right]$
–20	58.7148	60.1209	60.1029	270.4712	–237.7379
–15	59.4084	60.4744	60.4643	271.6263	–237.4425
–10	60.1137	60.8315	60.827	272.7727	–237.1437
–5	60.8298	61.1921	61.191	273.9109	–236.8414
0	61.5559	61.5559	61.5559	275.0414	–236.5354
5	62.2912	61.9229	61.9217	276.1646	–236.2258
10	63.035	62.2926	62.2882	277.281	–235.9125
15	63.7863	62.6651	62.6552	278.3909	–235.5955
20	64.5446	63.0401	63.0225	279.4947	–235.2746
25	65.309	63.4173	63.3899	280.5928	–234.95
50	69.1992	65.3317	65.2246	286.0053	–233.2688
100	77.0956	69.2399	68.835	296.5151	–229.6114
150	84.8097	73.1533	72.3008	306.6879	–225.5624

(continued)

Table B.25 (continued)

ϑ [°C]	C_p $\left[\frac{\text{J}}{\text{molK}}\right]$	$\overline{C_p}\Big\vert_{\vartheta_0}^{\vartheta}$ $\left[\frac{\text{J}}{\text{molK}}\right]$	$\overline{\overline{C_p}}\Big\vert_{\vartheta_0}^{\vartheta}$ $\left[\frac{\text{J}}{\text{molK}}\right]$	S_m $\left[\frac{\text{J}}{\text{molK}}\right]$	H_m $\left[\frac{\text{kJ}}{\text{mol}}\right]$
200	92.1092	76.9899	75.5801	316.5644	–221.1374
250	98.8948	80.7012	78.6578	326.1569	–216.3601
300	105.1476	84.2619	81.5348	335.469	–211.2569
350	110.894	87.6618	84.2208	344.5038	–205.8538
400	116.1833	90.9008	86.7297	353.267	–200.1751
450	121.0717	93.985	89.0773	361.7666	–194.2422
500	125.6132	96.9235	91.2792	370.013	–188.0737
600	133.8185	102.404	95.3031	385.7918	–175.093
700	140.9635	107.4149	98.8972	400.6919	–161.345
800	147.1451	112.0025	102.126	414.7837	–146.9335
900	152.5829	116.216	105.0455	428.1375	–131.941
1000	157.3368	120.0958	107.6993	440.8152	–116.4397
1100	161.4912	123.6744	110.1217	452.8705	–100.4936
1200	165.1264	126.9806	112.3412	464.3517	–84.1587
1300	168.3145	130.0401	114.3817	475.3025	–67.4832
1400	171.118	132.8763	116.2638	485.7628	–50.5086
1500	173.5906	135.5099	118.0049	495.7687	–33.2707
1600	175.7778	137.9596	119.6204	505.3532	–15.8
1700	177.7185	140.2424	121.1233	514.5463	1.8767
1800	179.4457	142.3733	122.525	523.3755	19.7366
1900	180.9873	144.3658	123.8357	531.8657	37.7597
2000	182.3674	146.232	125.0639	540.0395	55.9287
2100	183.6063	147.9828	126.2175	547.9176	74.2284
2200	184.7215	149.6278	127.3032	555.5191	92.6458
2300	185.7282	151.1759	128.3269	562.8613	111.1691
2400	186.6393	152.6349	129.2941	569.9601	129.7882
2500	187.4659	154.0118	130.2094	576.83	148.4942
2600	188.2179	155.3132	131.077	583.4844	167.2789
2700	188.9037	156.5448	131.9009	589.9358	186.1355
2800	189.5305	157.7118	132.6843	596.1954	205.0577
2900	190.1048	158.8191	133.4303	602.2737	224.0399
3000	190.6322	159.8708	134.1417	608.1806	243.0771
3100	191.1176	160.8711	134.821	613.9249	262.1649
3200	191.5652	161.8234	135.4704	619.5149	281.2994
3300	191.9789	162.731	136.092	624.9586	300.4769
3400	192.3619	163.5969	136.6876	630.2629	319.6941

(continued)

Table B.25 (continued)

ϑ [°C]	C_p $\left[\frac{J}{molK}\right]$	$\overline{C_p}\Big\vert_{\vartheta_0}^{\vartheta}$ $\left[\frac{J}{molK}\right]$	$\overline{\overline{C_p}}\Big\vert_{\vartheta_0}^{\vartheta}$ $\left[\frac{J}{molK}\right]$	S_m $\left[\frac{J}{molK}\right]$	H_m $\left[\frac{kJ}{mol}\right]$
3500	192.7173	164.4239	137.259	635.4347	338.9483
3600	193.0475	165.2145	137.8077	640.4801	358.2368
3700	193.3548	165.9709	138.3351	645.405	377.557
3800	193.6413	166.6954	138.8426	650.2149	396.907
3900	193.9088	167.3898	139.3313	654.9149	416.2847
4000	194.1589	168.0559	139.8023	659.5096	435.6882
4500	195.1934	171.0167	141.9255	681.0535	533.0399
5000	195.9568	173.4744	143.7313	700.5382	630.8367

Table B.26 Absolute molar specific enthalpy and entropy of C_2H_5OH(liq). The dependency of the enthalpy on the pressure is supposed to be insignificant. Liquid ethanol is treated as an incompressible fluid, i.e. the entropy does not need to be corrected. Reference temperature for averaged heat capacities is $\vartheta_0 = 0\,°C$

ϑ [°C]	C $\left[\frac{J}{molK}\right]$	$\overline{C}\Big\vert_{\vartheta_0}^{\vartheta}$ $\left[\frac{J}{molK}\right]$	$\overline{\overline{C}}\Big\vert_{\vartheta_0}^{\vartheta}$ $\left[\frac{J}{molK}\right]$	S_m $\left[\frac{J}{molK}\right]$	H_m $\left[\frac{kJ}{mol}\right]$
–40	94.2886	98.5383	98.4156	135.0857	–284.1454
–35	95.1866	99.0821	98.9859	137.0958	–283.6718
–30	96.157	99.6515	99.5791	139.0835	–283.1934
–25	97.2024	100.2471	100.1957	141.0512	–282.7101
–20	98.3255	100.8695	100.8359	143.0013	–282.2213
–15	99.5284	101.5193	101.5	144.936	–281.7267
–10	100.8135	102.1969	102.1882	146.8575	–281.2258
–5	102.1826	102.9029	102.9007	148.7677	–280.7184
0	103.6376	103.6376	103.6376	150.6688	–280.2039
5	105.1801	104.4015	104.3992	152.5625	–279.6819
10	106.8114	105.1949	105.1854	154.4508	–279.1519
15	108.5328	106.0181	105.9963	156.3353	–278.6136
20	110.3454	106.8715	106.832	158.2179	–278.0665
25	112.25	107.7552	107.6924	160.1	–277.51
50	123.1721	112.6355	112.3629	169.5565	–274.5721
75	136.4333	118.2934	117.6347	179.2084	–271.3319
100	151.9748	124.7247	123.4797	189.1894	–267.7314
116	163.0598	129.2379	127.5041	195.7981	–265.2123

Table B.27 Absolute molar specific enthalpy and entropy of CH_3OH at $p_0 = 1\,\text{bar}$. Reference temperature for averaged heat capacities is $\vartheta_0 = 0\,°\text{C}$

ϑ [°C]	C_p $\left[\frac{\text{J}}{\text{molK}}\right]$	$\overline{C_p}\big\vert_{\vartheta_0}^{\vartheta}$ $\left[\frac{\text{J}}{\text{molK}}\right]$	$\overline{\overline{C_p}}\big\vert_{\vartheta_0}^{\vartheta}$ $\left[\frac{\text{J}}{\text{molK}}\right]$	S_m $\left[\frac{\text{J}}{\text{molK}}\right]$	H_m $\left[\frac{\text{kJ}}{\text{mol}}\right]$
–20	41.5565	42.0448	42.0385	232.8241	–202.8629
–15	41.7917	42.1687	42.1651	233.6392	–202.6546
–10	42.0378	42.2962	42.2945	234.4432	–202.445
–5	42.2944	42.4271	42.4267	235.2368	–202.2342
0	42.5615	42.5615	42.5615	236.0207	–202.022
5	42.8385	42.6992	42.6988	236.7952	–201.8085
10	43.1253	42.8401	42.8385	237.5609	–201.5936
15	43.4213	42.9843	42.9804	238.3184	–201.3773
20	43.7261	43.1314	43.1246	239.068	–201.1594
25	44.0395	43.2816	43.2708	239.8101	–200.94
50	45.7184	44.073	44.0287	243.4217	–199.8184
100	49.4777	45.8184	45.6382	250.2579	–197.4402
150	53.4971	47.7046	47.3035	256.7257	–194.8664
200	57.5467	49.66	48.9674	262.9229	–192.09
250	61.4901	51.6342	50.5953	268.8999	–189.1135
300	65.2553	53.5934	52.1675	274.6834	–185.944
350	68.8119	55.516	53.6744	280.2895	–182.5914
400	72.1547	57.3891	55.1124	285.7291	–179.0664
450	75.2924	59.2059	56.4819	291.0111	–175.3794
500	78.2398	60.9634	57.7852	296.1434	–171.5403
600	83.6226	64.3004	60.2078	305.9873	–163.4418
700	88.3737	67.4076	62.4097	315.3132	–154.8367
800	92.5326	70.2934	64.4149	324.1616	–145.7873
900	96.2113	72.9731	66.2469	332.5707	–136.3462
1000	99.4471	75.4622	67.926	340.5743	–126.5598
1100	102.2884	77.7746	69.4691	348.2022	–116.47
1200	104.7846	79.9237	70.8907	355.4812	–106.1136
1300	106.9811	81.9223	72.2038	362.436	–95.523
1400	108.9183	83.7829	73.4196	369.0894	–84.726
1500	110.6312	85.5168	74.5481	375.4623	–73.7468
1600	112.1499	87.1349	75.5982	381.5741	–62.6063
1700	113.5003	88.6468	76.5775	387.4424	–51.3225
1800	114.7043	90.0616	77.493	393.0837	–39.9111
1900	115.7808	91.3875	78.3506	398.5129	–28.3858
2000	116.746	92.6317	79.1557	403.7437	–16.7586

(continued)

Table B.27 (continued)

ϑ [°C]	C_p $\left[\frac{\text{J}}{\text{molK}}\right]$	$\overline{C_p}\big\vert_{\vartheta_0}^{\vartheta}$ $\left[\frac{\text{J}}{\text{molK}}\right]$	$\overline{\overline{C_p}}\big\vert_{\vartheta_0}^{\vartheta}$ $\left[\frac{\text{J}}{\text{molK}}\right]$	S_m $\left[\frac{\text{J}}{\text{molK}}\right]$	H_m $\left[\frac{\text{kJ}}{\text{mol}}\right]$
2100	117.6136	93.801	79.913	408.7887	–5.0399
2200	118.3956	94.9015	80.6266	413.6594	6.7612
2300	119.1023	95.9386	81.3004	418.3665	18.6367
2400	119.7425	96.9173	81.9377	422.9199	30.5795
2500	120.324	97.8421	82.5414	427.3283	42.5833
2600	120.8533	98.7171	83.1142	431.6003	54.6425
2700	121.3364	99.5461	83.6585	435.7434	66.7524
2800	121.7782	100.3323	84.1765	439.7647	78.9084
2900	122.1832	101.0789	84.6702	443.6708	91.1068
3000	122.5553	101.7887	85.1412	447.4677	103.344
3100	122.8979	102.4642	85.5912	451.1611	115.6169
3200	123.214	103.1077	86.0217	454.7562	127.9227
3300	123.5062	103.7215	86.4339	458.2579	140.2589
3400	123.7768	104.3074	86.8291	461.6707	152.6232
3500	124.0279	104.8673	87.2084	464.9988	165.0136
3600	124.2613	105.4029	87.5727	468.2462	177.4282
3700	124.4787	105.9155	87.9231	471.4166	189.8653
3800	124.6813	106.4067	88.2603	474.5133	202.3235
3900	124.8706	106.8777	88.5852	477.5397	214.8012
4000	125.0476	107.3298	88.8984	480.4988	227.2972
4500	125.7806	109.3413	90.3113	494.3779	290.0138
5000	126.3225	111.0136	91.5143	506.9362	353.0461

Table B.28 Absolute molar specific enthalpy and entropy of CH_3OH(liq). The dependency of the enthalpy on the pressure is supposed to be insignificant. Liquid methanol is treated as an incompressible fluid, i.e. the entropy does not need to be corrected. Reference temperature for averaged heat capacities is $\vartheta_0 = 0\,°\text{C}$

ϑ [°C]	C $\left[\frac{\text{J}}{\text{molK}}\right]$	$\overline{C}\big\vert_{\vartheta_0}^{\vartheta}$ $\left[\frac{\text{J}}{\text{molK}}\right]$	$\overline{\overline{C}}\big\vert_{\vartheta_0}^{\vartheta}$ $\left[\frac{\text{J}}{\text{molK}}\right]$	S_m $\left[\frac{\text{J}}{\text{molK}}\right]$	H_m $\left[\frac{\text{kJ}}{\text{mol}}\right]$
–40	72.5503	74.5339	74.476	108.5685	–243.8644
–35	72.958	74.7886	74.7431	110.1121	–243.5007
–30	73.4066	75.0569	75.0226	111.6326	–243.1348
–25	73.8958	75.3387	75.3143	113.1317	–242.7665
–20	74.4254	75.634	75.618	114.611	–242.3958
–15	74.9956	75.943	75.9338	116.0722	–242.0222
–10	75.6065	76.2657	76.2615	117.5166	–241.6457
–5	76.2587	76.6022	76.6011	118.9458	–241.2661

(continued)

Table B.28 (continued)

ϑ [°C]	C $\left[\frac{\text{J}}{\text{molK}}\right]$	$\overline{C}\Big\vert_{\vartheta_0}^{\vartheta}$ $\left[\frac{\text{J}}{\text{molK}}\right]$	$\overline{\overline{C}}\Big\vert_{\vartheta_0}^{\vartheta}$ $\left[\frac{\text{J}}{\text{molK}}\right]$	S_m $\left[\frac{\text{J}}{\text{molK}}\right]$	H_m $\left[\frac{\text{kJ}}{\text{mol}}\right]$
0	76.9527	76.9527	76.9527	120.3609	–240.8831
5	77.6894	77.3175	77.3164	121.7634	–240.4965
10	78.4694	77.6966	77.692	123.1544	–240.1061
15	79.2938	78.0904	78.0799	124.5351	–239.7117
20	80.1637	78.499	78.4801	125.9066	–239.3131
25	81.08	78.9228	78.8927	127.27	–238.91
50	86.3963	81.2782	81.1466	134.0013	–236.8192
75	93.0188	84.0505	83.7289	140.6746	–234.5793
100	100.9939	87.2616	86.6454	147.3908	–232.1569
116	106.7768	89.5487	88.683	151.7497	–230.4954

Table B.29 Absolute molar specific enthalpy and entropy of air at $p_0 = 1\,\text{bar}$. Reference temperature for averaged heat capacities is $\vartheta_0 = 0\,°\text{C}$

ϑ [°C]	C_p $\left[\frac{\text{J}}{\text{molK}}\right]$	$\overline{C_p}\Big\vert_{\vartheta_0}^{\vartheta}$ $\left[\frac{\text{J}}{\text{molK}}\right]$	$\overline{\overline{C_p}}\Big\vert_{\vartheta_0}^{\vartheta}$ $\left[\frac{\text{J}}{\text{molK}}\right]$	S_m $\left[\frac{\text{J}}{\text{molK}}\right]$	H_m $\left[\frac{\text{kJ}}{\text{mol}}\right]$
–20	29.0566	29.0643	29.0642	194.0651	–1.434
–15	29.0602	29.0663	29.0662	194.6335	–1.2887
–10	29.0641	29.0684	29.0684	195.191	–1.1434
–5	29.0683	29.0706	29.0706	195.7381	–0.99805
0	29.0729	29.0729	29.0729	196.2752	–0.85269
5	29.0779	29.0754	29.0754	196.8026	–0.70732
10	29.0833	29.078	29.0779	197.3207	–0.56191
15	29.0891	29.0807	29.0806	197.8298	–0.41648
20	29.0953	29.0836	29.0834	198.3303	–0.27102
25	29.102	29.0866	29.0864	198.8224	–0.12553
50	29.1429	29.104	29.103	201.1672	0.60251
100	29.2654	29.1517	29.1468	205.3678	2.0625
150	29.4447	29.218	29.2048	209.0582	3.53
200	29.6769	29.3026	29.2759	212.3591	5.0078
250	29.9538	29.4045	29.3584	215.3536	6.4984
300	30.265	29.5216	29.4503	218.1015	8.0038
350	30.5994	29.6515	29.5495	220.6466	9.5253
400	30.9465	29.7916	29.6539	223.0214	11.0639
450	31.297	29.9394	29.7618	225.2511	12.62
500	31.643	30.0925	29.8715	227.355	14.1936
600	32.3004	30.4065	30.0911	231.2436	17.3912
700	32.8981	30.7204	30.3051	234.7783	20.6516

(continued)

Table B.29 (continued)

ϑ [°C]	C_p $\left[\frac{\text{J}}{\text{molK}}\right]$	$\overline{C_p}\Big\|_{\vartheta_0}^{\vartheta}$ $\left[\frac{\text{J}}{\text{molK}}\right]$	$\overline{\overline{C_p}}\Big\|_{\vartheta_0}^{\vartheta}$ $\left[\frac{\text{J}}{\text{molK}}\right]$	S_m $\left[\frac{\text{J}}{\text{molK}}\right]$	H_m $\left[\frac{\text{kJ}}{\text{mol}}\right]$
800	33.438	31.0271	30.5099	238.0228	23.969
900	33.9076	31.3217	30.7034	241.023	27.3368
1000	34.3176	31.6012	30.8846	243.8137	30.7485
1100	34.6778	31.8649	31.0539	246.4222	34.1987
1200	34.9967	32.1129	31.2118	248.8712	37.6827
1300	35.281	32.3458	31.3591	251.1791	41.1969
1400	35.536	32.5647	31.4968	253.3613	44.738
1500	35.7664	32.7706	31.6258	255.4309	48.3033
1600	35.9757	32.9645	31.7468	257.399	51.8905
1700	36.167	33.1474	31.8605	259.2751	55.4978
1800	36.3427	33.3201	31.9677	261.0675	59.1234
1900	36.5049	33.4835	32.069	262.7834	62.7659
2000	36.6553	33.6383	32.1647	264.4291	66.424
2100	36.7955	33.7854	32.2556	266.0102	70.0966
2200	36.9265	33.9252	32.3419	267.5316	73.7828
2300	37.0494	34.0584	32.424	268.9977	77.4816
2400	37.1651	34.1855	32.5023	270.4125	81.1924
2500	37.2744	34.3069	32.577	271.7795	84.9144
2600	37.3779	34.423	32.6485	273.1018	88.6471
2700	37.4761	34.5343	32.717	274.3823	92.3899
2800	37.5696	34.641	32.7827	275.6236	96.1422
2900	37.6587	34.7436	32.8458	276.828	99.9036
3000	37.7439	34.8422	32.9064	277.9978	103.6738
3100	37.8254	34.9371	32.9648	279.1349	107.4523
3200	37.9036	35.0286	33.0211	280.2412	111.2388
3300	37.9787	35.1169	33.0754	281.3181	115.0329
3400	38.051	35.2021	33.1279	282.3674	118.8344
3500	38.1208	35.2845	33.1786	283.3904	122.643
3600	38.1881	35.3642	33.2277	284.3885	126.4585
3700	38.2533	35.4414	33.2753	285.3628	130.2806
3800	38.3165	35.5163	33.3213	286.3144	134.1091
3900	38.3778	35.5888	33.366	287.2445	137.9438
4000	38.4376	35.6593	33.4095	288.154	141.7846
4500	38.7187	35.9839	33.6094	292.4228	161.0747
5000	38.9875	36.2708	33.7858	296.2932	180.5013

Appendix C
Caloric State Diagrams

C.1 Water

See (Figs. C.1, C.2 and C.3).

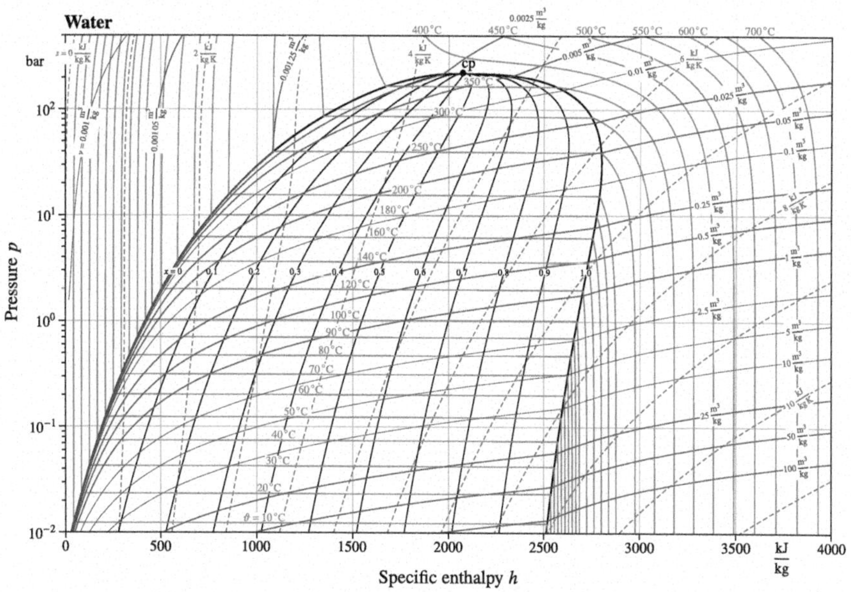

Fig. C.1 log p, h-diagram of water, generated with CoolProp, see [6]

A. Schmidt, *Technical Thermodynamics Workbook for Engineers*,
https://doi.org/10.1007/978-3-031-50172-2

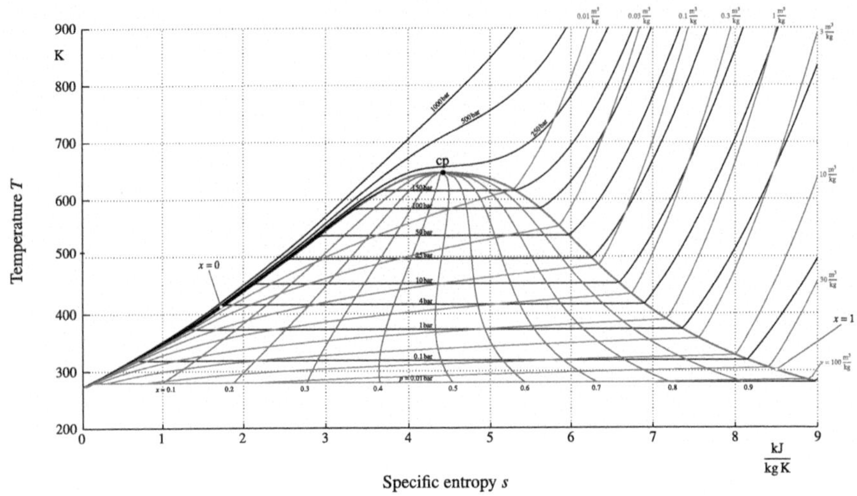

Fig. C.2 T, s-diagram of water, generated with XSteam, see [4]

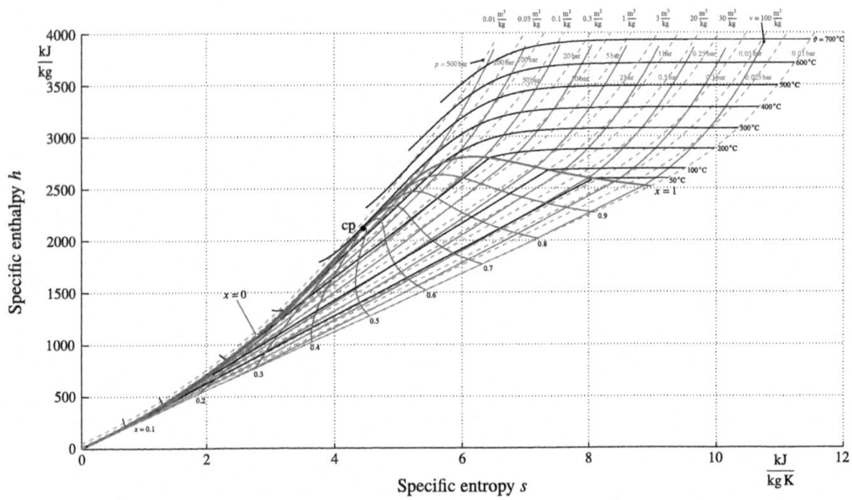

Fig. C.3 h, s-diagram of water, generated with XSteam, see [4]

C.2 Refrigerants

See (Figs. C.4, C.5, C.6,C.7 and C.8).

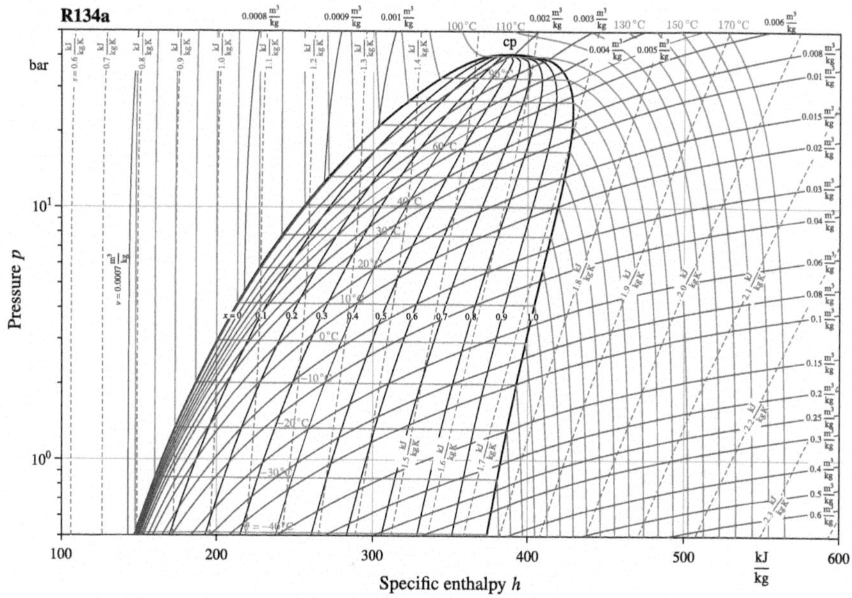

Fig. C.4 log p, h-diagram of R134a, generated with CoolProp, see [6]

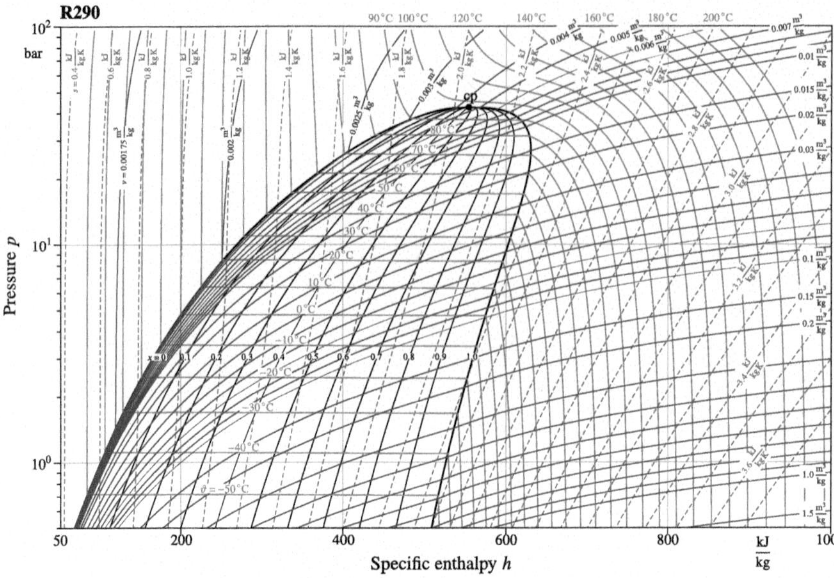

Fig. C.5 log p, h-diagram of R290, generated with CoolProp, see [6]

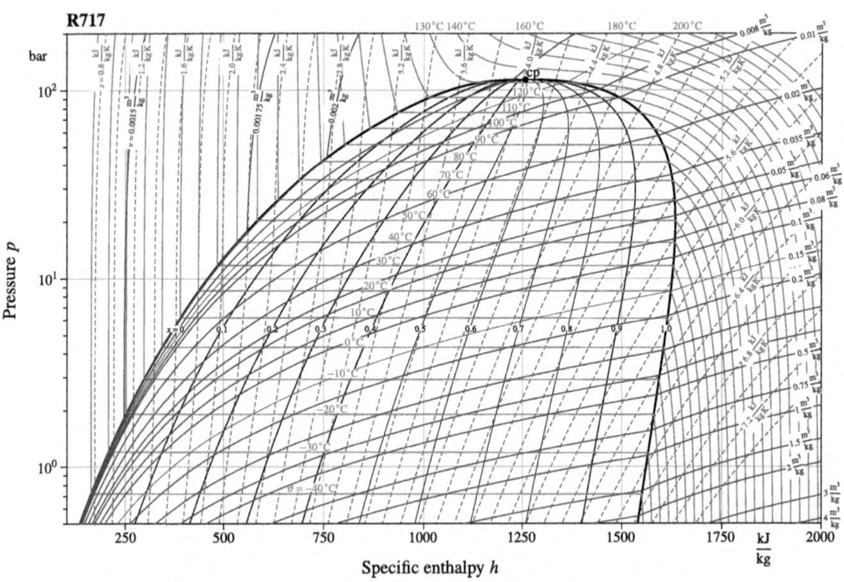

Fig. C.6 log p, h-diagram of R717, generated with CoolProp, see [6]

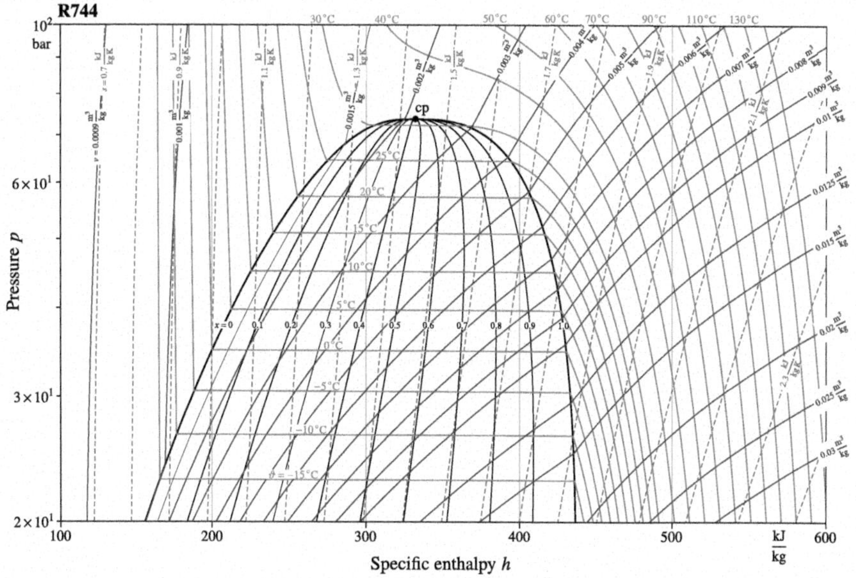

Fig. C.7 log p, h-diagram of R744, generated with CoolProp, see [6]

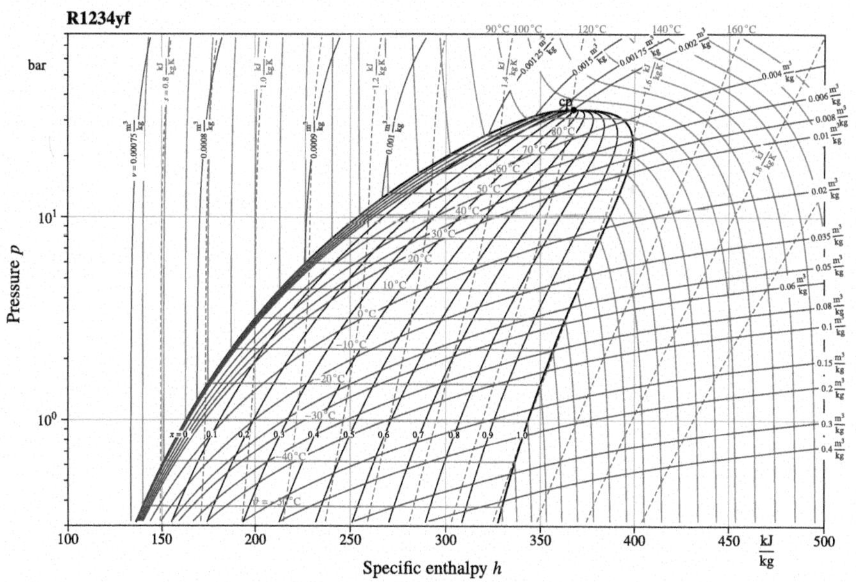

Fig. C.8 log p, h-diagram of R1234yf, generated with CoolProp, see [6]

Appendix D
The h_{1+x}, x-Diagram

See (Figs. D.1 and D.2).

A. Schmidt, *Technical Thermodynamics Workbook for Engineers*,
https://doi.org/10.1007/978-3-031-50172-2

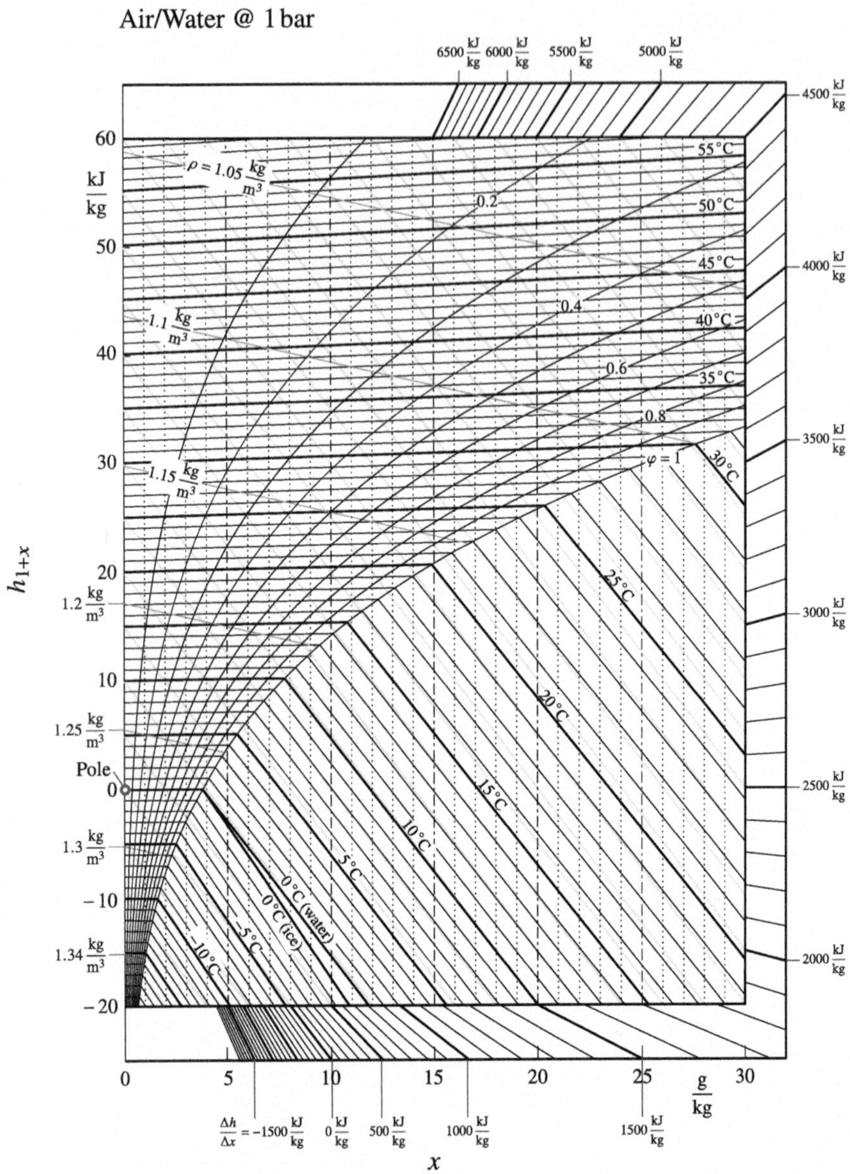

Fig. D.1 h_{1+x}, x-diagram (atmospheric air + water)

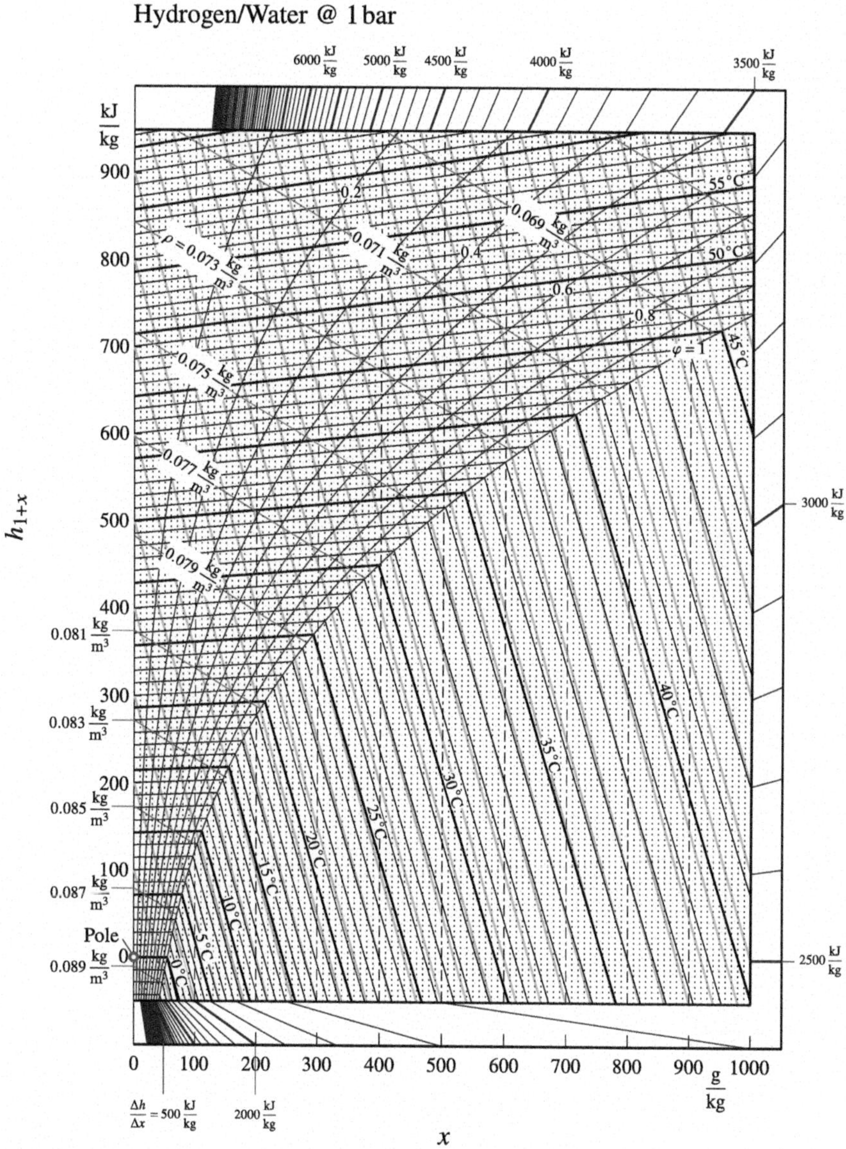

Fig. D.2 h_{1+x}, x-diagram (hydrogen + water)

Appendix E
Formulary

The following pages show the collection of formulae used to solve the tasks. A detailed derivation is given in [1].

A. Schmidt, *Technical Thermodynamics Workbook for Engineers*,
https://doi.org/10.1007/978-3-031-50172-2

Technical Thermodynamics

State values

$z=\frac{Z}{m}\qquad h=u+pv\qquad \mathrm{d}s=\frac{\delta q}{T}+\frac{\delta\psi}{T}=\delta s_a+\delta s_i$

$\Delta e_{a,12}=\frac{1}{2}\left(c_2^2-c_1^2\right)+g\left(z_2-z_1\right)$

Equations of state (ideal gas)

$pV=mRT$ resp. $pv=RT$

$\mathrm{d}u=c_v\,\mathrm{d}T\xrightarrow{c_v=\text{const.}}u_2-u_1=c_v\left(T_2-T_1\right)$

$\mathrm{d}h=c_p\,\mathrm{d}T\xrightarrow{c_p=\text{const.}}h_2-h_1=c_p\left(T_2-T_1\right)$

$\mathrm{d}s=\frac{\mathrm{d}u}{T}+\frac{p\,\mathrm{d}v}{T}=\frac{\mathrm{d}h}{T}-\frac{v\,\mathrm{d}p}{T}$

$\xrightarrow{c_v,c_p=\text{const.}}s_2-s_1=c_p\ln\frac{T_2}{T_1}-R\ln\frac{p_2}{p_1}=c_v\ln\frac{T_2}{T_1}+R\ln\frac{v_2}{v_1}$

$c_p=c_v+R\qquad \kappa=\frac{c_p}{c_v}\qquad \bar{c}\Big|_{\vartheta_1}^{\vartheta_2}=\frac{\bar{c}\Big|_{0^\circ\mathrm{C}}^{\vartheta_2}\vartheta_2-\bar{c}\Big|_{0^\circ\mathrm{C}}^{\vartheta_1}\vartheta_1}{\vartheta_2-\vartheta_1}$

$\overline{\overline{c_v}}\Big|_{\vartheta_1}^{\vartheta_2}=\frac{\overline{\overline{c_v}}\Big|_{\vartheta_0}^{\vartheta_2}\ln\frac{T_2}{T_0}-\overline{\overline{c_v}}\Big|_{\vartheta_0}^{\vartheta_1}\ln\frac{T_1}{T_0}}{\ln\frac{T_2}{T_1}}$

$R_M=R\cdot M=8.3143\,\frac{\mathrm{kJ}}{\mathrm{kmol\,K}}\qquad M=\frac{m}{n}$

Equations of state (incompressible liquids)

$\mathrm{d}u=c\,\mathrm{d}T\xrightarrow{c=\text{const.}}u_2-u_1=c\left(T_2-T_1\right)$

$\mathrm{d}h=c\,\mathrm{d}T+v\,\mathrm{d}p$

$\xrightarrow{c=\text{const.}}h_2-h_1=c\left(T_2-T_1\right)+v\left(p_2-p_1\right)$

$\mathrm{d}s=\frac{\mathrm{d}u}{T}\xrightarrow{c=\text{const.}}s_2-s_1=c\ln\frac{T_2}{T_1}$

Thermodynamic mean temperature

$T_m=\frac{q_{12}+\psi_{12}}{s_2-s_1}\qquad T_m=\frac{q_{12}}{s_{a,12}}\qquad T_m=\frac{\psi_{12}}{s_{i,12}}$

Closed system

1. Law

$W_{12}+Q_{12}=U_2-U_1+m\left[\frac{1}{2}\left(c_2^2-c_1^2\right)+g\left(z_2-z_1\right)\right]$

with

$W_{12}=W_{12,V}+W_{12,\mathrm{mech}}+\Psi_{12}$ and $W_{12,V}=-\int_1^2 p\,\mathrm{d}V$

$W_{\mathrm{eff}}=W_{12}+p_{\mathrm{env}}\left(V_2-V_1\right)$

2. Law

$S_2-S_1=m\left(s_2-s_1\right)=S_{a,12}+S_{i,12}=\sum\frac{Q_{12}}{T_m}+S_{i,12}$

Exergy

$E_{x,2}=E_{x,1}+W_{\mathrm{eff}}+E_{x,Q}-\Delta E_{x,V}$ with $\Delta E_{x,V}=T_{\mathrm{env}}S_{i,12}$

$E_{x,Q}=Q_{12}\left(1-\frac{T_{\mathrm{env}}}{T_m}\right)=Q_{12}-T_{\mathrm{env}}\left(S_2-S_1\right)+T_{\mathrm{env}}S_{i,12}$

$E_{x,i}=m_i\left(u_i-u_{\mathrm{env}}\right)+$

$+m_i\left[\frac{1}{2}c_i^2+gz_i+T_{\mathrm{env}}\left(s_{\mathrm{env}}-s_i\right)-p_{\mathrm{env}}\left(v_{\mathrm{env}}-v_i\right)\right]$

Open system - steady state flow process

Continuity

$\sum_{i,\mathrm{in}}\dot{m}_i=\sum_{j,\mathrm{out}}\dot{m}_j$ with $\dot{m}=\rho cA=\frac{cA}{v}$

1. Law

$\dot{Q}+P_t+\sum_{i,\mathrm{in}}\dot{m}_i\left[h_i+\frac{1}{2}c_i^2+gz_i\right]=\sum_{j,\mathrm{out}}\dot{m}_j\left[h_j+\frac{1}{2}c_j^2+gz_j\right]$

with

$w_{t,12}=\frac{P_t}{\dot{m}}=\underbrace{\int_1^2 v\,\mathrm{d}p+\psi_{12}}_{=y_{12}}+\left[\frac{1}{2}\left(c_2^2-c_1^2\right)+g\left(z_2-z_1\right)\right]$

2. Law

$\dot{S}_i+\dot{S}_a+\sum_{i,\mathrm{in}}\dot{m}_i s_i=\sum_{j,\mathrm{out}}\dot{m}_j s_j$ with $\dot{S}_a=\sum\frac{\dot{Q}}{T_m}$

Exergy

$\sum_{i,\mathrm{in}}\dot{m}_i e_{x,S,i}+\dot{E}_{x,Q}+P_t=\Delta\dot{E}_{x,V}+\sum_{j,\mathrm{out}}\dot{m}_j e_{x,S,j}$

$\dot{E}_{x,S,i}=\dot{m}e_{x,S,i}$

$=\dot{m}\left[h_i-h_{\mathrm{env}}+\frac{1}{2}c_i^2+gz_i+T_{\mathrm{env}}\left(s_{\mathrm{env}}-s_i\right)\right]$

$\Delta\dot{E}_{x,V}=T_{\mathrm{env}}\dot{S}_i$ and $\dot{E}_{x,Q}=\dot{Q}_{12}\left(1-\frac{T_{\mathrm{env}}}{T_m}\right)$

Open system - unsteady system

1. Law

$m_{\mathrm{Sys},\Delta t}\left(u_{\mathrm{Sys},\Delta t}+\frac{c_{\mathrm{Sys},\Delta t}^2}{2}+gz_{\mathrm{Sys},\Delta t}\right)+$

$-m_{\mathrm{Sys},0}\left(u_{\mathrm{Sys},0}+\frac{c_{\mathrm{Sys},0}^2}{2}+gz_{\mathrm{Sys},0}\right)=$

$\int_0^{\Delta t}\left(h_1+\frac{c_1^2}{2}+gz_1\right)\dot{m}_1\,\mathrm{d}t+$

$-\int_0^{\Delta t}\left(h_2+\frac{c_2^2}{2}+gz_2\right)\dot{m}_2\,\mathrm{d}t+\int_0^{\Delta t}\dot{Q}\,\mathrm{d}t+\int_0^{\Delta t}P_t\,\mathrm{d}t$

Thermal flow machines

$\eta_{s,T}=\frac{h_2-h_1}{h_{2s}-h_1}$ and $\eta_{s,V}=\frac{h_{2s}-h_1}{h_2-h_1}$

$T_{2s}=T_1\left(\frac{p_2}{p_1}\right)^{\frac{\kappa-1}{\kappa}}$

Thermodynamic cycle

$\oint\mathrm{d}Z=0$ and $\sum Q+\sum W=0$

$\eta_C=1-\frac{T_{\min}}{T_{\max}}$ and $\eta_{\mathrm{ex}}=\frac{\text{ex. benefit}}{\text{ex. effort}}$

$\eta_{\mathrm{th}}=\eta_{\mathrm{ex}}\eta_C$

$\eta_{\mathrm{th,TE}}=\frac{\text{benefit}}{\text{effort}}=\frac{P}{\dot{Q}_{\mathrm{in}}}$

$\varepsilon_{\mathrm{CM}}=\frac{\text{benefit}}{\text{effort}}=\frac{\dot{Q}_{\mathrm{in}}}{P}\qquad \varepsilon_{\mathrm{CM,C}}=\frac{T_{\min}}{T_{\max}-T_{\min}}$

$\varepsilon_{\mathrm{HP}}=\frac{\text{benefit}}{\text{effort}}=\frac{\dot{Q}_{\mathrm{out}}}{P}\qquad \varepsilon_{\mathrm{HP,C}}=\frac{T_{\max}}{T_{\max}-T_{\min}}$

$\varepsilon_{\mathrm{HP}}=\varepsilon_{\mathrm{CM}}+1$

Technical Thermodynamics

	isochoric	isobaric	isothermal	isentropic	polytropic
p,v	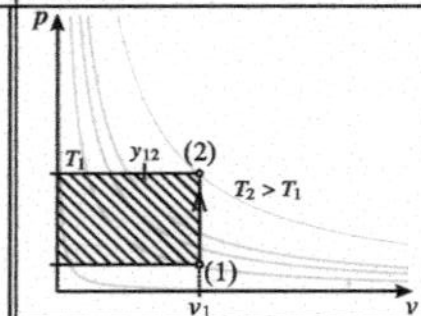	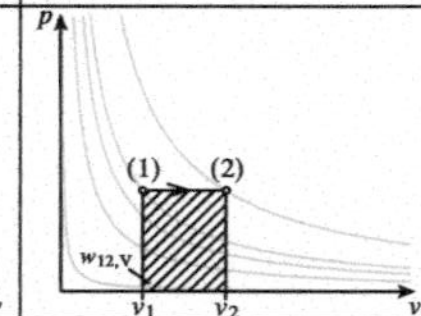	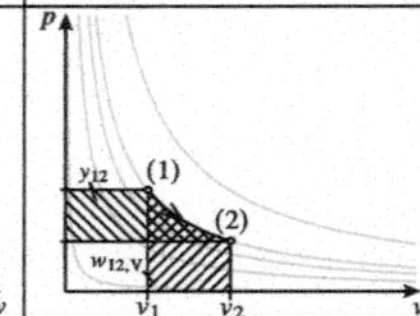	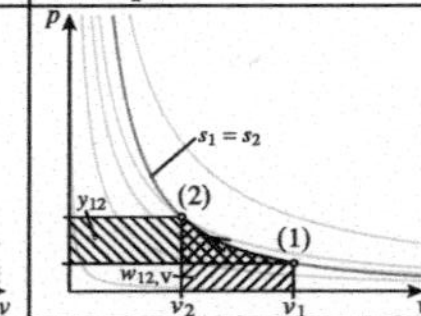	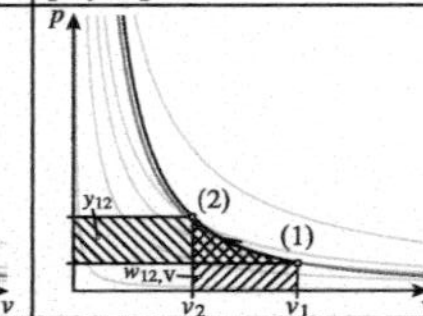
T,s	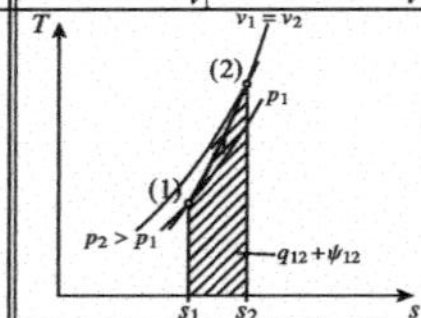	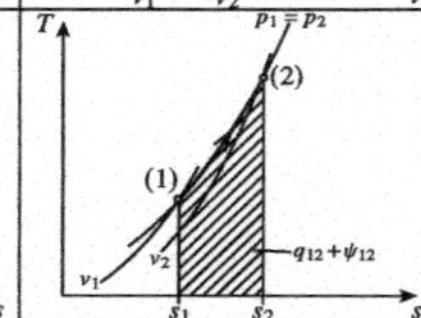	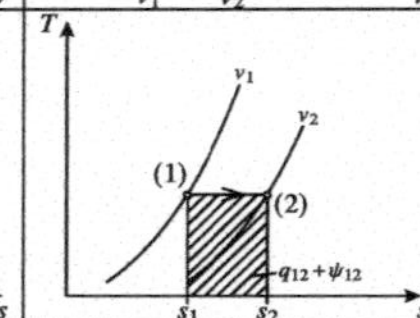	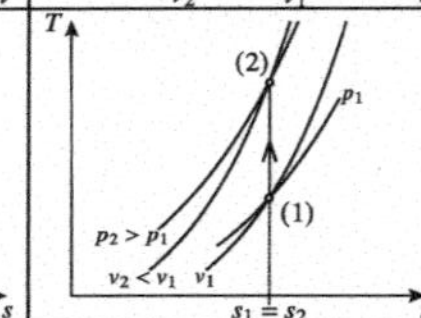	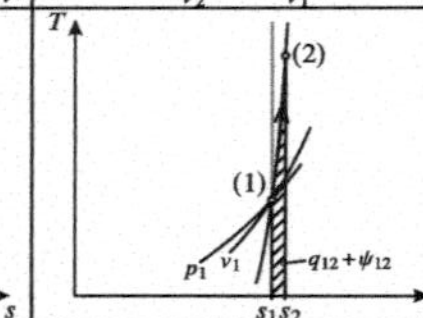
p,v,T	$\frac{p_1}{T_1}=\frac{p_2}{T_2}$	$\frac{v_1}{T_1}=\frac{v_2}{T_2}$	$p_1v_1=p_2v_2$	$p_1v_1^{\kappa}=p_2v_2^{\kappa}$ $\frac{T_2}{T_1}=\left(\frac{p_2}{p_1}\right)^{\frac{\kappa-1}{\kappa}}$	$p_1v_1^{n}=p_2v_2^{n}$ $\frac{T_2}{T_1}=\left(\frac{p_2}{p_1}\right)^{\frac{n-1}{n}}$
s_2-s_1	$s_2-s_1=c_v\ln\frac{T_2}{T_1}$ $s_2-s_1=c_v\ln\frac{p_2}{p_1}$	$s_2-s_1=c_p\ln\frac{T_2}{T_1}$ $s_2-s_1=c_p\ln\frac{v_2}{v_1}$	$s_2-s_1=R\ln\frac{v_2}{v_1}$ $s_2-s_1=R\ln\frac{p_1}{p_2}$	$s_2-s_1=0$	$s_2-s_1=c_v\frac{n-\kappa}{n-1}\ln\frac{T_2}{T_1}$
$w_{12,V}$	$w_{12,V}=-\int_1^2 p\,dv=0$	$w_{12,V}=-p\,(v_2-v_1)$ $w_{12,V}=-R\,(T_2-T_1)$	$w_{12,V}=-RT\ln\frac{v_2}{v_1}$ $w_{12,V}=-RT\ln\frac{p_1}{p_2}$ $w_{12,V}=-p_1v_1\ln\frac{p_1}{p_2}$ $w_{12,V}=-p_2v_2\ln\frac{p_1}{p_2}$	$w_{12,V}=\frac{p_1v_1}{\kappa-1}\left[\left(\frac{v_1}{v_2}\right)^{\kappa-1}-1\right]$ $w_{12,V}=\frac{RT_1}{\kappa-1}\left[\frac{T_2}{T_1}-1\right]$ $w_{12,V}=c_v\,(T_2-T_1)$ $w_{12,V}=\frac{RT_1}{\kappa-1}\left[\left(\frac{p_2}{p_1}\right)^{\frac{\kappa-1}{\kappa}}-1\right]$	$w_{12,V}=\frac{p_1v_1}{n-1}\left[\left(\frac{v_1}{v_2}\right)^{n-1}-1\right]$ $w_{12,V}=\frac{RT_1}{n-1}\left[\frac{T_2}{T_1}-1\right]$ $w_{12,V}=\frac{RT_1}{n-1}\left[\left(\frac{p_2}{p_1}\right)^{\frac{n-1}{n}}-1\right]$
y_{12}	$y_{12}=\int_1^2 v\,dp=v\,(p_2-p_1)$	$y_{12}=0$	$y_{12}=w_{12,V}$	$y_{12}=\kappa w_{12,V}$	$y_{12}=nw_{12,V}$

Technical Thermodynamics

Real fluids

$$x=\frac{\text{mass vavpour}}{\text{mass total}}=\frac{m_V}{m_V+m_L}$$

$$x=\frac{v-v'}{v''-v'}=\frac{u-u'}{u''-u'}=\frac{h-h'}{h''-h'}=\frac{s-s'}{s''-s'}$$

$$\Delta h_v=h''-h'$$

Clausius-Clapeyron

$$\frac{dp}{dT}=\frac{s''-s'}{v''-v'}=\frac{\Delta h_v}{T\,(v''-v')}$$

Joule-Thomson

$$\left(\frac{\partial T}{\partial p}\right)_h=-\frac{\delta_T}{c_p}=\delta_h$$

Mixture of ideal gases

$$\xi_i=\frac{m_i}{m_{\text{total}}}\quad x_i=\frac{n_i}{n_{\text{total}}}\quad \sigma_i=\frac{V_i}{V_{\text{total}}}\quad \pi_i=\frac{p_i}{p_{\text{total}}}$$

$$\sum_i\xi_i=1\quad \sum_i x_i=1\quad \sum_i\sigma_i=1\quad \sum_i\pi_i=1$$

$$\pi_i=\sigma_i=\pi_i$$

$$p_{\text{total}}=p_1+p_2+\dots+p_n=\sum_i p_i$$

$$p_iV_{\text{total}}=m_iR_iT\quad p_{\text{total}}V_{\text{total}}=m_{\text{total}}R_{\text{total}}T$$

$$R_{\text{total}}=\sum_i\xi_iR_i\quad M_{\text{total}}=\sum_i x_iM_i$$

$$R_{\text{total}}M_{\text{total}}=R_M\quad M_{\text{total}}=\frac{x_i}{\xi_i}M_i$$

$$h_{\text{total}}=\sum_i\xi_ih_i\quad u_{\text{total}}=\sum_i\xi_iu_i$$

$$c_{p,\text{total}}=\sum_i\xi_ic_{p,i}\quad c_{v,\text{total}}=\sum_i\xi_ic_{v,i}$$

$$S_{\text{total}}=\sum_i m_is_i(p_i,T)\quad s_i-s_0=c_{p,i}\ln\frac{T_i}{T_0}-R_i\ln\frac{p_i}{p_0}$$

Adiabatic mixing @ $p_{\text{total}}=\text{const.}, T=\text{const.}$

$$S_{i,12}=m_1R_1\ln\left(\frac{p_{\text{total}}}{p_1}\right)+m_2R_2\ln\left(\frac{p_{\text{total}}}{p_2}\right)$$

Humid air

$$x=\frac{m_w}{m_a}\quad x_v=\frac{m_v}{m_a}=\frac{R_a}{R_v}\frac{p_v}{p-p_v}=0.622\frac{p_v}{p-p_v}$$

$$z_{1+x}=\frac{Z}{m_a}$$

$$\varphi=\frac{p_v}{p'_v}=\frac{x_v}{x_v+0.622}\cdot\frac{p}{p'_v}\quad \varphi_{1,\text{ref}}=\frac{p_{\text{ref}}}{p_1}\cdot\varphi_1$$

$$v_{1+x}=\frac{V}{m_a}=\frac{R_aT}{p}\left[1+\frac{R_v}{R_a}\cdot x_v\right]=\frac{R_aT}{p}\cdot(1+1.608\cdot x_v)$$

$$v_{1+x}=(1+x_v)\cdot v$$

Reference level

$$T_{0,a}=273.15\,\text{K}\quad p_{0,a}=1\,\text{bar}$$

$$T_{0,w}=273.16\,\text{K}\quad p_{0,w}=0.006\,117\,\text{bar}$$

unsaturated

$$h_{1+x}=h_a+x\cdot(\Delta h_{v,0}+c_{p,v}\cdot\vartheta)$$

$$h_a=c_{p,a}\cdot\vartheta$$

$$s_{1+x}=c_{p,a}\ln\frac{T}{T_{0,a}}-R_a\ln\frac{p-p_v}{p_{0,a}}+x_v\left[c_{p,v}\ln\frac{T}{T_{0,w}}-R_v\ln\frac{p_v}{p_{0,w}}+\frac{\Delta h_{v,0}}{T_{0,a}}\right]$$

$$s'_{1+x}=c_{p,a}\ln\frac{T}{T_{0,a}}-R_a\ln\frac{p-p'_v}{p_{0,a}}+x'\left[c_{p,v}\ln\frac{T}{T_{0,w}}-R_v\ln\frac{p'_v}{p_{0,w}}+\frac{\Delta h_{v,0}}{T_{0,a}}\right]$$

fog region

$$h_{1+x}=h_a+x'\cdot h_v+(x-x')\cdot h_{\text{liq}}$$

$$h_a=c_{p,a}\cdot\vartheta\quad h_v=\Delta h_{v,0}+c_{p,v}\cdot\vartheta\quad h_{\text{liq}}=c_{\text{liq}}\cdot\vartheta$$

$$s_{1+x}=s'_{1+x}+(x-x')\cdot c_{\text{liq}}\ln\frac{T}{T_{0,w}}$$

freezing fog region

$$h_{1+x}=h_a+x'\cdot h_v+(x-x')\cdot h_{\text{ice}}$$

$$h_a=c_{p,a}\cdot\vartheta\quad h_v=\Delta h_{v,0}+c_{p,v}\cdot\vartheta\quad h_{\text{ice}}=\Delta h_{m,0}+c_{\text{ice}}\cdot\vartheta$$

$$s_{1+x}=s'_{1+x}+(x-x')\cdot\left[\frac{\Delta h_{m,0}}{T_{0,w}}+c_{\text{ice}}\ln\frac{T}{T_{0,w}}\right]$$

$$R_a=287.1\,\frac{\text{J}}{\text{kg K}}\quad R_v=461.5\,\frac{\text{J}}{\text{kg K}}$$

$$c_{p,a}=1.004\,\frac{\text{kJ}}{\text{kg K}}\quad c_{p,v}=1.86\,\frac{\text{kJ}}{\text{kg K}}$$

$$c_{\text{liq}}=4.19\,\frac{\text{kJ}}{\text{kg K}}\quad c_{\text{ice}}=2.05\,\frac{\text{kJ}}{\text{kg K}}$$

$$\Delta h_{v,0}=2500\,\frac{\text{kJ}}{\text{kg}}\quad \Delta h_{m,0}=-333\,\frac{\text{kJ}}{\text{kg}}$$

Steady state flow processes

incompressible flow

$$\left(p+\frac{\rho}{2}c^2+g\rho z\right)_2-\left(p+\frac{\rho}{2}c^2+g\rho z\right)_1=-\rho\psi_{12}<0$$

adiabatic, frictional tube flow

$$h+\frac{1}{2}v^2\left(\frac{\dot m}{A}\right)^2=\text{const.}=h_1^+$$

Supersonic flow

$$a=\sqrt{\left(\frac{\partial p}{\partial\rho}\right)_s}=\sqrt{\kappa pv}=\sqrt{\kappa RT}$$

$$\text{Ma}=\frac{c}{a}$$

Laval nozzle

$$\frac{dA}{A}=\left(\frac{1}{c^2}-\frac{1}{a^2}\right)v\,dp$$

$$\frac{dc}{c}=-\frac{dA}{A\,(1-\text{Ma}^2)}$$

adiabatic diffuser

$$\eta_{sD}=\frac{\Delta h_s}{\Delta h}=\frac{h_{2s}-h_1}{h_2-h_1}=\frac{c_1^2-c_{2s}^2}{c_1^2-c_2^2}$$

adiabatic nozzle

$$\eta_{sN}=\frac{\Delta h}{\Delta h_s}=\frac{h_2-h_1}{h_{2s}-h_1}=\frac{c_1^2-c_2^2}{c_1^2-c_{2s}^2}$$

Technical Thermodynamics

Technical combustion

Stoichiometry

$$\xi_C = \frac{m_C}{F} = c \quad \xi_{H_2} = \frac{m_{H_2}}{F} = h \quad \xi_{O_2} = \frac{m_{O_2}}{F} = o$$

$$\xi_S = \frac{m_S}{F} = s \quad \xi_{N_2} = \frac{m_{N_2}}{F} = n \quad \xi_W = \frac{m_W}{F} = w$$

$$\xi_{Ash} = \frac{m_{Ash}}{F} = a$$

$$c + h + s + o + n + a + w = 1$$

$$M_H = 1\,\frac{kg}{kmol} \quad M_C = 12\,\frac{kg}{kmol} \quad M_O = 16\,\frac{kg}{kmol}$$

$$M_N = 14\,\frac{kg}{kmol} \quad M_S = 32\,\frac{kg}{kmol}$$

$$\frac{m_{O_2,min}}{m_F} = o_{min} = 2.667 \cdot c + 8 \cdot h + s - o$$

$$l_{min} = \frac{m_{a,\,min}}{m_F} = \frac{1}{\xi_{O_2,dry\,air}} \cdot o_{min}$$

$$\lambda = \frac{m_a}{m_{a,min}} = \frac{l}{l_{min}}$$

$$m_A = m_a + m_v = m_a + x m_a = m_a(1+x) = \lambda m_{a,min}(1+x)$$

$$\mu_{EG,CO_2} = \frac{m_{EG,CO_2}}{m_F} = 3.667 \cdot \frac{m_C}{m_F} = 3.667 \cdot c$$

$$\mu_{EG,SO_2} = \frac{m_{EG,SO_2}}{m_F} = 2 \cdot \frac{m_S}{m_F} = 2 \cdot s$$

$$\mu_{EG,H_2O} = \frac{m_{EG,H_2O}}{m_F} = 9 \cdot h + w + \lambda \cdot l_{min} \cdot x$$

$$\mu_{EG,N_2} = \frac{m_{EG,N_2}}{m_F} = n + \lambda \cdot l_{min} \cdot \xi_{N_2,dry\,air}$$

$$\mu_{EG,O_2} = \frac{m_{EG,O_2}}{m_F} = l_{min} \cdot \xi_{O_2,dry\,air} \cdot (\lambda - 1) = o_{min} \cdot (\lambda - 1)$$

$$\sum_i \mu_{EG,i} = (1-a) + \lambda \cdot l_{min} \cdot (1+x)$$

$$\xi_{EG,i} = \frac{m_{EG,i}}{m_{EG}} = \frac{\mu_{EG,i}}{\sum_i \mu_{EG,i}}$$

$$\sigma_k = \frac{V_k}{V_F} \quad \sigma_k = x_k = \frac{n_k}{n_F}$$

$$O_{min} = 0.5 \cdot x_{H_2} + 0.5 \cdot x_{CO} + \sum_i \left(a_i + \frac{b_i}{4} - \frac{z_i}{2} + r_i\right) \cdot x_{C_aH_bO_zN_pS_r,i} - x_{O_2}$$

$$L_{min} = \frac{n_{a,min}}{n_F} = \frac{1}{x_{O_2,dry\,air}} \cdot O_{min}$$

$$\lambda = \frac{m_a}{m_{a,min}} = \frac{n_a M_a}{n_{a,min} M_a} = \frac{n_a}{n_{a,min}} = \frac{L}{L_{min}}$$

$$n_A = \lambda \cdot n_{a,min}(1 + 1.608 \cdot x)$$

$$\nu_{EG,CO_2} = \frac{n_{EG,CO_2}}{n_F} = x_{CO} + \sum_i a_i \cdot x_{C_aH_bO_zN_pS_r,i} + x_{CO_2}$$

$$\nu_{EG,SO_2} = \frac{n_{EG,SO_2}}{n_F} = \sum_i r_i \cdot x_{C_aH_bO_zN_pS_r,i}$$

$$\nu_{EG,H_2O} = \frac{n_{EG,H_2O}}{n_F} = x_{H_2} + \sum_i \frac{b_i}{2} \cdot x_{C_aH_bO_zN_pS_r,i} + 1.608 \cdot \lambda \cdot L_{min} \cdot x$$

$$\nu_{EG,N_2} = \frac{n_{EG,N_2}}{n_F} = x_{N_2} + \sum_i \frac{p_i}{2} \cdot x_{C_aH_bO_zN_pS_r,i} + \lambda \cdot L_{min} \cdot x_{N_2,dry\,air}$$

$$\nu_{EG,O_2} = \frac{n_{EG,O_2}}{n_F} = L_{min} \cdot x_{O_2,dry\,air} \cdot (\lambda - 1) = O_{min} \cdot (\lambda - 1)$$

$$x_{EG,i} = \frac{n_{EG,i}}{n_{EG}} = \frac{\nu_{EG,i}}{\nu_{EG}} = \frac{\nu_{EG,i}}{\sum_i \nu_{EG,i}} \quad \mu_{EG,i} = \nu_{EG,i} \cdot \frac{M_{EG,i}}{M_F}$$

$$\dot{n}_{liq} = \frac{\dot{n}_{EG,H_2O} - x_s \dot{n}_{EG}}{1 - x_s}$$

Energetics

For $\vartheta_{EG} > \vartheta_\tau$

$$q = \mu_{EG} \cdot \overline{c_{p,EG}}\big|_{\vartheta_0}^{\vartheta_{EG}} (\vartheta_{EG} - \vartheta_0) + a \cdot \overline{c_{Ash}}\big|_{\vartheta_0}^{\vartheta_{EG}} (\vartheta_{EG} - \vartheta_0) + {} - H_U(T_0) - \overline{c_{p,F}}\big|_{\vartheta_0}^{\vartheta_F} (\vartheta_F - \vartheta_0) - \frac{\dot{m}_A}{\dot{m}_F} \cdot \sum_{i=1}^{n} \xi_{A,i} \cdot \overline{c_{p,i}}\big|_{\vartheta_0}^{\vartheta_a} (\vartheta_a - \vartheta_0)$$

$$\vartheta_{ad} = \vartheta_0 + \frac{H_U(T_0) + \overline{c_{p,F}}\big|_{\vartheta_0}^{\vartheta_F} (\vartheta_F - \vartheta_0) + \lambda l_{min} \overline{c_{p,a}}\big|_{\vartheta_0}^{\vartheta_a} (\vartheta_a - \vartheta_0)}{\mu_{EG} \cdot \overline{c_{p,EG}}\big|_{\vartheta_0}^{\vartheta_{ad}} + a \cdot \overline{c_{Ash}}\big|_{\vartheta_0}^{\vartheta_{ad}}}$$

$$H_0(T_0) = H_U(T_0) + \mu_{EG,H_2O} \cdot \Delta h_v(T_0)$$

$$H_{UM}(T_0) = \sum_i \frac{n_i}{n_F} \cdot H_{UM,i}(T_0)$$

$$H_{0M}(T_0) = \sum_i \frac{n_i}{n_F} \cdot H_{0M,i}(T_0)$$

$$\eta_K = \frac{|q_E|}{H_U} \quad \eta_F = \frac{|q_E + q_{iso}|}{H_U}$$

$$\Delta h_{EG} = \sum_i \mu_{EG,i} \cdot [h_{EG}(\varepsilon_G) - h_{EG}(T_0)]$$

Chemical reactions

$$H_{m,i}(T) = \underbrace{\Delta_B^0 H_{m,i}(T_0)}_{\text{chemical}} + \underbrace{\overline{C_p}\Big|_{T_0}^{T} (T - T_0)}_{\text{caloric}}$$

$$\dot{Q} = \sum_{i,Products} \dot{n}_i H_{m,i}(T_i) - \sum_{j,Educts} \dot{n}_j H_{m,j}(T_j)$$

$$\Delta_R^0 H_m(T_0) = \sum_{i,Products} \frac{\dot{n}_i}{\dot{n}_F} \Delta_B^0 H_{m,i}(T_0) - \sum_{j,Educts} \frac{\dot{n}_j}{\dot{n}_F} \Delta_B^0 H_{m,j}(T_0) = -H_{UM}(T_0)$$

$$S_m(T, p_i) = S_m(T, p_0) - R_m \ln \frac{p_i}{p_0}.$$

$$\dot{S}_{II} - \dot{S}_I = \sum_{i,Products} \dot{n}_i S_{m,i}(T_i, p_i) - \sum_{j,Educts} \dot{n}_j S_{m,j}(T_j, p_j).$$

References

1. Schmidt A (2022) Technical Thermodynamics for Engineers: Basics and Applications, 2nd edn. Springer Nature
2. Baehr HD, Kabelac S (2012) Thermodynamik: Grundlagen und technische Anwendungen. Springer, Wiesbaden
3. Lucas K (2008) Thermodynamik: Die Grundgesetze der Energie- und Stoffumwandlungen. Springer, Berlin
4. Holmgren M, Steam X (2018) Thermodynamic properties of water and steam. https://de.mathworks.com/matlabcentral/fileexchange/9817-x-steam--thermodynamic-properties-of-water-and-steam. Accessed 01 June 2018
5. McBride BJ, Zehe MJ, Gordon S (2002) NASA Glenn Coefficients for Calculating Thermodynamic Properties of Individual Species. Glenn Research Center, NASA/TP-2002-211556
6. Bell IH, Wronski J, Quoilin S, Lemort V (2014) Pure and Pseudo-pure Fluid Thermophysical Property Evaluation and the Open-Source Thermophysical Property Library CoolProp. Ind Eng Chem Res 53(6):2498–2508

A. Schmidt, *Technical Thermodynamics Workbook for Engineers*,
https://doi.org/10.1007/978-3-031-50172-2

Index

A. Schmidt, *Technical Thermodynamics Workbook for Engineers*,
https://doi.org/10.1007/978-3-031-50172-2

The manufacturer's authorised representative in the EU is Springer Nature Customer Service Centre GmbH, Europaplatz 3, 69115 Heidelberg, Germany. If you have any concerns regarding our products, please contact ProductSafety@springernature.com

Printed and bound by CPI Group (UK) Ltd, Croydon, CR0 4YY

31/07/2026

02182909-0001